Edexcel
ASChemistry

2nd Edition

Edexcel

George Facer
ASChemistry

2nd Edition

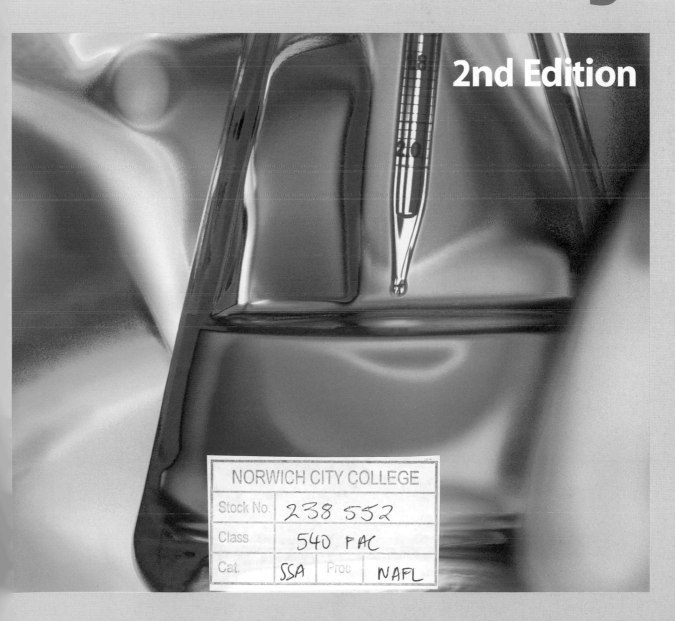

This book is dedicated to the memory of my father, John Facer, who inspired me, and many others, in the love of chemistry when he was a teacher at Epsom College.

Philip Allan Updates, an imprint of Hodder Education, an Hachette UK company, Market Place, Deddington, Oxfordshire OX15 0SE

Orders
Bookpoint Ltd, 130 Milton Park, Abingdon, Oxfordshire OX14 4SB
tel: 01235 827720   fax: 01235 400454
e-mail: uk.orders@bookpoint.co.uk
Lines are open 9.00 a.m.–5.00 p.m., Monday to Saturday, with a 24-hour message answering service. You can also order through the Philip Allan Updates website: www.philipallan.co.uk

© Philip Allan Updates 2008

ISBN 978-0-340-95760-8

First printed 2008
Impression number   7   6   5   4
Year   2013   2012   2011   2010   2009

This textbook has been written specifically to support students studying Edexcel AS Chemistry. The content has been neither approved nor endorsed by Edexcel and remains the sole responsibility of the author.

All efforts have been made to trace copyright on items used.
Front cover photograph reproduced by permission of Science Photo Library

Design by Juha Sorsa
Printed in Italy

Hachette UK's policy is to use papers that are natural, renewable and recyclable products and made from wood grown in sustainable forests. The logging and manufacturing processes are expected to conform to the environmental regulations of the country of origin.

Contents

Introduction

About this book

This textbook covers the Edexcel specification for AS chemistry. Generally, the order follows that of the specification except that atomic structure has been placed before energetics.

The first seven chapters describe and explain the material of **Unit 1: The core principles of chemistry**. The following nine chapters cover the material of **Unit 2: Application of core principles of chemistry**. There are slight changes to the specification order. For instance, the greenhouse effect is included with infrared spectra in Chapter 15. Mechanisms are incorporated into Chapter 14, which covers further organic chemistry. Chapter 17 is designed to help you prepare for internally assessed practical work. It should also help in answering practical based questions in Unit Tests 1 and 2.

Margin comments are provided throughout the book. These comprise valuable reminders and snippets of information and include examiner's tips (indicated by an ⓔ symbol), which clarify what you need to know and common sources of confusion.

Each chapter is preceded by information about the element that has the same atomic number as the chapter number. The intention is to give an insight into some of the interesting aspects of the chemistry of the elements, some of which are not covered by the AS specification.

This book is not a guide to the practical chemistry that all AS candidates will study. However, many of the reactions that will be met in the laboratory are detailed throughout the book.

At the back of the book (page 340), there is a periodic table that gives relative atomic masses to one decimal place. This should be referred to for atomic numbers, atomic masses and symbols of the elements. The table is similar to the one printed on the back of the examination papers.

Online resources
The online resources (**www.hodderplus/philipallan**) have three components:
- end-of-chapter multiple-choice tests for students to print off and complete
- practice exam-style unit tests
- exam-style questions with graded student answers, supported by examiner's comments, written by Rod Beavon, author of the Philip Allan Updates Edexcel AS Unit Guides

Required previous knowledge and skills

It is assumed that all AS chemistry students have covered at least Double Award Science to GCSE level. All students should be:

- familiar with the use of a calculator
- able to change the subject of an algebraic equation
- able to draw straight-line and curved graphs from supplied data and to extrapolate graphs
- confident in the use of scientific (standard) notation, e.g. that the number 1234 can be written as 1.234×10^3 and that 1.234×10^{-3} is the same as 0.001234

Scheme of assessment

Assessment objectives

There are three assessment objectives tested at AS:

- **AO1** Knowledge and understanding of science and of how science works — 38.5% of the AS mark.
- **AO2** Application of knowledge and understanding of science and of how science works — 38.5% of the AS mark.
- **AO3** How science works — 23% of the AS mark.

'How science works' is now an integral part of the course and the examination system. Do not expect questions such as 'Give an example of how science works'. The theme is part of the underlying chemistry in the specification and the exam questions. Examples of 'How science works' include:

- knowledge and understanding of the development of chemical theories — for example, atomic theory from Dalton to the Rutherford/Bohr theory of the atom
- applications of theory to modern-day use
- the use of instruments such as the mass spectrometer in industrial and research applications
- a question that starts with enthalpy of formation data and then asks for the amount of a chemical needed to heat up the contents of a can of self-heating soup
- comparison of lattice energy values obtained from Born–Haber data with those obtained from a theoretical model, leading to a realisation that ionic and covalent bonding are the two ends of a continuum

The unit tests

The AS course consists of three units. The material in Unit 1 is tested by a written examination of 1 hour 15 minutes. The exam is in two sections. Section A contains only objective test questions (multiple choice); Section B contains some short-answer questions and some questions requiring longer answers. There will also be questions on the analysis and evaluation of practical work.

Unit Test 2 follows a similar pattern, but contains a third section comprising one extended-answer question based on one or more contemporary issues. Quality of written communication is evaluated in this paper, but there are no marks set aside for use of good English. Rather, you will be judged on your ability to give a coherent and logical explanation of a piece of chemistry.

Unit 3 measures ability in the laboratory. Except for a few overseas candidates, this is assessed internally:

- Activity A is an assessment of general practical competence. This is measured over at least eight experiments, chosen by your centre. The evidence that the centre must provide is the date, field and title of each experiment.
- Activities B and C are qualitative observation and quantitative measurement respectively.
- Activity D is a preparation, which can be either inorganic or organic.

The marks for the three units are converted to 'uniform marks', the total of which determines the AS grade awarded.

Examination technique

Mark allocation

In both AS papers the marks for each part of the question are given in brackets. This is a much better guide as to how much to write than the number of dotted lines provided for the answer. If there are 2 marks, two statements must be made. For example, if the question asks for the conditions for a particular reaction and there are 2 marks available, there must be two different conditions given, such as solvent, temperature or catalyst.

Alternative answers

Do *not* give alternative answers. If one of them is wrong, the examiner will not award any marks for this part of the question. If both answers are correct, you *would* score the mark. However, there is no point in risking one answer being wrong. Beware also of contradictions, such as giving the reagent as concentrated sulfuric acid and then writing $H_2SO_4(aq)$ in an equation.

Writing your answers

In Edexcel AS chemistry exams the answers are written in the spaces on the question paper. If part of your answer is written elsewhere on the page, alert the examiner by writing, for example, 'see below' or 'continued on page 5'. Exam papers will be marked online, so question papers and answers will be electronically scanned. For this reason, it is essential *not to write outside the borders* marked on the page, unless you indicate that you have done so.

For the multiple-choice questions, you must put a cross in the box corresponding to your chosen answer. If you change your mind, put a horizontal line through the cross. Then, mark your new choice with a cross. The multiple-choice questions contain four possible answers, labelled A, B, C and D. Only one of these is the correct answer to the question. Be aware that some questions are expressed as negatives — for example, 'Which of the following species is **not** polar?'

Correction fluid and coloured pens

Do not use either of these. Mistakes should be crossed out neatly before writing the new answer. Red or other coloured inks will not show up as being different from black or blue ink when the paper is scanned ready for online marking. If you cross something out and then wish you had not done so, add the words 'please ignore crossing out'. The examiner will then read and mark the crossed-out material.

Command words

It is important that you respond correctly to key words or phrases in the question.

- **Define** — definitions of important terms such as relative atomic mass or standard enthalpy of formation are frequently asked for. You *must* know these definitions. They are printed in red in this book.
- **Name** — give the full name of the substance, *not* its formula.
- **Identify** — give either the name or the formula.
- **Write the formula** — a molecular formula, such as C_2H_5Cl, will suffice, as long as it is unambiguous. It is no use writing C_2H_4O for the formula of ethanal, or C_3H_7Br for the formula for 2-bromopropane. This also applies to equations. For example, the equation $C_2H_4 + Br_2 \rightarrow C_2H_4Br_2$ would not score a mark, because the formula $C_2H_4Br_2$ is ambiguous.
- **Draw or write the structural formula** — this must clearly show the bonding and the position of the functional group. All double bonds must be shown. For example, an acceptable structural formula of but-1-ene is $H_2C=CHCH_2CH_3$.
- **Draw the displayed formula or the full structural formula** — all the atoms and all the bonds in the molecule *must* be shown. For example, the displayed formula of ethanoic acid is:

- **State** — give the answer without any explanation. For example, if asked to state in which direction the position of equilibrium moves, the answer is simply 'to the left' or 'to the right'.
- **State, giving your reasons** — this is a difficult type of question. First, look at the mark allocation. Then state the answer (usually 1 mark unless the answer is one of only two possibilities — a so-called 50/50 question) and follow this with an explanation containing enough chemical points to score the remaining marks.
- **Explain or justify your answer** — look at the mark allocation and then give the same number of pieces of chemical explanation, or even one extra. For example, in answer to the question 'explain why but-2-ene has two geometric isomers (2)', the first point is that there is restricted rotation about the double bond and the second point is that there are two different groups on each double-bonded carbon atom.
- **Deduce** — the data supplied in the question, or an answer from a previous part of the question, are used to work out the answer. The data could be numerical or they could be the results of qualitative tests on an unknown substance.

Alternatively, knowledge from another part of the specification may be needed to answer a question about a related topic or similar substance.

- **Suggest** — candidates are not expected to know the answer. Knowledge and understanding of similar compounds have to be used to work out (deduce) the answer. For example, the shape of SF_6 is covered in the specification, so students should be able to deduce the shape of the PCl_6^- ion. Alternatively, the question might ask candidates to suggest the identity of an organic compound because there are not sufficient data to decide between two possible isomers.

- **Compare** *or* **explain the difference between** — valid points must be made about *both* substances. For example, if the question asks for an explanation for the difference in the boiling temperatures of hydrogen fluoride and hydrogen chloride, the different types and strengths of intermolecular forces in *both* substances must be described, together with an explanation of what causes these differences.

- **Calculate** — it is essential to show all working. For example, if the question asks for an empirical formula to be derived from % mass data, it must be clear to the examiner that the candidate has first divided by the relative atomic mass and then by the smallest answer. Work should always be set out so that if a mistake is made and the wrong answer calculated, the examiner can identify the mistake and award marks for the consequential steps in the calculation. An answer without working will score a maximum 1 mark. Always give your final answer to the number of significant figures justified by the number of significant figures in the data. Never round to one or two significant figures in the intermediate steps of a calculation.

- **Identify the reagent** — give the *full* name or formula. Answers such as 'acidified dichromate' or 'OH^- ions' may not score full marks. The name of a reagent is the name on the bottle.

- **State the conditions** — do not automatically write down 'heat under reflux'. The answer might be 'at room temperature' or you might be expected to know the necessary solvent (e.g. ethanol as a solvent for the elimination of HBr from bromoalkanes) or a specific catalyst (e.g. platinum or nickel in the addition of hydrogen to alkenes). If a concentrated acid is needed in a reaction, this must be stated. In the absence of any knowledge of the reaction, then try 'heat under reflux' — it might be correct!

Equations

- Equations must always be balanced. Word equations never score any marks.
- Ionic equations and half-equations must also balance for charge.
- State symbols must be included:
  - if the question asks for them
  - in all thermochemical equations
  - if a precipitate or a gas is produced
- The use of the symbols [O] and [H] in organic oxidation and reduction reactions, respectively, is acceptable. Equations using these symbols must still be properly balanced.
- Organic formulae used in equations must be written in such a way that their structures are unambiguous.

Stability

'Alkanes are stable' has no meaning. 'Stability' must only be used when comparing two states or two sets of compounds. Alkanes may be unreactive, but a mixture of methane and air is thermodynamically unstable. This means that methane and oxygen are unstable compared with their combustion products, carbon dioxide and water.

Graphs

Normally, there is a mark for labelling the axes. When sketching a graph, make sure that any numbers are on a linear scale. The graph should start at the right place, have the correct shape and end at the right place. An example is the Maxwell–Boltzmann distribution, which starts at the origin, rises in a curve to a maximum and tails off as an asymptote to the x-axis.

Diagrams of apparatus

Make sure that a flask and condenser are not drawn as one continuous piece of glassware. The apparatus must work. Be particularly careful when drawing a condenser. There must be an outlet to the air somewhere in the apparatus. In distillation, the top should be closed and the outlet should be at the end of the condenser. For heating under reflux, the top of the condenser must be open. It is always safer to draw an electrical heater, in case one of the reagents is flammable.

Doom and gloom

Avoid apocalyptic environmental predictions. The ozone layer is only damaged by CFCs, not by other chlorine compounds. Discarded plastic does not decimate all wildlife. Inorganic fertilisers do not wipe out all aquatic life.

Read the question

In one paper I was marking, the candidates had been asked to draw the isomer of 2-bromopropane. About half failed to read the question properly and drew the structure of 2-bromopropane, not its isomer.

Quality of written communication (QWC)

The most important thing is to convey the meaning clearly, accurately and in a logical order. Minor spelling errors do not matter, unless they cause ambiguity or imply a different group in a molecule. You could spell phenolphthalein in a variety of ways without being penalised, but phenylphthalein will not earn a mark because 'phenyl' is a specific group and is different from phenol.

George Facer

Chapter 1
Fundamental concepts

Hydrogen: fuel for space and the future

Atomic number:	1
Electron configuration:	$1s^1$
Symbol:	H
Isotopes:	^1H (99.985%), ^2H (deuterium 0.015%) and ^3H (tritium)
Abundance:	0.88% by mass in the Earth's crust

Hydrogen fuel sending the shuttle into space

COREL

The only product of the combustion of hydrogen is water. The problems of hydrogen as a fuel are its manufacture and its storage.

Manufacture

Hydrogen is obtained from methane (a non-renewable fossil fuel) or made by the electrolysis of water, which requires a large amount of electrical energy. This energy could be produced:

- from solar power — very inefficient at present
- by wind farms — energy costs of building these are high
- by nuclear power

Storage

Hydrogen is a gas that cannot be liquefied above −240°C. Even then, the liquid has a very low density — the energy released from the combustion of 1 cm³ of liquid hydrogen is four times less than that of petrol. However, the energy released *per gram* is higher than other fuels, so liquid hydrogen is used as the fuel in the booster rockets of the space shuttle.

Research is taking place into other methods of storing hydrogen. Many metals, particularly transition metals, absorb the gas. The gas is pumped into a tank containing the metal and the hydrogen is released when the metal is warmed. A problem here is the weight of the solid metal of the tank. Other absorption materials, such as alkaline-doped carbon nanotubes, are being tried. So far, no economic, safe storage material has been discovered.

Fuel cells

An internal combustion engine has a low efficiency — less than 30% of the chemical energy is converted into the work of propelling a car. A more efficient way is to make electricity using the reaction of hydrogen with oxygen. This is done in a fuel cell. The anode is made of porous carbon impregnated with nickel, which acts as a catalyst for the oxidation of hydrogen to H^+ ions. Oxygen from the air is reduced at the cathode and forms OH^- ions, which then react with the H^+ ions produced at the anode to form water.

Fundamental concepts

Chemistry is the study of matter and the changes that can be made to matter.

The term 'matter' means anything that has mass and takes up space. This includes metals, plastics and fertilisers, but not light or magnetism, or abstract ideas such as health or beauty.

Chemistry is concerned with the properties of matter — what happens when it is heated, when it is mixed with other substances, when electricity is passed through it — and the reasons for the changes that occur in the different circumstances.

Substances have two types of property:
- **Physical properties** are those that can be observed without changing the identity of the substance, for example by melting, compressing, magnetising or bending the substance. At the end of a physical change, the substance is still there, in a different shape or state. The physical states are solid, liquid and gas:
  - A solid has a fixed volume (at a given temperature) and shape.
  - A liquid flows and takes the shape of the part of the container that it fills. It has a fixed volume at a given temperature.
  - A gas completely fills the container in which it is placed. It can easily be compressed to fit in a container of a smaller volume.
- **Chemical properties** are those that are observed when the substance changes its identity, for example on the addition of acid or on burning in air. After a chemical change (chemical reaction) has taken place, one or more new substances are formed. Each substance has its own unique chemical properties.

Chemistry as a spectacle. (Inset) The Chinese characters for chemistry, which mean 'the study of change'.

Elements

An element is a substance that cannot be broken down into two or more different substances.

An atom is the smallest uncharged particle of an element.

A compound is made of two or more elements chemically joined together.

Elements are the building blocks from which all compounds are made.

The ancient Greeks thought that all matter was made from four elements — fire, earth, air and water. As scientists began to experiment and apply scientific method to analysing results, this idea was gradually abandoned. The big step forward came in 1869 when Dmitri Mendeléev arranged the elements in order of atomic mass and produced a periodic table similar to that in use today.

One hundred and twelve elements have been positively identified, although only 90 occur naturally on earth. The others have either been made in nuclear reactors or been found in the debris from a nuclear explosion.

All the elements originally came from the stars. Inside a star, including our sun, hydrogen atoms are converted into atoms of other elements by **nuclear fusion**. A small amount of mass is lost and this is turned into energy — the sunlight and starlight that we can see. The creation of elements from hydrogen in the stars is taking place continually. However, as a star ages, its outer layer begins to collapse, which causes the temperature of the star to increase further. Some of the atoms of elements already created fuse together and form still heavier atoms, such as uranium, lead and gold. The temperature increase causes a huge explosion and the outer layers are thrown into space. This is what happens in a supernova. The debris from supernovae will eventually condense under the gravity of another star and form planets. Thus the elements formed in the stars form the rocks, the seas and even our own bodies (Table 1.1).

SCIENCE PHOTO LIBRARY

Dmitri Mendeléev

Earth's crust			Human body		
Rank	Element	% by mass	Rank	Element	% by mass
1st	Oxygen	49.5	1st	Oxygen	65
2nd	Silicon	25.7	2nd	Carbon	18
3rd	Aluminium	7.5	3rd	Hydrogen	10
4th	Iron	4.7	4th	Nitrogen	3
5th	Calcium	3.4	5th	Calcium	1.5
6th	Sodium	2.6	6th	Phosphorus	1.2
7th	Potassium	2.4	7th	Potassium	0.2
8th	Magnesium	1.9	8th	Sulfur	0.2
9th	Hydrogen	0.88	9th	Chlorine	0.2
10th	Titanium	0.58	10th	Sodium	0.1
11th	Chlorine	0.19	11th	Magnesium	0.05
12th	Phosphorus	0.12	12th	Iron	0.04

Table 1.1 The abundance by mass of the most common elements in the Earth's crust and in the human body

The periodic table

The modern form of the periodic table has the elements arranged in atomic number order (in order of the number of protons in the nucleus).

It is shown in Table 1.2 and also, in more detail, at the back of this book (page 340). It is printed on the back page of the question booklet in all AS and A2 examinations.

	1	2											3	4	5	6	7	0
1	H																	He
2	Li	Be											B	C	N	O	F	Ne
3	Na	Mg											Al	Si	P	S	Cl	Ar
4	K	Ca	Sc	Ti	V	Cr	Mn	Fe	Co	Ni	Cu	Zn	Ga	Ge	As	Se	Br	Kr
5	Rb	Sr	Y	Zr	Nb	Mo	Tc	Ru	Rh	Pd	Ag	Cd	In	Sn	Sb	Te	I	Xe
6	Cs	Ba	La*	Hf	Ta	W	Re	Os	Ir	Pt	Au	Hg	Tl	Pb	Bi	Po	At	Rn
7	Fr	Ra	Ac*	Rf														

Table 1.2 The periodic table

*There are 14 f-block rare-earth elements between lanthanum and hafnium, and 14 f-block actinide elements between actinium and rutherfordium.

- The vertical columns are called groups. The numbers in brown are the group numbers.
- The elements in a group, such as in groups 1, 2 or 7, show similar, but steadily changing, physical and chemical properties. In group 7, for example, fluorine and chlorine are gases; bromine is a liquid and iodine is a solid. All four react with hydrogen to form acids.
- The horizontal rows are called periods. The numbers in violet are the period numbers.
- The elements in a period change from metals on the left to non-metals on the right. The noble gases are in group 0.
  - The elements in red are metals.
  - The elements in green are metalloids (semi-metals).
  - The elements in blue are non-metals.

Important points about the periodic table

Metals
The metals are on the left and in the middle of the table.

The physical properties of metals are that:
- they conduct electricity when solid and when liquid
- they are malleable and ductile
- apart from mercury, they are all solids at room temperature

e You need to know the names and symbols of at least the first 20 elements (hydrogen to calcium). You must also know to which group each belongs.

ANDREW LAMBERT PHOTOGRAPHY/SCIENCE PHOTO LIBRARY

Three well-known metals: (a) copper, (b) mercury and (c) magnesium

The chemical properties of metals are that:

- they form positive ions (**cations**) in their compounds; for example, sodium forms Na^+ ions and magnesium forms Mg^{2+} ions
- the more reactive metals react with an acid to give a salt and hydrogen; for example, magnesium reacts with dilute sulfuric acid to form magnesium sulfate and hydrogen gas
- the most reactive metals react with water to give a hydroxide and hydrogen; for example, sodium reacts with water to form sodium hydroxide and hydrogen gas
- they react with cations of less reactive metals in displacement reactions; for example, zinc reacts with a solution of copper sulfate to form zinc sulfate and a residue of copper metal
- they become more reactive *down* a group in the periodic table; for example, potassium is more reactive than sodium
- their oxides and hydroxides are bases; for example, magnesium oxide is a base and reacts with nitric acid to form magnesium nitrate and water

Non-metals

Non-metals are on the right of the table.

The chemical properties of non-metals are that:

- they form negatively charged ions (**anions**) in many compounds with metals; for example, chlorine forms Cl^- ions and oxygen forms O^{2-} ions
- they form covalent bonds with other non-metals
- apart from in group 0, they become more reactive *up* a group in the periodic table; for example, chlorine is more reactive than bromine, which is more reactive than iodine

Organic chemistry is about compounds in which carbon is covalently bonded to other carbon atoms, and to other elements such as hydrogen, oxygen and nitrogen.

- their oxides are acidic, reacting with water to form a solution of an acid; for example, sulfur dioxide reacts with water to form sulfurous acid, H_2SO_3, which ionises to form $H^+(aq)$ ions, making the solution acidic

Metalloids

Between the metals and non-metals there are some elements that do not fit easily into either category. These elements are called semi-metals or metalloids. The properties of metalloids are that:

- they do not form ions
- they are semiconductors of electricity, particularly when a small amount of impurity is added
- their oxides are weakly acidic; for example, silicon dioxide reacts with the strong alkali sodium hydroxide to form sodium silicate

Silicon is used as a semiconductor in computers.

Force and energy

Force

There are four types of force:

- strong nuclear
- weak nuclear
- gravitational
- electromagnetic

Strong and weak nuclear forces are only important inside the nucleus of an atom. Gravitational forces are so weak that they are only effective if at least one of the objects is astronomically large, such as the sun or the Earth. The only forces that affect chemistry are the electromagnetic forces.

A positively charged object attracts a negatively charged object; there is a force of repulsion between two objects that have the same charge.

The strength of the force depends on:

- the size of the charges — the *bigger* the charges, the stronger is the force acting between them
- the distance between the centres of the two objects — the smaller the distance between the centres, the stronger is the force between them

In chemistry, the most common charged particles are ions.

An ion is a charged atom or group of atoms.

- A **cation** is an atom or group of atoms that has lost one or more electrons and so is *positively* charged. Metals form cations. Na^+ and Mg^{2+} are examples.
- An **anion** is an atom or group of atoms that has gained one or more electrons and so is *negatively* charged. Non-metals form anions. Cl^-, OH^- and O^{2-} are examples.

e Remember this as 'unlike charges attract; like charges repel'.

> *e* Remember that metal ions and the ammonium ion, NH_4^+, are positively charged. Anions, such as chloride and sulfate, are negatively charged. The nucleus contains positively charged protons, which are surrounded by negatively charged electrons.

Energy

If there is a force of attraction between two particles, energy has to be supplied to separate them. The amount of energy required depends on the strength of the force between the particles, which in turn depends upon the size of the charges and how close the two centres are. This means that more energy is required to separate small, highly charged ions than less highly charged or bigger ions.

Conversely, if two oppositely charged particles are brought closer together, energy is released.

Energy can neither be created nor destroyed. However, one type of energy can be converted into other types. For instance, in a car engine, the chemical energy lost by the combustion of petrol is converted into the kinetic energy of the car moving along the road, the potential energy gained if the car climbs a hill, and heat and sound energy.

In an **exothermic** reaction, chemical energy is converted to heat energy. This means that the temperature increases as heat is produced. The combustion of petrol is an example of an **exothermic** reaction.

If heat is absorbed, the chemical reaction is **endothermic**, and the temperature falls as heat energy is converted into chemical energy.

Chemical energy is measured as a quantity called **enthalpy**, H. In an exothermic reaction, chemical energy is lost as it is changed into heat energy. Therefore, the value for the *change* in enthalpy, ΔH, is negative. In an endothermic reaction, the value of ΔH is positive.

◀ The unit of energy is the joule, J. However, this is so small that chemists normally use the kilojoule, kJ.

◀ The Greek letter Δ means a change, so ΔH means a change in the quantity H (enthalpy or chemical energy available as heat).

Amount of substance

When we go shopping we buy cheese by weight, drinks by volume and cans of baked beans by number. Scientists measure substances by mass or volume.

Mass

The unit of mass normally used is the gram. However, other mass units, such as the kilogram, kg, milligram, mg, nanogram, ng, or tonne are sometimes used.

$1\,g = 1000\,mg$ or $1\,mg = 10^{-3}\,g$

$1\,g = 1\,000\,000\,000\,ng$ or $1\,ng = 10^{-9}\,g$

$1\,kg = 1000\,g$

$1\,tonne = 1000\,kg$

Volume

The common units of volume are the cm^3 and the bigger unit dm^3 (also called the litre, l).

$1\,cm^3 = 0.001\,dm^3$

$1\,dm^3 = 1000\,cm^3$

Worked example 1
A liquid has a volume of 23.75 cm^3. Convert this to a volume in dm^3.

Answer

$$\text{volume} = \frac{23.75}{1000} = 0.02375 \, \text{dm}^3$$

e Remember that a volume in cm^3 converts into a much smaller volume in dm^3.

Worked example 2
A gas has a volume of 24 dm^3. Convert this into a volume in cm^3.

Answer
$$\text{volume} = 24 \times 1000 = 24\,000 \, \text{cm}^3$$

e Remember that a volume in dm^3 converts into a much larger volume in cm^3.

The mole

Although substances are measured out according to their mass or volume, chemicals react by numbers of particles.

The equation for the formation of water from hydrogen and oxygen is:
$$2H_2 + O_2 \rightarrow 2H_2O$$

This means that hydrogen and oxygen react in the ratio of two molecules of hydrogen to one molecule of oxygen.

Atoms and molecules are so small that when two chemicals are mixed and react, billions and billions of molecules are involved.

The mass of one molecule of water is 2.99×10^{-23} g and in 1 g of water there are 3.34×10^{22} molecules, or 33 400 000 000 000 000 000 000 molecules. Counting in such large numbers is almost impossible. To get round this problem the concept of the **mole** was introduced. The mole is just a very large number of particles.

◀ The symbol for the mole is mol.

The mole is defined as the number of carbon atoms in exactly 12 g of the carbon-12 isotope. This number has been calculated to be 6.02×10^{23} carbon atoms. It is called the **Avogadro constant** and has the symbol, L or N_A.

Avogadro constant $= 6.02 \times 10^{23}$ mol^{-1}

One mole is the amount of substance containing the Avogadro number of atoms, molecules or groups of ions.

For example:
- 1 mol of sodium, Na, contains 6.02×10^{23} sodium atoms.
- 1 mol of water, H_2O, contains 6.02×10^{23} water molecules.
- 1 mol of sodium chloride, Na^+Cl^-, contains 6.02×10^{23} pairs of Na^+ and Cl^- ions.

Chemicals react in a ratio by moles. The energy released in an exothermic reaction depends on the number of moles that react. Therefore, ΔH has units of kJ mol^{-1}. An understanding of the mole is essential to A-level chemistry.

Mole ratios

Consider the equation:

$$H_2SO_4 + 2NaOH \rightarrow Na_2SO_4 + 2H_2O$$

Sulfuric acid and sodium hydroxide react in the **molar ratio** of 1:2. This means that 0.111 mol of sulfuric acid reacts with 0.222 mol of sodium hydroxide. The equation also shows that 0.111 mol of sodium sulfate and 0.222 mol of water are formed.

Worked example 1

Calculate the number of moles of silver nitrate that react with 1.23 mol of sodium chloride, according to the equation:

$$AgNO_3 + NaCl \rightarrow AgCl + NaNO_3$$

Answer

moles of $AgNO_3$:moles of NaCl = 1:1
moles of silver nitrate = moles of sodium chloride
$\qquad\qquad = 1.23$ mol

Worked example 2

Calculate the number of moles of phosphoric acid, H_3PO_4, that react with 0.0132 mol of potassium hydroxide, according to the equation:

$$H_3PO_4 + 3KOH \rightarrow K_3PO_4 + 3H_2O$$

Answer

moles of H_3PO_4:moles of KOH = 1:3
moles of H_3PO_4 = $\frac{1}{3} \times$ moles of KOH
$\qquad = \frac{1}{3} \times 0.0132$
$\qquad = 0.0044$ mol

Worked example 3

Calculate the number of moles of hydrogen gas, H_2, formed when 0.0246 mol of aluminium reacts with excess sulfuric acid, according to the equation:

$$2Al + 3H_2SO_4 \rightarrow Al_2(SO_4)_3 + 3H_2$$

Answer

moles of H_2:moles of Al = 3:2
moles of H_2 = $\frac{3}{2} \times$ moles of Al
$\qquad = \frac{3}{2} \times 0.0246$
$\qquad = 0.0369$ mol

The mole concept is further developed in Chapter 3.

Significant figures

In all calculations, the answer should be given to an appropriate number of significant figures.

e Remember that the number of moles is a measure of the number of particles, and that mole ratios are found from the balanced chemical equation for the reaction.

The method of working out the number of significant figures is based on these rules:

■ For whole numbers and decimals of numbers *greater* than 1, all the figures, *including any zeros at the end*, are counted. For example:
   – the number 102 is written to three significant figures (s.f.)
   – the number 1.02 is also written to 3 s.f.
   – the number 1.020 is written to 4 s.f.
   – the number 10.20 is also written to 4 s.f.
   – the measurement 100 cm^3 is expressed to 3 s.f.

> **e** Do not confuse decimal places with significant figures. The number of decimal places is determined by the number of digits (including all zeros) that occur *after* the decimal point. A numerical value written to three decimal places has three numbers after the decimal point.
>
> ■ 0.123 is written to three decimal places, and is also written to three significant figures.
> ■ 1.23 is written to two decimal places, expressed to three significant figures.
> ■ 0.0123 is written to four decimal places, expressed to three significant figures.
> ■ 0.010 is written to three decimal places, but only two significant figures.

■ For decimals in numbers *less* than 1, count all the numbers, except the zeros *before* and *immediately after* the decimal point. For example:
   – the number 0.202 is written to 3 s.f.
   – the number 0.2020 is written to 4 s.f.
   – the number 0.00202 is written to 3 s.f., not 5 s.f.
   – the number 0.02200 is written to 4 s.f.

■ For values written in scientific notation, all the numbers before and after the decimal place are counted as significant figures. The power of 10 is not counted. For example, the number 1.02×10^{-3} is written to 3 s.f.

Rounding numbers

Questions often ask for numbers to be rounded to a certain number of significant figures.

If you have to round a number to *three* significant figures, look at the *fourth* significant figure. If it is less than 5, do not alter the third significant figure. For example, 6.372 becomes 6.37, because the fourth significant figure, 2, is less than 5.

If the fourth significant figure is 5 or more, increase the third significant figure by 1. For example, 6.379 becomes 6.38, because the fourth significant figure, 9, is greater than 5. The number 2.365 is rounded up to 2.37, because the fourth significant figure is 5.

Do not round up in stages. For example, 6.3749 rounds to 6.37, because the *fourth* figure is 4, which is less than 5. Do not round 6.3749 to 6.375 and then to 6.38.

Questions

1 Identify the element in group 5 and period 3 of the periodic table.

2 Identify the most reactive metal in group 2 of the periodic table.

3 Name two elements that are liquids at room temperature.

4 An element X loses three electrons. What is the charge on the ion formed?

5 Convert 12 kg into grams.

6 Convert 1.2×10^6 g into tonnes.

7 Convert 0.0234 g into milligrams (mg).

8 Convert 22.4 dm^3 into a volume in cm^3.

9 Convert a volume of 23.7 cm^3 into a volume in dm^3.

10 Calculate:
   a the number of atoms in 1 mol of CO_2
   b the number of chloride ions in 1 mol of calcium chloride, $CaCl_2$
   (The Avogadro constant, $N_A = 6.02 \times 10^{23}$ mol^{-1})

11 Nitrogen reacts with hydrogen to form ammonia, according to the equation:
   $$N_2 + 3H_2 \rightarrow 2NH_3$$
   Calculate:
   a the number of moles of nitrogen, N_2, that react with 0.246 mol of hydrogen, H_2
   b the number of moles of ammonia produced

12 Nitric acid reacts with calcium hydroxide according to the equation:
   $$Ca(OH)_2 + 2HNO_3 \rightarrow Ca(NO_3)_2 + 2H_2O$$
   Calculate the number of moles of calcium hydroxide that react with 0.0642 mol of nitric acid.

13 Write down the number of significant figures in the following:
   a 1.202   c 0.002220
   b 3.30   d 2.34×10^{-3}

14 Round the following to three significant figures:
   a 2.3447   c 4.872499
   b 0.04375   d 1.524×10^{-5}

Chapter 2
Atomic structure and the periodic table

Helium: balloons, diving and superconductors

Atomic number: 2

Electron configuration: $1s^2$

Symbol: He

The word helium comes from the Greek word *helios* (the sun). It was so-named in 1868 when dark lines in the spectrum of light from the sun showed the existence of this new element. Twenty-seven years later, Sir William Ramsey isolated it from uranium ore.

Occurrence

Helium is thought to be the second most abundant element in the universe, hydrogen being the most abundant. It is chemically completely inert, but is too light to remain in the Earth's atmosphere. Radioisotopes, such as uranium-238, steadily emit alpha-rays, which are helium nuclei. These pick up electrons and become helium atoms. Helium is found in natural gas deposits, mainly in the USA, Canada and the Sahara.

Balloons

Helium has the second lowest density of all gases. It does not burn, so is used to fill balloons and airships. Airships were originally filled with hydrogen, which has half the density of helium, but burns in air. In 1937, the German airship *Hindenburg* was destroyed in a spectacular fire. It was filled with hydrogen and a spark caused by a collision with its mooring gantry set it alight with the loss of many lives.

Helox

Divers take a supply of compressed air with them. If a diver stays below 5 metres, the nitrogen in the pressurised air starts to dissolve in the blood. If the diver comes to the surface too quickly, nitrogen comes out of solution and small bubbles of the gas are released into the bloodstream. This is called the 'bends', as the pain can make the diver double up in agony. Severe cases can cause death. Also, when the pressure exceeds about 6 atm, nitrogen becomes poisonous. The diver suffers from nitrogen narcosis, which can be fatal.

Commercial deep-sea divers breathe a mixture of helium and oxygen, called helox, and so can live and work at extreme depths for several days without suffering from nitrogen narcosis. The low density of this gaseous mixture causes the divers to speak in a squeaky high-pitched voice.

Superconductors

Many metals lose their electrical resistance at very low temperatures. Liquid helium, which has a boiling point of 4 K (–269°C), is used to cool the metals in the very powerful electromagnets that are used in MRI (magnetic resonance imaging) machines in hospitals.

Modern airships are filled with helium

COREL

Atomic structure and the periodic table

Dalton's atomic theory

The father of modern atomic theory was John Dalton, a schoolteacher who was born in 1776.

He used measurements of the masses in which elements combine to propose his hypothesis, which was that:

- all matter is made of atoms
- all atoms of a given element are identical, with the same mass, but have different masses from the atoms of other elements
- a compound is a combination of the atoms of two or more elements in a specific ratio
- atoms can neither be created nor destroyed
- in a chemical reaction, atoms in the reactants are rearranged to give the products of the reaction

John Dalton

Dalton thought of an atom as a spherical hard object, rather like a marble. The word atom is derived from Greek and means 'uncuttable'. Dalton's theory formed the basis for understanding chemical reactions, formulae and chemical equations and allowed huge advances to be made in the nineteenth century. The needs of the industrial revolution led to advances in industrial chemistry. New manufacturing processes for the production of, for example, dyes, bleaches, acids and alkalis were developed.

The nuclear atom

Doubts about the accuracy of Dalton's atomic theory began to form around the beginning of the twentieth century. J. J. Thomson discovered that all metals, when heated, give off identical, tiny, negatively charged particles and he was able to measure their mass-to-charge ratio (m/e). Millikan measured the charge on these particles and hence the mass. He found that they had a mass nearly two thousand times smaller than that of a hydrogen atom. These particles are called **electrons**.

If atoms contain negatively charged electrons, they must also contain positively charged particles. The first idea was that the electrons were somehow embedded in the positive matter of an atom. The discovery of radioactivity by Becquerel and the separation of radium from the uranium ore pitchblende by Marie Curie led to the knowledge that alpha particles are helium atoms that have lost two electrons. Geiger and Marsden, working on suggestions by Rutherford, bombarded gold foil with alpha particles. Their results led Rutherford to propose the nuclear theory of the atom:

- An atom contains a small, central, positively charged nucleus.
- The diameter of the nucleus is about $\frac{1}{10\,000}$ that of the atom. This means that if an atom were the size of a football, the nucleus would be smaller than a full stop on this page. Alternatively, if the nucleus were the size of a golf ball, the atom would have a diameter of about 400 metres.
- Almost all the mass of the atom is concentrated in the nucleus.
- The electrons orbit the nucleus.

Aston invented the **mass spectrometer** to measure the masses of atoms precisely. He made the startling discovery that not all the atoms of an element have the same mass. In a sample of neon, 91% of the atoms have a mass of 3.32×10^{-23} g, which is 20 times heavier than a hydrogen atom. However, there are atoms of neon with masses 21 and 22 times heavier than hydrogen atoms. Other elements were also found to have atoms of different mass but identical chemical properties. These different atoms of the same element are called **isotopes**.

This immediately caused a problem with the simple nuclear theory. How could the positive centre of one atom of an element differ from that of another atom of the same element?

There was a further long-standing problem associated with the periodic table. This is an arrangement of the elements in increasing atomic mass devised by Mendeléev. Argon has a relative atomic mass of 40, which is one greater than that of potassium. This means that potassium, on the basis of mass only, should be placed in group 0 with the noble gases, and argon in group 1 with the alkali metals. A similar problem arises with tellurium, Te, and iodine. This is clearly absurd. No-one could solve this riddle, until Moseley devised an experiment that measured the positive charge in the nucleus. This positive charge is called the atomic number. A simultaneous discovery of the proton led to the theory that the atomic number is the number of protons in the nucleus. This value is the same for all isotopes of an element. Thus the elements in the periodic table used today are arranged in order of *increasing atomic number* and not according to their atomic mass.

A third subatomic particle was then suggested — this is the **neutron**.

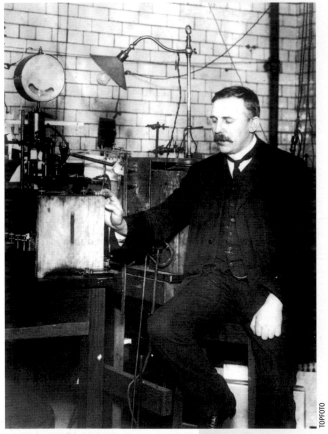

Ernest Rutherford

The final Rutherford atomic theory is as follows:

- The nucleus of an atom of a particular element contains a fixed number of positively charged protons. This number is called the **atomic number**.
- The nucleus also contains a number of neutrons. The number varies from one isotope to another. Neutrons are not charged (neutral).
- The **mass number** of an isotope is the sum of the number of protons and neutrons in the nucleus of an atom of that isotope.
- The number of electrons in a neutral atom of an element is equal to the number of protons in the nucleus.
- The electrons orbit the nucleus.

The masses and charges of these three subatomic particles are shown in Table 2.1.

Particle	Symbol	Charge/C	Relative charge	Mass/g	Relative mass
Proton	p	$+1.60 \times 10^{-19}$	+1	1.67×10^{-24}	1
Neutron	n	0	0	1.67×10^{-24}	1
Electron	e^-	-1.60×10^{-19}	-1	9.11×10^{-28}	0.00055 (which is usually ignored)

Table 2.1 Masses and charges of subatomic particles

During the last 50 years, many more subatomic particles, such as mesons and neutrinos, have been discovered. However, this is beyond the scope of A-level chemistry.

Never state that the mass of an electron is zero. You do not need to know the charges on subatomic particles in coulombs or their masses in grams. It is the *relative* values that you must know.

Definitions

The atomic number (Z) of an element is the number of protons in the nucleus of an atom of that element.

All atoms and ions of a given element have the same atomic number, which is different from the atomic numbers of other elements.

The relative atomic mass (r.a.m.) of an element is the weighted average mass of an atom of that element divided by $\frac{1}{12}$ the mass of a carbon-12 atom.

These values were often approximated to whole numbers, e.g. the r.a.m. of magnesium was quoted as 24, but to three significant figures it is 24.3. The relative atomic masses in the periodic table used by Edexcel are given to one decimal place. The mass of an atom of the carbon-12 isotope is defined as being exactly 12 atomic mass units and is the reference used for all relative atomic and isotopic masses.

The relative molecular mass of an element (such as O_2) or a compound (such as H_2O) is the average mass of a molecule of that element or compound divided by $\frac{1}{12}$ the mass of a carbon-12 atom.

The mass number of an isotope of an element is the sum of the number of protons and neutrons in the nucleus of an atom of that isotope.

Mass numbers apply only to *isotopes* and are always whole numbers.

The development of the understanding of the atom, from Dalton to Rutherford and beyond, is an example of how science works. A theory is suggested, based on experimental evidence, and that theory is modified steadily as more experimental evidence is obtained.

The relative isotopic mass is the mass of an atom of that isotope divided by $\frac{1}{12}$ the mass of a carbon-12 atom.

Isotopes are atoms of the same element that have the same number of protons but different numbers of neutrons.

Isotopes apply to a single element. They have the same number of protons in the nucleus, so they have the same atomic number. However, they have different numbers of neutrons in the nucleus, so the mass numbers are different. Isotopes of the same element have identical chemical properties because they have identical electron configurations.

Most elements have more than one stable and naturally occurring isotope. Some of these are shown in Table 2.2.

e Do not confuse relative atomic mass, which is an *average*, with relative isotopic mass, which refers to a single type of atom.

Element	Name of isotope	Symbol	Atomic number	Number of neutrons	Mass number	Relative isotopic mass	%
Hydrogen	Protium	^1H	1	0	1	1.008	99.99
	Deuterium	^2H	1	1	2	2.014	0.01
Lithium	Lithium-6	^6Li	3	3	6	6.015	7.42
	Lithium-7	^7Li	3	4	7	7.016	92.58
Carbon	Carbon-12	^{12}C	6	6	12	12 exactly	98.89
	Carbon-13	^{13}C	6	7	13	13.003	1.11
	Carbon-14	^{14}C	6	8	14	14.003	<0.01

Table 2.2 Some naturally occurring isotopes

Mass spectrometer

The masses of atoms, molecules and fragments of molecules can be measured using a mass spectrometer.

1 All the air in the spectrometer is first pumped out.

2 A sample of the element or compound in gaseous form is injected into the mass spectrometer (Figure 2.1, compartment A) and is bombarded with high-energy electrons from an electron gun. If the substance is a solid, it has to be heated to produce a sufficient number of gaseous particles. The energy from the bombarding electrons strips electrons from the atoms or molecules, forming positive ions:

$$A(g) + energy \rightarrow A^+(g) + e^- \quad \text{where A represents an atom}$$
$$M(g) + energy \rightarrow M^+(g) + e^- \quad \text{where M represents a molecule}$$

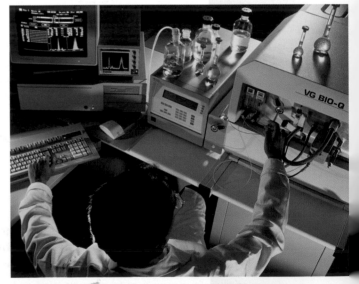

A mass spectrometer

The positive ions formed from molecules can then fragment into two or more particles, one positive and the other neutral. For instance, from chlorine:

$$Cl_2(g) \rightarrow Cl_2^+(g) \rightarrow Cl^+(g) + Cl(g)$$

and from ethanol:

$$C_2H_5OH(g) \rightarrow C_2H_5OH^+(g) \rightarrow C_2H_5^+(g) + OH(g)$$
$$C_2H_5OH(g) \rightarrow C_2H_5OH^+(g) \rightarrow C_2H_5(g) + OH^+(g)$$

3 The positive ions are accelerated by the high electric potential (Figure 2.1, compartment A), pass through slits and emerge as a parallel beam of ions. All the ions have the same energy, so those of a particular mass have the same speed.

4 The ions are then deflected by a powerful magnetic field (Figure 2.1, compartment B).Those with a greater mass are deflected less than those of a smaller mass. Therefore, the ions are sorted according to mass. The angle deflected depends upon the strength of the magnetic field and the mass-to-charge ratio (m/e) of the ion.

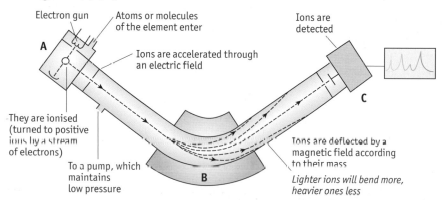

Figure 2.1
A representation of a mass spectrometer

5 The ions of lowest mass are detected first (Figure 2.1, compartment C). The magnetic field is gradually increased and the ions of greater mass are then detected. The detector is coupled to a computer that calculates the m/e ratio of each positive ion and its relative abundance compared with the most abundant particle. This is then converted into a percentage for each positive ion.

To summarise, the stages in mass spectrometry are:
- **bombardment** by high-energy electrons to create positive ions
- **acceleration** of positive ions by an electric potential
- **deflection** of positive ions by a magnetic field
- **detection** of positive ions

ⓔ Remember that these positive ions could be positively charged atoms, positively charged molecules or positively charged fragments of molecules.

Uses of mass spectrometers

Determination of relative atomic mass of an element
Data that have been obtained from a mass spectrometer, such as those in Figure 2.2, can be used to calculate the relative atomic mass of an element.

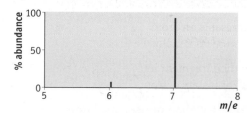

m/e	%
6.015	7.42
7.016	92.58

Figure 2.2 The signal and data from a mass spectrum of lithium

Worked example 1

Lithium has two isotopes, ^6Li and ^7Li. Use the data in Figure 2.2 to calculate the relative atomic mass of lithium.

Answer

$$r.a.m. = \frac{\left(m/e\ ^6Li \times \%\,^6Li\right) + \left(m/e\ ^7Li \times \%\,^7Li\right)}{100}$$

$$= \frac{\left(6.015 \times 7.42\right) + \left(7.016 \times 92.58\right)}{100}$$

$$= 6.94$$

Worked example 2

Phosphorus, relative atomic mass 31.0, consists of one type of atom only. Solid phosphorus has the formula P_4. A mass spectrum of phosphorus showed peaks at 124, 62 and 31. Identify the species that caused these peaks.

Answer

The peak at 124 is caused by the molecular ion $^{31}P_4{}^+$, the peak at 62 by the fragment $^{31}P_2{}^+$ and the peak at 31 by the fragment $^{31}P^+$.

ⓔ Do not forget that the species detected are positively charged.

Detection of drugs in urine samples

Most drugs, for example the anabolic steroid nandrolone, have a unique molar mass. A urine sample is taken from an athlete, placed in a low-energy mass spectrometer and a pulse of energy from a laser is fired in. This vaporises the solution and the large steroid molecules become charged. The spectrum is then analysed. If there is an m/e peak equivalent to that of the molecular ion of nandrolone, then the athlete is shown to have used that steroid.

In the pharmaceutical industry

The identity of a new compound can be determined by measuring the m/e value of the peak caused by the molecular ion. To prevent the decomposition of the fragile large molecule, a special low-energy mass spectrometer is used. Either a laser is used (as described above) or a solution of the substance is sprayed through a needle at high electric potential. A fog of charged particles is produced and these are analysed in the mass spectrometer for their m/e values.

Carbon-14 dating

The relative amounts of the isotopes carbon-12 and carbon-14 in a sample can be found using a mass spectrometer. Carbon-14 is radioactive and its activity halves every 5730 years as it changes to nitrogen-14.

When cosmic rays collide with nuclei in the Earth's upper atmosphere, neutrons are produced. If a neutron collides with a nitrogen nucleus, an atom of carbon-14 and a proton are formed. The ratio of carbon-14 to carbon-12 remained fairly constant until the Industrial Revolution, when 'old' carbon in coal was burnt releasing carbon dioxide with a very low carbon-14 content. The difference between the historical ratio and that measured in the mass spectrum of a carbon-containing object is the basis of carbon-14 dating.

Some rocks can be dated by a similar method. Uranium-238 decays eventually to lead-206. If the ratio of these two isotopes is 1:1, it means that half the uranium has decayed. As its half-life is about 4×10^9 years, this would mean that the rock was formed 4×10^9 years ago.

Electron configuration

Rutherford tried to explain the arrangements of electrons around the nucleus in terms of classical physics — Newton's laws of motion — but this did not work. An orbiting electron should spiral towards the nucleus and become absorbed by it and it was calculated that an atom would exist for less than a nanosecond! This problem was solved by quantum theory. When it was proposed by Max Planck in 1900, quantum theory shook the foundations of the scientific world. However, it became gradually accepted over the next few years. Planck suggested that energy was not continuous, but came in tiny packets called 'quanta'. In terms of electron config-uration, this means that an electron can only have certain discrete levels of energy. For instance, in a hydrogen atom, an electron can have specific energies, such as 5.45×10^{-19} J or 2.43×10^{-19} J, but not any value in between.

Carbon-14 dating suggests that the Turin shroud is only around 500 years old

Niels Bohr understood the importance of quantum theory and used it to explain the electromagnetic radiation produced when elements are excited when heated to high temperatures or by being placed in discharge tubes.

When hydrogen is energised in a discharge tube, spectral lines of different frequencies are emitted. One series, the Lyman series, is found in the ultraviolet region and another, the Balmer series, in the visible part of the electromagnetic spectrum. Each series has lines with frequencies that converge on a single value.

The spectra of other elements and their compounds are more complex. The visible spectrum of energised lithium or sodium vapour contains many lines; that of sodium also includes a pair of lines very close together (a doublet).

Bohr suggested that the electron in hydrogen can exist in definite circular **orbits** around the nucleus. The first orbit is the closest to the nucleus, the second is further away and so on. The first orbit has the lowest energy. The electron can only have the energy associated with a particular orbit, which means that an electron can be in a certain orbit or in a different orbit but not in between the two. These orbits are sometimes called **quantum shells**.

A quantum shell defines the energy level of the electrons. All electrons in a given quantum shell have similar, but not identical energies and electrons in the first quantum shell are in the lowest energy level for that element. A simplified view of a quantum shell is that it is the region around the nucleus in which an electron is found. The first quantum shell or orbit is in the region closest to the nucleus.

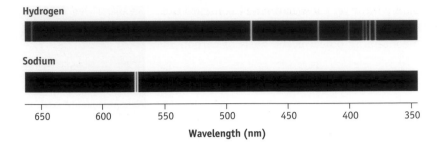

Hydrogen

Sodium

| 650 | 600 | 550 | 500 | 450 | 400 | 350 |

Wavelength (nm)

Emission spectra of hydrogen and sodium

When hydrogen atoms are heated, electrons are promoted from the '**ground state**' (their most stable or lowest energy state) to a higher, excited state. In terms of electron orbits, this means that the electron is promoted from the first orbit to an outer orbit. This is not a stable state and the electron drops back, giving out energy in the form of light. The energy given out is the difference between the energy of the electron in its outer orbit and the energy of the orbit into which it drops.

- The Lyman series is obtained when an electron in a hydrogen atom drops from an excited state back to the ground state, which is when it is in the first orbit.
- The Balmer series is caused by the electron dropping back to the second orbit.

The energy levels in a hydrogen atom and the transitions that cause the spectral lines are shown in Figure 2.3. The orbit number (principal quantum number) is represented by n.

Unfortunately, this simple picture did not explain the complex spectra of other elements. The next step was to suggest that the orbits could be divided into sub-orbits consisting of different **orbitals.**

The **first orbit** is not divided. Its orbital is spherical and is designated by 1s, where 1 means the first orbit and the s means that it is an s-type orbital.

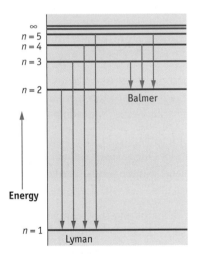

Figure 2.3 The energy transitions in a hydrogen atom

The **second orbit** is divided into two types of subshell, s and p. The 2s-orbital is the same shape as the 1s-orbital but has a larger radius (Figure 2.4).

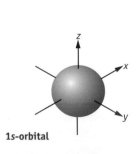

1s-orbital

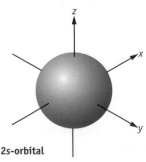

2s-orbital

Figure 2.4 The shape of 1s- and 2s-orbitals

There are three 2p-orbitals. The shape of each can be considered to be like two pears stuck together. The three orbitals point along the three axes, x, y and z (Figure 2.5).

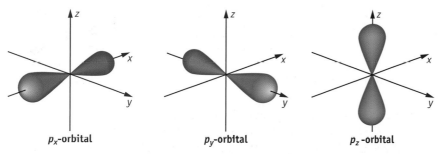

p_x-orbital p_y-orbital p_z-orbital

Figure 2.5 *The shape of 2p-orbitals*

The size and shape of an orbital is drawn so that there is a 90% probability of finding the electron within its boundary. This fits with the modern theory that describes electrons in atoms as having the properties of standing waves. The standing wave for any s-electron is such that it can be found much closer to the nucleus than a p-electron in the same shell. This is called **penetrating** towards the nucleus. Because of this, the energy level of the 2s-orbital is slightly lower than that of the 2p-orbital. However, the energies of the three 2p-orbitals are the same.

The **third orbit** consists of three subshells with different types of orbital.

■ One subshell consists of a 3s-orbital, which is a spherical orbital with a slightly bigger radius than that of the 2s-orbital.

■ The second subshell has three 3p-orbitals, which are the same shape as 2p-orbitals but bigger. Electrons in these 3p-orbitals penetrate less towards the nucleus than do the 3s-electrons.

■ The third subshell has five 3d-orbitals, the shapes of which are shown in Figure 2.6. Electrons in the 3d-orbitals penetrate towards the nucleus even less than electrons in the 3p-orbitals.

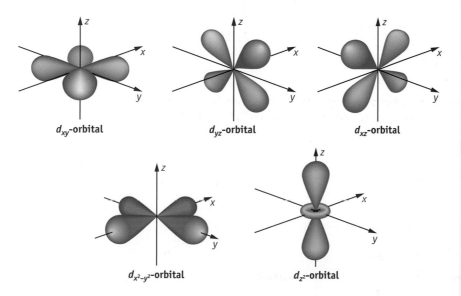

d_{xy}-orbital d_{yz}-orbital d_{xz}-orbital

$d_{x^2-y^2}$-orbital d_{z^2}-orbital

Figure 2.6 *The shapes of d-orbitals*

The energies of the subshells (s-, p- and d-orbitals) differ from each other because of the effect of penetrating towards the nucleus. The energy level of the 2s-orbital is lower than that of the 2p-orbitals. The energy level of the 3s-orbital is lower than that of the 3p-orbitals, which is lower than that of the 3d-orbitals. In a particular element, all 3p-orbitals have the same energy as each other and all 3d-orbitals have the same energy as each other, which is higher than that of the 3p-orbitals.

The **fourth orbit** contains one 4s-, three 4p-, five 4d- and seven 4f-orbitals.

The extent to which the energies of the subshells are split from each other depends upon the number of protons in the nucleus. In hydrogen (one proton) there is no splitting. If there are more than 18 protons, the splitting is so great that the energy level of the 4s-orbital becomes lower than that of the 3d-orbital. This has a significant effect on the electronic structure of elements beyond argon.

Detailed electron configuration

The exact electronic structure of an isolated atom in the ground state can be predicted using the energy level diagram (Figure 2.7) and the following rules:

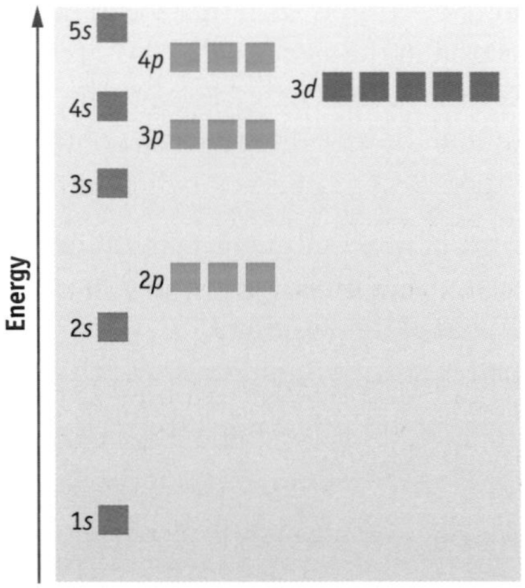

Figure 2.7 Relative energy levels of orbitals in an atom with atomic number greater than 18

- An orbital can hold a maximum of two electrons, one with a spin in the clockwise direction and one in the counter-clockwise direction. Electrons with opposite spins are represented by ↑ and ↓, or ↿ and ⇂.
- The aufbau (building up) principle states that electrons go into an atom starting in the lowest energy orbital, then to the next lowest and so on until the correct number of electrons has been added.

The ground state of an atom is where the electrons occupy the lowest possible energy levels.

- Hund's rule states that if there is more than one orbital in a subshell, electrons are initially added to the orbitals so that the electrons have parallel spins.
- Pauli's exclusion principle states that all electrons in an atom must be in different orbitals or have different spins.

Overall, this means that:
- A single orbital can hold a maximum of two electrons, with opposite spins.
- The first orbit can hold a maximum of two electrons — both in the 1s-orbital, but with opposite spins.
- The second orbit can hold a maximum of eight electrons — two in the 2s- and two in each of the three 2p-orbitals, i.e. $(1 \times 2) + (3 \times 2) = 8$.
- The third orbit can hold a maximum of 18 electrons — two in the 3s-, two in each of the three 3p- and two in each of the five 3d-orbitals, i.e. $2 + 6 + 10 = 18$.
- The fourth orbit can hold a maximum of 32 electrons, i.e. $2 + 6 + 10 + 14 = 32$.

The electron configurations of magnesium and vanadium are shown in Figure 2.8.

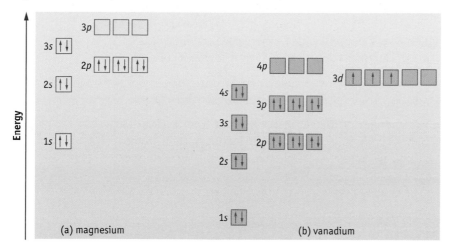

Figure 2.8
The electron
configuration of (a) a
magnesium atom and
(b) a vanadium atom

The electron configuration of magnesium can be written as $1s^2\,2s^2\,2p^6\,3s^2$.

The electron configuration of vanadium can be written as $1s^2\,2s^2\,2p^6\,3s^2\,3p^6\,3d^3\,4s^2$. Note that the three $3d$ electrons are in different orbitals and have parallel spins.

A full list of the electron configurations of the first 36 elements is given in Table 2.3.

A short way of writing an electron configuration is to give the symbol of the preceding noble gas followed by the detail of the electrons added subsequently. For example, the electron configuration of manganese is $[Ar]\,3d^5\,4s^2$ and that of bromine is $[Ar]\,3d^{10}\,4s^2\,4p^5$, where $[Ar]$ means $1s^2\,2s^2\,2p^6\,3s^2\,3p^6$.

The order of filling orbitals is $1s$, $2s$, $2p$, $3s$, $3p$, $4s$, $3d$, $4p$, $5s$, $4d$, $5p$, $6s$... This is shown diagrammatically in Figure 2.9.

Figure 2.9 The order
in which orbitals are
filled

Note that the d-orbitals fill 'late', after the s-orbital of the next orbit (or shell) has received electrons. However, there are two slight variations in the order of filling:

- chromium is $[Ar]\,3d^5\,4s^1$
  not $[Ar]\,3d^4\,4s^2$
- copper is $[Ar]\,3d^{10}\,4s^1$
  not $[Ar]\,3d^9\,4s^2$

This is because an atom is more stable if it has a half-filled or filled set of $3d$-orbitals and a single electron in the $4s$-orbital, rather than four or nine $3d$-electrons and two in the $4s$-orbital.

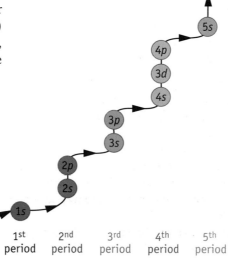

Atomic number	Symbol	1s	2s	2p	3s	3p	3d	4s	4p	
1	H	1								
2	He	2								
3	Li	2	1							s-block elements
4	Be	2	2							
5	B	2	2	1						p-block elements
6	C	2	2	2						
7	N	2	2	3						
8	O	2	2	4						
9	F	2	2	5						
10	Ne	2	2	6						
11	Na	2	2	6	1					s-block elements
12	Mg	2	2	6	2					
13	Al	2	2	6	2	1				p-block elements
14	Si	2	2	6	2	2				
15	P	2	2	6	2	3				
16	S	2	2	6	2	4				
17	Cl	2	2	6	2	5				
18	Ar	2	2	6	2	6				
19	K	2	2	6	2	6	0	1		s-block elements
20	Ca	2	2	6	2	6	0	2		
21	Sc	2	2	6	2	6	1	2		d-block elements
22	Ti	2	2	6	2	6	2	2		
23	V	2	2	6	2	6	3	2		
24	Cr	2	2	6	2	6	5	1		
25	Mn	2	2	6	2	6	5	2		
26	Fe	2	2	6	2	6	6	2		
27	Co	2	2	6	2	6	7	2		
28	Ni	2	2	6	2	6	8	2		
29	Cu	2	2	6	2	6	10	1		
30	Zn	2	2	6	2	6	10	2		
31	Ga	2	2	6	2	6	10	2	1	p-block elements
32	Ge	2	2	6	2	6	10	2	2	
33	As	2	2	6	2	6	10	2	3	
34	Se	2	2	6	2	6	10	2	4	
35	Br	2	2	6	2	6	10	2	5	
36	Kr	2	2	6	2	6	10	2	6	

Table 2.3 Electron configuration of the first 36 elements

The periodic table

Since the time of Dalton, chemists have tried to bring order out of the chaos of the chemical and physical properties of the elements. A number of attempts were made. For example, Newland suggested that there was a repeating pattern of properties every eight elements — his 'Law of Octaves'. Newland's contribution to modern understanding was the idea of **periodicity**. This is that there is a pattern of regularly repeating physical and chemical properties. It was Mendeléev who put this on a sound basis and produced the periodic table, the understanding of which is fundamental to chemistry. Mendeléev's original table has subsequently been modified and the modern form is provided at the back of this book (page 340). You should use it whenever you need relative atomic mass values, as well as when you are studying the chemistry of the elements and their compounds.

Periodic properties

Melting points

The melting points of the elements lithium to neon increase steadily for the first four elements and then fall dramatically at the fifth. This pattern is repeated with the elements sodium to argon (Table 2.4).

Element	Melting point/K	Element	Melting point/K
Lithium	454	Sodium	371
Beryllium	1551	Magnesium	922
Boron	2573	Aluminium	933
Carbon	3925	Silicon	1683
Nitrogen	63	Phosphorus	317
Oxygen	55	Sulfur	386
Fluorine	53	Chlorine	172
Neon	25	Argon	84

Table 2.4 Melting points of the elements in period 2 and period 3

First ionisation energies

The repeating pattern of the first ionisation energies of the elements is shown and explained later in this chapter (pages 34–37).

Valency

The **valency** of an element can be defined as the number of electrons that it uses in bonding. Thus sodium (electron configuration 2,8,1) loses one electron when it forms an Na^+ ion and so has a valency of 1. Likewise chlorine (2,8,7) gains an electron to form the Cl^- ion, so it also has a valency of 1. The valencies of the elements in the period lithium to fluorine are in the order 1, 2, 3, 4, 3, 2, 1. This pattern is repeated exactly for the lowest valencies of the elements sodium to chlorine in the next period.

Division of the periodic table into groups, periods and blocks

In the periodic table, the vertical columns are called **groups**. All the elements in a group contain the same number of electrons in their outer orbit. For example, all the elements in group 1 have one electron in their outer orbit, which is in an *s*-orbital. All the elements in group 7 have seven electrons in their outer orbit, arranged $s^2 p^5$.

The horizontal rows are called **periods**. All elements in the same period have the same number of orbits containing electrons. For example, all the elements in the third period, Na to Ar, have electrons in the first, second and third orbits.

The table is also divided into **blocks**:

■ The *s*-block consists of the elements in groups 1 and 2. An *s*-block element has its highest-energy electron in an *s*-orbital, i.e. the last electron added goes into an *s*-orbital.

- The **p-block** contains the elements in groups 3 to 7 and group 0 (the noble gases). A p-block element has its highest-energy electron in a p-orbital, i.e. the last electron added goes into a p-orbital.
- The **d-block** contains the elements scandium to zinc in period 4 and those elements below them. A d-block element cannot be defined in terms of energy because the energy level of the d-orbitals is altered by the presence of electrons in the outer s-orbital. It can only be defined in aufbau terms, i.e. the last electron added to a d-block element goes into a d-orbital.
- The **f-block** contains the 14 rare-earth elements (cerium to lutetium), which fit in the periodic table between lanthanum ($Z = 57$) and hafnium ($Z = 72$), and the 14 actinide elements (thorium to lawrencium), which come after actinium ($Z = 89$). The last electron added to an f-block element goes into an f-orbital.

The electronegativity of the elements increases from left to right in the s- and p-blocks (except for the noble gases, to which the term electronegativity does not apply). Electronegativity decreases down a group.

> The electronegativity of an element is the extent to which it attracts a pair of electrons in a covalent bond towards itself.

The elements become less metallic going from left to right across a period and more metallic going down a group. There is no exact definition of a metal, but the typical properties of a metal are as follows:

- A metallic element conducts electricity.
- A metal is malleable and ductile. Malleable means that it can be hammered or pressed into a different shape; ductile means that it can be drawn out into wires.
- A metal forms positive ions (cations) in compounds. When a metal atom forms a positive ion, it loses one or more electrons.
- Metals react with oxygen to form oxides that are bases. (A base is a substance that accepts H^+ ions from an acid.)
- The more reactive metals react with acids to form a salt and hydrogen gas.

The periodic table can be used to predict the electron configuration of an element because it indicates the order in which orbitals are filled (Figure 2.10).

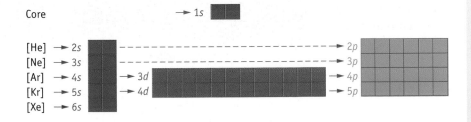

Figure 2.10
The periodic table and the order of filling orbitals

The forces within an atom

Inside an atom, there are forces that act inside the nucleus and forces that act on the electrons.

Z is the atomic number of an element. It is the number of *protons* in the nucleus and so equals the charge on the nucleus.

The most electronegative element is fluorine (top right); the least electronegative element is caesium (bottom left).

Metals and graphite are the only elements that conduct electricity when solid.

Inside the nucleus

Inside the nucleus there are strong and weak nuclear forces, which are responsible for holding the nucleus together. Whenever there is a change in nuclear structure, very large amounts of energy are involved.

Radioactivity

The nuclei of some isotopes, such as carbon-14 and all the isotopes of radium and uranium, are unstable and spontaneously emit radiation. The level of radioactivity given off by a radioisotope is determined by its half-life. A short half-life means that the isotope is very radioactive but the radioactivity decreases rapidly over a short time. Some radioisotopes have long half-lives, such as the 4 billion years of the uranium-238 isotope. A long half-life means that exposure to the radioisotope for a short period is not very dangerous. However, its radio-activity will not diminish noticeably. The most dangerous radioisotopes are those with half-lives from several days to a few years.

The radiation is of three main types:

- α-**rays** — alpha-rays are fast-moving helium nuclei. A helium nucleus consists of two protons and two neutrons. Alpha-rays are stopped by a few centimetres of air or by a sheet of paper.
- β-**rays** — beta-rays are very fast-moving electrons. They are stopped by a thin sheet of metal.
- γ-**rays** — gamma-rays are similar to X-rays but have an even higher frequency and hence energy. They need a thick layer of lead or concrete to absorb them.

It is dangerous to breathe in or eat any material containing α or β emitters (e.g. polonium-210, which is an α-emitter). This is because the radiation will destroy cells that absorb the chemical containing the isotope. However, use can be made of this. Iodine is concentrated in the thyroid gland. Cancers of the thyroid are treated with minute doses of the radioactive isotope, iodine-131. Its half-life is 8 days, so the radioactivity rapidly decreases to a negligible level.

A nuclear reactor

Gamma-rays can be absorbed by human tissue, causing changes to the cells. High doses will kill cells and lower doses may cause mutations. Cancer cells are more likely to be destroyed than healthy cells. Radiotherapy, with controlled doses from cobalt-60, is used to treat many forms of cancer. Gamma-rays also blacken photographic film.

Nuclear fission

Some heavy nuclei, such as uranium-235 and plutonium-239, are **fissile**. This means that when the nucleus absorbs a neutron, it becomes so unstable that it breaks into several smaller pieces.

Huge amounts of energy are released. A typical fission of uranium-235 produces strontium-90, xenon-143 and three neutrons. Uranium-235 absorbs slow (thermal) neutrons more efficiently than it absorbs fast neutrons. Therefore, if the neutrons produced in the fission of one nucleus are slowed down (by a moderator such as graphite), another fission reaction will occur. This is known as a **chain reaction**, and is the principle behind a nuclear reactor.

Nuclear fission can be regarded as an environmentally friendly way of generating electricity. It does not produce any greenhouse gases and, as so little is used to produce a year's supply of electricity, there is a virtually unlimited source of uranium. One gram of uranium-235 produces as much energy by fission as burning over 2 tonnes of coal. All developed and developing countries require vast amounts of electrical energy. Traditionally, this has been provided by power stations fired by fossil fuels. However, these produce huge amounts of carbon dioxide, a gas that contributes to global warming. In the short term, nuclear fission is the only answer to the threat of climate change. The disadvantage of nuclear power is that the fission products and the materials with which the reactor is built become dangerously radioactive, and have to be safely stored for thousands of years. Also, the technology of nuclear power can be adapted to make nuclear weapons.

An atomic bomb explosion

Accidents, such as that at Chernobyl, can cause many deaths. However, so can accidents in coalmines, and death can result from breathing the pollution from fossil-fuel power stations and from wars over limited oil resources. If the sea levels rise significantly as a result of global warming caused by burning fossil fuels, millions in low-lying countries will die as a result of extensive flooding.

Nuclear fusion

When the nuclei of light isotopes such as hydrogen-1 or hydrogen-2 (deuterium) fuse together to make isotopes such as helium-4, enormous amounts of energy are released. This is the source of the energy that drives the universe. All stars were originally made of hydrogen. Under the enormous pressures and temperatures of these massive astronomical bodies, fusion takes place. Energy is released as electromagnetic radiation, in the form of radio waves, visible light and X-rays.

In a hydrogen bomb, a fusion reaction is triggered by the extreme conditions created by the explosion of a uranium fission device. When this reaction is harnessed, it will be possible to generate power without the problems associated with nuclear fission reactors.

Forces that act on electrons in an atom

The forces that act on electrons in an atom are electromagnetic. They obey Coulomb's law, which states that the magnitude of the force between two charged objects is directly proportional to the product of their charges and inversely proportional to the square of the distance between their centres. This is expressed mathematically as:

$$\text{force} \propto \frac{q_+ q_-}{r^2}$$

where q_+ and q_- are the charges on the two objects and r is the distance between their centres.

The rules are:
- opposite charges attract; like charges repel
- the bigger the charges, the stronger is the force
- the larger the sum of the radii of the particles, the weaker is the force

This means that electrons are attracted towards the nucleus. The greater the atomic number, the stronger is the force of attraction. Electrons that are further away from the nucleus are attracted less than those closer to the nucleus.

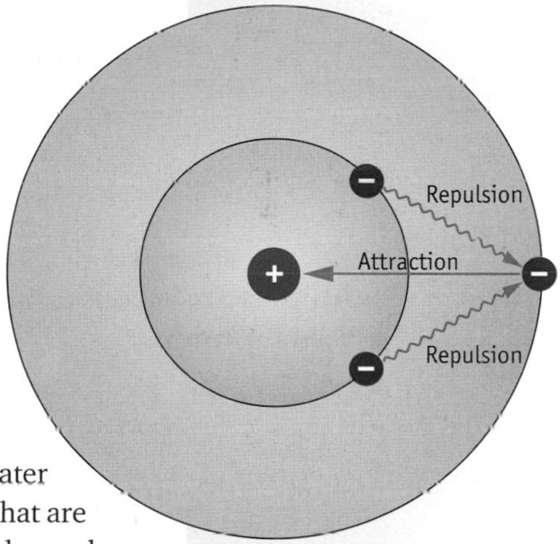

In addition, because they have the same charge, electrons repel each other. Because they are more densely packed, the inner electrons repel outer electrons much more than the other outer electrons do.

Figure 2.11 The forces acting on an electron in an atom

Shielding and the effective nuclear charge

Effective nuclear charge across period 1
In a hydrogen atom, the nucleus has a charge of +1 and there is one electron in a 1s-orbital. There are no forces of repulsion, so the electron feels the full force of attraction of a +1 charge.

In a helium atom, the nucleus has a charge of +2 and there are two electrons in the 1s-orbital. These electrons repel each other slightly. The result is that the net force of attraction between the nucleus and each electron is slightly less than that between a +2 charge and one electron. Therefore, the helium nucleus is said to have an **effective nuclear charge** of slightly less than 2.

Effective nuclear charge across period 2
The situation becomes more complicated for lithium and the remaining elements. In lithium, the nucleus has a charge of +3 and the outer 2s electron is strongly repelled by the two inner 1s electrons. The nucleus is **shielded** by the inner electrons and the effective nuclear charge is approximately +1. This is the +3 nuclear charge, minus the effect of two negatively charged screening electrons.

The effective nuclear charge is the net charge on the nucleus, after allowing for the electrons in orbit around the nucleus shielding its full charge.

The next element is beryllium. The nucleus has a charge of $+4$, there are two $1s$ electrons shielding the nucleus and two $2s$ electrons that also repel each other slightly. Therefore, the effective nuclear charge is not exactly $+2$ ($+4$ nuclear charge minus the effect of the two negative inner electrons) — it is slightly less than $+2$ because of the extra repulsion by the two electrons in the outer orbit.

The situation is slightly more complicated with the next element, boron. The atomic number of boron is five (a nuclear charge of $+5$). There are two $1s$ electrons that shield the outer electrons from the nucleus. The two $2s$ electrons are closer to the nucleus than the single $2p$ electron and they repel it. Therefore, the effective nuclear charge is significantly less than the $+3$ value predicted by the simplified idea that effective nuclear charge is equal to the atomic number of the element minus the number of inner-shell electrons.

Similar arguments apply to other periods — the effective nuclear charge increases across a period, but does not increase by as much as $+1$ between successive elements.

Effective nuclear charge in a group
The effective nuclear charge acting on the outer electrons of the elements in the same group of the periodic table hardly varies. This is because the nuclear charge and the number of inner shielding electrons both increase by the same amount.

For instance, sodium has 11 protons and, therefore, a nuclear charge of $+11$. It has two electrons in the first shell and eight in the second shell. These ten electrons shield the outer, third-shell electron very efficiently and the effective nuclear charge is close to $+1$ ($+11 - 2 - 8 = +1$).

Potassium has 19 protons and, therefore, a nuclear charge of $+19$. The outer, fourth-shell electron is shielded from the nucleus by two electrons in the first shell, eight electrons in the second shell and eight electrons in the third shell, making a total of 18 inner shielding electrons. From sodium to potassium, the number of protons has increased by eight and the number of shielding electrons has also increased by eight. Therefore, potassium also has an effective nuclear charge close to $+1$.

Similar arguments apply to other groups — the effective nuclear charge is approximately the same for all members of a group in the periodic table.

Shielding in the d-block elements
The increase in atomic number and hence in the nuclear charge is matched by the increase in the number of inner shielding electrons. Therefore, the effective nuclear charge of the d-block elements in the same period hardly alters.
- Scandium has 21 protons. Its electron configuration is $1s^2\ 2s^2\ 2p^6\ 3s^2\ 3p^6\ 3d^1\ 4s^2$. Therefore, it has 19 inner electrons shielding the 21 protons and has an effective nuclear charge of approximately $+2$.

e Do not state that there is more shielding in potassium than in sodium. This fails to make the important point that there is more to shield because of the increase in nuclear charge.

- Titanium has 22 protons. Its electron configuration is $1s^2\,2s^2\,2p^6\,3s^2\,3p^6\,3d^2\,4s^2$. Therefore, it has 20 inner electrons shielding the 22 protons and has an effective nuclear charge of approximately +2.
- Vanadium has 23 protons. Its electron configuration is $1s^2\,2s^2\,2p^6\,3s^2\,3p^6\,3d^3\,4s^2$. Therefore, it has 21 inner electrons shielding the 23 protons and has an effective nuclear charge of approximately +2.

Size of atoms

Across a period, the trend is for atomic radii to *decrease*. This is because the greater nuclear charge increases the force of attraction on the electrons, drawing them closer to the nucleus. This is slightly offset by the increased repulsion of the electrons from each other. However, the net effect is a decrease in radius (Table 2.5).

	Group 1	Group 2	Group 3	Group 4	Group 5	Group 6	Group 7
Element	Li	Be	B	C	N	O	F
Radius/pm	150	112	112	70	65	60	50
Element	Na	Mg	Al	Si	P		Cl
Radius/pm	186	160	143	110	100		100
Element	K	Ca					Br
Radius/pm	230	200					115
Element	Rb						I
Radius/pm	240						140

Table 2.5 Trends in atomic radius

Down a group, the trend is for atomic radii to steadily increase. This is caused by the increase in the number of occupied shells. The electron in the third shell is further from the nucleus than the electron in the second shell.

- Lithium: electron structure 2,1; two occupied shells.
- Sodium: electron structure 2,8,1; three occupied shells.
- Potassium: electron structure 2,8,8,1; four occupied shells.
- Rubidium: electron structure 2,8,18,8,1; five occupied shells.

Size of positive ions

A positive ion is always *smaller* than its neutral atom. If an atom loses all its outer electrons, the radius of the resulting ion is much smaller than the atomic radius. This is because:

- there is one fewer shell of electrons
- there are fewer electrons in the positive ion than in the atom, so the electron–electron repulsion is less, causing a further reduction in the radius

For ions with the same electron configuration (e.g. Na^+, Mg^{2+} and Al^{3+}), the ion with the greatest charge will have the smallest radius.

The three positive ions in Table 2.6 have the same number of electrons (ten) arranged in the same way ($1s^2\,2s^2\,2p^6$). Therefore, the electron–electron repulsion is the same.

	Group 1	Group 2	Group 3	Group 4	Group 5	Group 6	Group 7
Atomic radius/pm	Na 186	Mg 160	Al 143	C 70	N 65	O 60	F 50
Ionic radius/pm	Na^+ 95	Mg^{2+} 65	Al^{3+} 50		N^{3-} 171	O^{2-} 140	F^- 136

Table 2.6 Atomic and ionic radii

The nuclear charge increases from 11 to 12 to 13. The force of attraction between the nucleus of the aluminium ion (13 protons) and its ten electrons is, therefore, greater than that between the nucleus of the other ions and their ten electrons. This causes the Al^{3+} ion to have the smallest radius.

Size of negative ions

A negative ion is always *larger* than its neutral atom. To form a negative ion, an atom must gain one or more electrons. The atom and the ion have the same atomic number, so forces of attraction remain the same. However, there is extra repulsion due to the increased number of electrons in the same shell. This causes the ion to expand, moving the electrons further from the nucleus until, once again, there is a balance between the forces of attraction and the forces of repulsion.

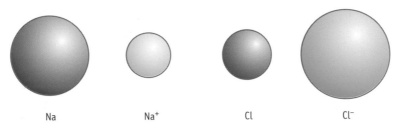

Na   Na⁺   Cl   Cl⁻

Figure 2.12 Relative sizes of atoms and ions

Ionisation energy (IE)

Ionisation energy is the energy needed to remove an electron from an atom, ion or molecule. The units are normally $kJ\ mol^{-1}$.

> The first ionisation energy of an element is the energy required to remove one electron from the ground state of each atom of a mole of *gaseous* atoms of that element.

The first ionisation energy is always a positive number (denoting an endothermic process). It can be represented by the equation:

$$A(g) \rightarrow A^+(g) + e^-$$

This means that a mole of gaseous positive ions is formed, regardless of whether the element is a metal or a non-metal. (See page 9 for an explanation of the term 'mole'.)

The first ionisation energy for hydrogen is represented in Figure 2.3 (page 22). It is the energy that is required to move the electron from the $n = 1$ orbit to the $n = \infty$ orbit. For chlorine, it is the energy change *per mole* for

$$Cl(g) \rightarrow Cl^+(g) + e^-$$

> The second ionisation energy is the energy required to remove one electron from each ion of a mole of gaseous singly charged positive ions of that element.

The second ionisation number is always a positive number (an endothermic process).

More energy is required to remove the second electron than to remove the first electron. This is because the second electron is being removed from a positive ion, which is smaller than the original atom and experiences a greater force of attraction. Even more energy would be required to remove the third electron, and so on.

The second ionisation energy can be represented by the equation:

$$A^+(g) \rightarrow A^{2+}(g) + e^-$$

For the element calcium, it is the energy change *per mole* for:

$$Ca^+(g) \rightarrow Ca^{2+}(g) + e^-$$

Successive ionisation energies

If the successive ionisation energies of an element are listed, it can be seen that there are steady increases and that big jumps occur at defined places. This is one piece of evidence for the existence of orbits or quantum shells.

Aluminium has 13 electrons. Its successive ionisation energies are shown in Table 2.7. The places where the ionisation energy has jumped to a proportionately much higher value are shown in red.

The energy required for the process $Ca(g) \rightarrow Ca^{2+}(g) + 2e^-$ is the *sum* of the first and second ionisation energies.

1st	2nd	3rd						
580	1820	2750						
4th	5th	6th	7th	8th	9th	10th	11th	
11 600	14 800	18 400	23 300	27 500	31 900	38 500	42 700	
12th	13th							
200 000	222 000							

Table 2.7 The successive ionisation energies of aluminium/kJ mol^{-1}

- The data in Table 2.7 show that the first three electrons are considerably easier to remove than the fourth, as there is a big jump from the third to the fourth ionisation energies. These first three electrons come from the outer shell.
- The last two electrons are very much harder to remove than the preceding eight, as there is a huge jump between the eleventh and twelfth ionisation energies. These electrons come from the inner shell.
- As the shells or orbits of the electrons get further from the nucleus, the energy level rises, so less energy is required to remove an electron from that shell. The values in Table 2.7, with the jumps after the third and eleventh ionisation energies, mean that an aluminium atom has three electrons in its outer orbit, eight nearer to the nucleus in an inner orbit and two electrons very close to the nucleus.

For Al, the first big jump occurs between the third and fourth ionisation energies, so Al has three outer electrons and is in group 3.

The group in which an element is found can be worked out by looking at where the first big jump in ionisation energy occurs. If it occurs between the fourth and fifth ionisation energies, then the element has four outer electrons and is in group 4.

Another way of presenting the data is in graphical form. As the variation between the first and last ionisation energies is so great, it is usual to plot the values as the logarithm of the ionisation energy. This is shown in Figure 2.13 for the element sodium ($Z = 11$).

It can be seen that there is a big jump between the first and second ionisation energies and another big jump between the ninth and tenth. This means that sodium has one electron in its outer orbit, eight in the next inner orbit and two in the orbit nearest to the nucleus. Thus, the electronic structure of sodium is 2, 8, 1.

Remember that the first electron to be removed comes from the outer orbit.

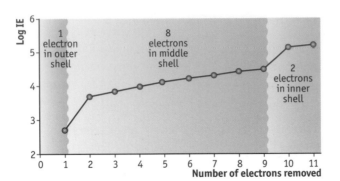

Figure 2.13 Successive ionisations of sodium

Worked example

The successive ionisation energies/kJ mol^{-1} of element X are listed below. Identify the group in the periodic table in which X occurs.

Ionisation energies of X:
1st 950; 2nd 1800; 3rd 2700; 4th 4800; 5th 6000; 6th 12 300; 7th 15 000

Answer

There is a big jump between the fifth and sixth ionisation energies. Therefore, the element has five electrons in its outer orbit and is in group 5.

Variation of first ionisation energies of the elements hydrogen to rubidium

The first ionisation energies of the elements show periodicity (see page 26). The pattern from lithium to neon is repeated exactly with the elements sodium to argon. Apart from the insertion of the *d*-block elements, this pattern is seen again from potassium to rubidium.

The general trend across a period is for the first ionisation energy to increase. However, there are a number of slight variations from this trend.

Figure 2.14 shows that there are maxima at each noble gas and minima at each group 1 metal. There are dips after the second and fifth elements in both periods 2 and 3. These factors can be explained as follows:

- Across a period, there is a general increase in the effective nuclear charge, which reaches a maximum at the noble gas in group 0 at the end of the period. This results in an increasing attraction for the highest energy electron and a general increase in the first ionisation energy.
- The group 1 element immediately after a noble gas has its outer electron in the next shell of higher energy. This electron is well shielded from the nucleus by the inner electrons. The effective nuclear charge is close to +1 and the atomic radius is greater than that of any atom in the preceding period. Therefore, the first ionisation energy of the group 1 element is lower than that of any of the previous elements.
- In period 2, the energy level of the 2*p*-electron is significantly higher than that of the 2*s*-electron, because it is partially shielded by the 2*s*-electrons (page 31). This means that less energy is required to remove the 2*p*-electron from a boron

This method of identifying the group to which an element belongs does not apply to the *d*-block elements because they all have two electrons in their outer orbit, apart from chromium and copper, which have only one outer electron.

atom than is required to remove the 2s-electron from a beryllium atom.

- The electron configuration of nitrogen is $1s^2\ 2s^2\ 2p_x^1\ 2p_y^1\ 2p_z^1$ and that of oxygen is $1s^2\ 2s^2\ 2p_x^2\ 2p_y^1\ 2p_z^1$. The electrons occupying different 2p-orbitals are further apart than two electrons occupying the same 2p-orbital. When an oxygen atom is ionised, it loses one of the two electrons in the $2p_x$-orbital. As there is more repulsion between the two electrons in this orbital than between electrons in different orbitals, it is easier to remove one of the paired 2p-electrons of oxygen than to remove one of the unpaired 2p-electrons of nitrogen.
- The same arguments explain the discrepancies in period 3, where the first ionisation energy of aluminium is lower than that of magnesium and the first ionisation energy of sulfur is lower than that of phosphorus.

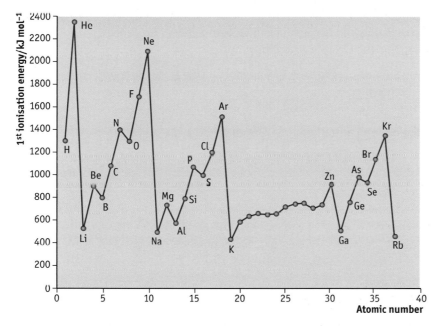

Figure 2.14 *Variation of the first ionisation energy with atomic number*

In period 4, there are dips after zinc and after arsenic. The electron configuration of zinc is $[Ar]\ 3d^{10}\ 4s^2$ and that of the next element, gallium, is $[Ar]\ 3d^{10}\ 4s^2\ 4p^1$. As the 4p-orbital is at a higher energy level than either the 4s- or the 3d-orbital, less energy is required to remove the 4p-electron in gallium than is required to remove one of the 4s-electrons in zinc.

The reason why the first ionisation energy of selenium (group 6) is lower than that of arsenic (group 5) is similar to the reason for the dip after nitrogen in period 2.

Electron affinity (EA)

The first electron affinity is the energy change when one electron is added to each atom in a mole of neutral gaseous atoms.

The values of first electron affinities are usually negative (exothermic).

Electron affinity can be represented by the equation:

$$A(g) + e^- \rightarrow A^-(g)$$

The negatively charged electron being added is brought towards the positively charged nucleus. There is a force of attraction between the two and, therefore, energy is released when the two are brought closer together.

The first electron affinity of oxygen is -142 kJ mol^{-1}. This is the energy change, per mole, for the process:

$$O(g) + e^- \rightarrow O^-(g)$$

> The second electron affinity is the energy change when one electron is added to each ion in a mole of singly charged gaseous negative ions.

The second electron affinity can be represented by the equation:

$$A^-(g) + e^- \rightarrow A^{2-}(g)$$

The second electron affinity of an element is always positive (endothermic) because energy is required to add an electron to an already negative ion. The incoming electron is repelled by the negative ion. Therefore, energy has to be supplied to bring the ion and the electron together.

The second electron affinity of oxygen is $+844$ kJ mol^{-1}. This is the energy change, per mole, for the process:

$$O^-(g) + e^- \rightarrow O^{2-}(g)$$

The variation of the first electron affinities of the elements in the second period is shown in Figure 2.15.

The second (and third) electron affinities are always positive numbers. They are endothermic reactions.

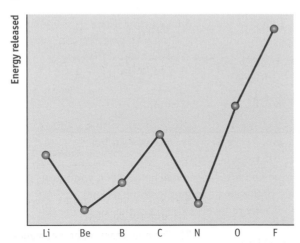

Figure 2.15 Energy released in the process $A(g) + e^- \rightarrow A^-(g)$

The electron is added to the outer orbit of the atom. As an electron is negatively charged and is being brought towards a positive nucleus, energy is released. The general trend is upwards because atomic radii decrease from lithium to fluorine, which causes the force of attraction to increase.

The dip at beryllium occurs because the added electron goes into an already singly occupied $2s$-orbital. Therefore, it experiences considerable repulsion and less energy is released.

The dip at nitrogen is caused by the extra repulsion of putting a second electron into the singly occupied $2p_x$-orbital.

Questions

1 Explain the difference between the terms 'relative isotopic mass' and 'relative atomic mass'.

2 Study the table below and answer the questions that follow.

a Identify the two elements of A–F that are isotopes.

Element	Number of protons	Number of neutrons	Number of electrons
A	19	20	18
B	17	18	17
C	12	12	10
D	17	20	16
E	35	44	36
F	18	22	18

b Identify which of A–F are neutral atoms.

c Identify which of A–F are cations.

d Identify which of A–F are anions.

e Identify which of A–F have the same electron configuration.

3 Explain why the relative atomic mass of most elements is not an exact whole number.

4 The relative atomic mass of copper is 63.5. Calculate the relative abundance of the two copper isotopes with relative isotopic masses of 63.0 and 65.0.

5 The data about silicon in the table below were obtained from a mass spectrometer.

m/e	% abundance
28	92.2
29	4.7
30	3.1

Calculate the relative atomic mass of silicon to 1 decimal place.

6 Identify the particle that has 13 protons, 14 neutrons and 10 electrons.

7 Bromine has two isotopes of mass numbers 79 and 81. The mass spectrum of a sample of bromine is shown below.

a Identify the particles responsible for the peaks.

b Deduce the relative abundance of the two isotopes.

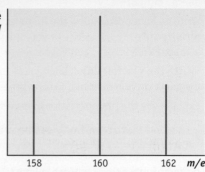

Relative intensity / m/e

c Small peaks are also seen at m/e values of 79 and 81. Write a chemical equation to show the formation of the particle that gives the peak at $m/e = 79$.

8 Explain how positive ions are formed in a mass spectrometer.

9 State the number of orbitals in the third shell of an atom.

10 Place the orbitals $4s$, $4p$, $4d$, $5s$, $5p$, $5d$, $6s$ in order of increasing energy levels.

11 Complete the electron configurations of atoms of the following elements:
Phosphorus: [Ne]...
Cobalt: [Ar]...

12 Write the full electron configuration of an excited sodium atom.

13 Fill in the outer electrons of a phosphorus atom in the boxes below.

14 Explain, with an example, the meaning of the term periodicity.

15 Consider atoms of boron, magnesium and bromine. In which is the effective nuclear charge the largest and in which is it the smallest?

16 List the particles Cl, Cl$^-$ and K$^+$ in order of increasing radius.

17 Write an equation to represent the first ionisation of:

a potassium

b argon

c bromine

18 Write an equation to show the fifth ionisation energy of fluorine.

19 Write an equation to show the first electron affinity of:

a magnesium   b nitrogen

20 Sketch the first ionisation energies of the elements sodium to argon. Explain the variation.

21 Explain why the first ionisation energies of the noble gases decrease from helium to krypton.

22 The successive ionisation energies of an element, X, are given in the table. To which group of the periodic table does element X belong?

Ionisation	Ionisation energy/kJ mol^{-1}
1st	1 000
2nd	2 260
3rd	3 390
4th	4 540
5th	6 990
6th	8 490
7th	27 100
8th	31 700
9th	36 600
10th	43 100

Chapter 3

Formulae, equations and moles

Lithium: medicine and nuclear fuel

Atomic number: 3

Electron configuration: $1s^2\,2s^1$

Symbol: Li

Natural isotopes: ^{6}Li (7.4%), ^{7}Li (92.6%)

Lithium is a soft, shiny metal in group 1 of the periodic table. It was first isolated from rock in 1817 — hence its name, from the Greek word *lithos* (a rock).

Lithium as a medicine

Lithium carbonate is used to treat some forms of mental illness, particularly manic-depressive psychosis.

Nuclear fusion

In the sun, at temperatures of around 15 000 000°C, hydrogen nuclei fuse together to form helium nuclei. The slight loss of mass during this process is converted into a huge amount of energy, according to Einstein's equation:

$$E = mc^2$$

where c is the speed of light.

Another fusion reaction is that between ^{6}Li nuclei and ^{2}H (deuterium) nuclei. If these are heated to millions of °C, they fuse together and then split, forming helium nuclei:

$$^6\text{Li} + {}^2\text{H} \rightarrow 2\,{}^4\text{He}$$

This reaction produces 4×10^8 kJ of energy per gram of lithium, which is similar to the amount of energy produced in 1 hour by the Hoover Dam hydroelectric power plant in the USA.

Fusion reactions are very difficult to achieve because of the extremely high temperatures necessary. Research is being carried out to create a controlled fusion reaction. If the energy created by controlled fusion could be harnessed, electricity could be generated without carbon dioxide emissions, while hydrogen could be obtained from water and used as a fuel for

Small-scale nuclear fusion

vehicles. In addition, there would be no radioactive waste. A hydrogen economy, fuelled by nuclear fusion, could eventually be the answer to the world's energy and environmental problems.

In a hydrogen bomb, the object is power without control. The lithium/deuterium mixture is heated under huge compression by the detonation of a fission nuclear explosion. The extreme conditions cause the nuclei to fuse, releasing awesome amounts of energy.

Formulae, equations and moles

Introduction

The ability to write correct chemical formulae is essential. Without this skill, equations and calculations become impossible.

- Many substances in the AS specification are either ionically bonded or are organic compounds. This chapter should help with working out the formulae of ionic compounds and Chapter 7 should help with organic compounds.
- Chemical equations are internationally understood and are central to the understanding of chemical reactions.
- Substances react in simple whole number ratios by moles. Therefore, the ability to convert masses and volumes into moles is an essential part of quantitative chemistry. Calculations involving moles occur in all unit tests, so it is vital to be confident and skilled in this area.

Formulae

Molecular formula

The molecular formula of a compound shows the number of atoms of each element in one molecule of the substance.

The molecular formula of glucose is $C_6H_{12}O_6$. One molecule of glucose contains six carbon atoms, 12 hydrogen atoms and six oxygen atoms.

Molar mass

Molar mass is the mass of 1 mol of the substance. The units are $g\ mol^{-1}$.

The value of the molar mass is found by adding the relative atomic masses of all the atoms present in the formula. The relative atomic masses can be found in the periodic table (page 340).

In order to calculate the molar mass of a substance (e.g. glucose, $C_6H_{12}O_6$), first use the periodic table to find the relative atomic masses. Here, the values needed are carbon = 12.0, hydrogen = 1.0, oxygen = 16.0.

$$\text{molar mass of glucose, } C_6H_{12}O_6 = (6 \times 12.0) + (12 \times 1.0) + (6 \times 16.0)$$
$$= 180.0\ g\ mol^{-1}$$

Empirical formula

The empirical formula of a substance is the simplest whole number ratio of the atoms of each element in the substance.

The empirical formula of glucose is CH_2O.

e Make sure that you use the atomic *mass* and not the atomic *number*. The atomic mass is the larger of the two numbers. The periodic table is printed on the back of all the unit test papers.

This is why glucose is classed as a carbohydrate, all of which have the formula $C_x(H_2O)_y$.

Experiments can be carried out to determine the mass of each element in a given mass of a substance. The absolute values are then converted to percentage by mass. The percentage values are used to determine the empirical formula.

This calculation is usually carried out in two steps.

Step 1: Divide each percentage by the relative atomic mass of the element, giving the answer to two decimal places.

This gives the number of moles of each element in 100 g of the compound.

Step 2: Divide the results of step 1 by the smallest value obtained in step 1.

This gives a simple ratio by moles of each element. It is usually a whole number ratio.

If the values obtained in step 2 are not integers or very close to integers, one more step must be carried out.

Step 3: Multiply the values obtained in step 2 by 2 so that they become integers. If this fails, multiply the values from step 2 by 3.

This converts values such as 2.5 or 2.33 into a whole number ratio.

> **ⓔ** You will not be asked to calculate the empirical formulae of compounds containing a large number of atoms. If you obtain a ratio such as 1:2:53, you have made a mistake. First check that you used atomic masses and not atomic numbers in step 1. Then check your arithmetic. It is easy to enter a wrong number into a calculator.

ⓔ Do not round values ending in .5, or .67 or .33 to the nearest whole number. Values such as 3.97 may be rounded up to 4.0.

Worked example

A compound contained 34.3% sodium, 17.9% carbon and 47.8% oxygen by mass. Calculate its empirical formula.

◀ This is best done in the form of a table.

Answer

Element	%	Divide by r.a.m.	Divide by the smallest	Ratio
Sodium	34.3	34.3/23.0 = 1.49	1.49/1.49 = 1	1
Carbon	17.9	17.9/12.0 = 1.49	1.49/1.49 = 1	1
Oxygen	47.8	47.8/16.0 = 2.99	2.99/1.49 = 2.01 ≈ 2	2

The empirical formula is $NaCO_2$.

Molecular formula from empirical formula

If the molar mass (or relative molecular mass) is known, the molecular formula can be derived from the empirical formula.

- First, the empirical mass is worked out. If this is the same numerical value as the molar mass, the molecular and empirical formulae are the same.
- If the empirical mass and molar mass are not the same, they will be in a simple ratio to each other. To find this ratio, divide the molar mass by the empirical

mass. If the answer is 2, then the molecular formula is twice the empirical formula. If the answer is 3, the molecular formula is three times the empirical formula.

In the worked example above, the empirical mass of $NaCO_2$ is $23.0 + 12.0 + (2 \times 16.0) = 67.0$. Given that the molar mass of the sodium compound is 134.0 g mol^{-1}, the molecular formula can be derived:

$$\frac{molar\ mass}{empirical\ mass} = \frac{134}{67} = 2$$

Hence, the molecular formula $= 2 \times NaCO_2 = Na_2C_2O_4$.

Worked example

A compound contains 38.7% calcium, 20.0% phosphorus and 41.3% oxygen by mass. The molar mass of the compound is 310 g mol^{-1}. Calculate:

a its empirical formula

b its molecular formula

Answer

a

Element	Divide by r.a.m.	Divide by smallest	Multiply by 2 to find the whole number ratio
Calcium	$38.7/40.1 = 0.97$	$0.97/0.65 = 1.5$	$1.5 \times 2 = 3$
Phosphorus	$20.0/31.0 = 0.65$	$0.65/0.65 = 1$	$1 \times 2 = 2$
Oxygen	$41.3/16.0 = 2.58$	$2.58/0.65 = 4$	$4 \times 2 = 8$

The empirical formula is $Ca_3P_2O_8$.

b Empirical mass $= (3 \times 40.1) + (2 \times 31.0) + (8 \times 16.0)$
$$= 310.3$$

This is numerically the same as the molar mass. Therefore, the molecular formula is $Ca_3P_2O_8$.

e In exams you must show your working. In this case, you must make it clear that because the molar and empirical masses are the same, the two formulae are the same.

Ionic compounds

- Ionic compounds are made up of cations and anions.
- Aqueous solutions of metal compounds are ionic, as are all ammonium compounds.

Cations

A cation is a positive ion formed by the loss of one of more electrons from an atom. Metals form cations. The charges on the cations met at AS are shown in Table 3.1.

The roman numerals in parentheses give the value of the positive charge on the cation in a compound. This is needed if the metal forms ions of more than one charge. For example, iron can lose two electrons and form Fe^{2+} ions or it can lose three electrons and form Fe^{3+} ions. Compounds containing Fe^{2+} ions are called iron(II) compounds; those containing Fe^{3+} ions are called iron(III) compounds.

Table 3.1 Cations and their charges

	1+ ions	2+ ions	3+ ions
Group 1	Li^+ Na^+ K^+ Rb^+ Cs^+		
Group 2		Be^{2+} Mg^{2+} Ca^{2+} Sr^{2+} Ba^{2+}	
Group 3			Al^{3+}
d-block	Ag^+ Cu^+ in copper(I) compounds	Zn^{2+} Mn^{2+} in manganese(II) compounds Fe^{2+} in iron(II) compounds Cu^{2+} in copper(II) compounds	Cr^{3+} in chromium(III) compounds Fe^{3+} in iron(III) compounds
Ammonium	NH_4^+		

NH_4^+ is a polyatomic ion as it contains more than one atom.

Anions

- An anion is a negative ion that has been formed by the gain of one or more electrons.
- Some non-metals form anions.
- **Polyatomic anions** contain more than one element. One of the elements is usually oxygen. For example, carbonates contain the CO_3^{2-} anion.

The anions that need to be known for AS are shown, with their charges, in Table 3.2.

Table 3.2 Anions and their charges

	Name	1– ions	2– ions	3– ions
Group 7	Fluoride	F^-		
	Chloride	Cl^-		
	Bromide	Br^-		
	Iodide	I^-		
Group 6	Oxide		O^{2-}	
	Sulfide		S^{2-}	
Group 5	Nitride			N^{3-}
Polyatomic ions	Hydroxide	OH^-		
	Nitrate	NO_3^-		
	Hydrogencarbonate	HCO_3^-		
	Manganate(VII)	MnO_4^-		
	Chlorate(I)	OCl^-		
	Chlorate(V)	ClO_3^-		
	Cyanide	CN^-		
	Superoxide	O_2^-		
	Peroxide		O_2^{2-}	
	Carbonate		CO_3^{2-}	
	Sulfate		SO_4^{2-}	
	Sulfite		SO_3^{2-}	
	Chromate(VI)		CrO_4^{2-}	
	Dichromate(VI)		$Cr_2O_7^{2-}$	
	Phosphate			PO_4^{3-}

How to work out the formula of an ionic compound

Ionic compounds consist of positive and negative ions held together in the solid by the attraction between their opposite charges. In solution, these ions are separated by water molecules.

The formula of an ionic compound shows the *ratio* of cations to anions. In sodium chloride, $NaCl$, there is one Na^+ ion for every Cl^- ion. In calcium chloride, $CaCl_2$, there are two Cl^- ions for every Ca^{2+} ion.

The ratio of ions in a particular compound can be worked out, based on the fact that the compound is neutral.

Rule 1: If the numerical charge on the ions is the same, the ratio is one cation to one anion.

Some examples of the application of rule 1 are shown below.

> *Worked example 1*
> What is the formula of sodium hydroxide?
>
> **Answer**
> The ions are Na^+ and OH^-.
> They have the same numerical charge.
> The formula of sodium hydroxide is $NaOH$.

> *Worked example 2*
> What is the formula of aluminium nitride?
>
> **Answer**
> The ions are Al^{3+} and N^{3-}.
> They have the same numerical charge.
> The formula of aluminium nitride is AlN.

> *Worked example 3*
> What is the formula of iron(II) sulfate?
>
> **Answer**
> The ions are Fe^{2+} and SO_4^{2-}.
> They have the same numerical charge.
> The formula of iron(II) sulfate is $FeSO_4$.

Rule 2: If the numerical charges are not the same, the numbers of each ion have to be worked out so that the total positive charge equals the total negative charge.

- Write down the formulae of the ions, including their charges.
- Find the lowest common multiple of the two charges. This equals the charge on the ion multiplied by the numbers of that ion in the formula.
- Work out how many of each ion is needed to give this total charge

Some examples of the application of rule 2 are shown below.

Worked example 1

Work out the formula of aluminium oxide.

Answer

The ions are Al^{3+} and O^{2-}.

The lowest common multiple of 3 and 2 is 6.

For a charge of +6, two Al^{3+} ions are needed; for a charge of –6, three O^{2-} ions are needed.

The formula of aluminium oxide is Al_2O_3.

Worked example 2

Work out the formula of calcium hydrogencarbonate.

Answer

The ions are Ca^{2+} and HCO_3^-.

The lowest common multiple of 2 and 1 is 2.

For a charge of +2, one Ca^{2+} ion is needed; for a charge of –2, two HCO_3^- ions are needed.

The formula of calcium hydrogencarbonate is $Ca(HCO_3)_2$.

---

e Parentheses are placed round the polyatomic HCO_3. This *must* be done if there is more than one polyatomic ion in the formula:

- calcium hydroxide is $Ca(OH)_2$, *not* $CaOH_2$
- iron(III) sulfate is $Fe_2(SO_4)_3$
- ammonium carbonate is $(NH_4)_2CO_3$

An alternative way to work out formulae using rule 2 is the 'swapping-over' method.

If the charges are not the same, the number of cations is equal to the value of the charge on the anion and the number of anions is equal to the value of the charge on the cation.

Some examples of the swapping-over method are shown below.

Worked example 1

What is the formula of sodium carbonate?

Answer

The ions are Na^+ and CO_3^{2-}.

Sodium is 1+, so there must be one carbonate ion. The carbonate ion is 2–, so there must be two sodium ions.

$$Na^{1+} \quad CO_3^{2-}$$

The formula of sodium carbonate is Na_2CO_3.

---

e You must not use the swapping-over method if the values of the charges are the same. It only applies when the values are different.

> **Worked example 2**
> What is the formula of aluminium sulfate?
>
> **Answer**
> The ions are Al^{3+} and SO_4^{2-}.
> Aluminium is 3+, so there must be three sulfate ions. The sulfate ion is 2−, so there must be two aluminium ions.
>
> $$Al^{3+} \quad SO_4^{2-}$$
>
> The formula of aluminium sulfate is $Al_2(SO_4)_3$.

e The ionic charges are left out when writing the formula of an ionic substance.

Hydrated ionic compounds

Many solid ionic compounds contain water of crystallisation. The number of water molecules in the formula of a hydrated compound is given by a full stop followed by the number of molecules of water. For instance, a formula unit of solid hydrated sodium carbonate contains ten molecules of water of crystallisation, so its formula is $Na_2CO_3.10H_2O$.

Some water molecules are dative-covalently bonded to the metal ion and some surround the anion (page 120).

Molar masses of ionic compounds

Molar masses of ionic compounds are calculated in the same way as molar masses of molecular substances. For example:

$$\text{molar mass of barium chloride, } BaCl_2 = 137.3 + (2 \times 35.5)$$
$$= 208.3 \text{ g mol}^{-1}$$

If an ionic compound contains water of crystallisation, the mass of water must also be taken into account. For example:

$$\text{molar mass of hydrated copper(II) sulfate, } CuSO_4.5H_2O$$
$$= 63.5 + 32.1 + (4 \times 16.0) + (5 \times 18.0)$$
$$= 249.6 \text{ g mol}^{-1}$$
$$\text{or } 250 \text{ g mol}^{-1} \text{ to three significant figures}$$

Note that the molar mass of water is 18 g mol^{-1}.

Dissolving ionic substances

Many ionic compounds are soluble in water. This is because of the strong forces of attraction between the positive cation and the slightly negative oxygen atom in a water molecule, and between the negative anion and the slightly positive hydrogen atom in a water molecule (page 166).

Equations representing the dissolving of ionic compounds have the formula of the ionic compound on the left and the *separate* ions on the right. State symbols must be included (see pages 51 and 52–54).

The equation for dissolving sodium chloride in water is:

$$NaCl(s) + aq \rightarrow Na^+(aq) + Cl^-(aq)$$

The equation for dissolving iron(III) sulfate in water is:

$$Fe_2(SO_4)_3(s) + aq \rightarrow 2Fe^{3+}(aq) + 3SO_4^{2-}(aq)$$

The symbol 'aq' stands for an unspecified number of water molecules. It comes from *aqua*, the Latin for water.

Chemical equations

- Chemical equations show the formulae of the reactants on the left-hand side and the formulae of the products on the right-hand side.
- The number of atoms of each element must be the same on the left as on the right.

Copper(II) oxide reacts with dilute hydrochloric acid to form copper(II) chloride and water. Therefore, in the equation, CuO and HCl are on the left and $CuCl_2$ and H_2O are on the right:

$$CuO + HCl \rightarrow CuCl_2 + H_2O$$

However, this equation does not balance. There is only one chlorine atom on the left, but two on the right. It also does not balance for hydrogen. To balance the equation a '2' must be put in front of the formula HCl.

The balanced equation is:

$$CuO + 2HCl \rightarrow CuCl_2 + H_2O$$

The equation means that 1 mol of copper(II) oxide reacts with 2 mol of hydrochloric acid to form 1 mol of copper(II) chloride and 1 mol of water.

The quantitative relationship in moles of the reactants and products is called the **stoichiometry** of the reaction.

Balancing equations

Some equations do not need balancing. For example, the equations for the combustion of sulfur and the reaction between magnesium oxide and sulfuric acid are already balanced:

$$S + O_2 \rightarrow SO_2$$
$$MgO + H_2SO_4 \rightarrow MgSO_4 + H_2O$$

There is no need to alter the coefficients in front of the formulae.

For equations that do need balancing, adopt the following strategy:

- Select the element that appears in the fewest number of species in the equation and balance it first.
- If there are two of these, start with the one with the least number of atoms in the formula.
- Then balance the next element, until all are balanced.
- Count groups such as sulfate, SO_4, and nitrate, NO_3, as entities rather than counting separate sulfur and oxygen atoms or nitrogen and oxygen atoms.

Example

Consider the combustion of propane, C_3H_8, which burns in excess air to form carbon dioxide and water. The unbalanced equation is:

$$C_3H_8 + O_2 \rightarrow CO_2 + H_2O$$

- Carbon and hydrogen both appear in only two species, but oxygen appears in three.

e Remember that a chemical equation not only identifies the products that are formed in the reaction, but also the molar ratio of the reactants and the products.

e Equations must always balance. *Never* write word equations. They will not score any marks.

e Never balance an equation by altering any formulae. Only the numbers in front of formulae may be altered.

- There are fewer carbon atoms than hydrogen atoms in propane, so start by balancing carbon. There are three carbon atoms on the left, so put a '3' in front of CO_2 to balance the carbon atoms.

$$C_3H_8 + O_2 \rightarrow 3CO_2 + H_2O$$

- Now balance for hydrogen. There are eight hydrogen atoms on the left, so there needs to be $4H_2O$ to get eight hydrogen atoms on the right.

$$C_3H_8 + O_2 \rightarrow 3CO_2 + 4H_2O$$

- Last, balance for oxygen. There are six oxygen atoms in $3CO_2$ and four oxygen atoms in $4H_2O$, so there has to be ten oxygen atoms or $5O_2$ on the left.

The balanced equation is:

$$C_3H_8 + 5O_2 \rightarrow 3CO_2 + 4H_2O$$

Example

A more difficult example is the combustion of ethane, C_2H_6, to form carbon dioxide and water. The unbalanced equation is:

$$C_2H_6 + O_2 \rightarrow CO_2 + H_2O$$

- Start with carbon. Put '2' as the coefficient of CO_2, then balance for hydrogen by putting '3' in front of H_2O.
- The number of oxygen atoms on the right is four (in $2CO_2$) plus three (in $3H_2O$), which is seven. Unfortunately, oxygen is diatomic. You have a choice:
  - Obtain seven oxygen atoms by having $3\frac{1}{2}O_2$ on the left. This is acceptable, as the molar ratio of ethane to oxygen would be $1:3\frac{1}{2}$. The half means half a mole, not half a molecule. The equation is:

$$C_2H_6 + 3\tfrac{1}{2}O_2 \rightarrow 2CO_2 + 3H_2O$$

  - Solve the problem of the odd number of oxygen atoms needed on the left by doubling the equation, which then becomes:

$$2C_2H_6 + 7O_2 \rightarrow 4CO_2 + 6H_2O$$

The ratios in the two equations are the same — $1:3\frac{1}{2}$ is the same ratio as $2:7$.

Example

An example where it is much easier if groups rather than individual atoms are counted is the reaction between aluminium and dilute sulfuric acid to form aluminium sulfate and water. The unbalanced equation is:

$$Al + H_2SO_4 \rightarrow Al_2(SO_4)_3 + H_2$$

- Double the number of aluminium atoms on the left to give:

$$2Al + H_2SO_4 \rightarrow Al_2(SO_4)_3 + H_2$$

This equation still does not balance.

- There are three SO_4 groups on the right, so there must be three on the left. Therefore, $3H_2SO_4$ is needed and $3H_2$ is produced. The balanced equation is:

$$2Al + 3H_2SO_4 \rightarrow Al_2(SO_4)_3 + 3H_2$$

Full equations have the full formulae for both molecular and ionic substances. The equations in the section above are all examples of full equations.

State symbols

These show the physical state of the substances in an equation. They are listed in Table 3.3.

Physical state	State symbol
Solid	(s)
Liquid	(l)
Gas	(g)
Aqueous solution	(aq)

Table 3.3 State symbols

State symbols should be added to an equation:
- when a gas is produced from a solution or from a liquid
- when a precipitate is formed
- if it is a thermochemical equation (see Chapter 5)
- if it is an ionic equation

For example, the equation for the reaction of solid calcium carbonate with dilute nitric acid to give calcium nitrate, carbon dioxide and water is:

$$CaCO_3(s) + 2HNO_3(aq) \rightarrow Ca(NO_3)_2(aq) + CO_2(g) + H_2O(l)$$

All nitrates are soluble in water.

The equation for the precipitation of copper(II) hydroxide from solutions of copper(II) sulfate and sodium hydroxide is:

$$CuSO_4(aq) + 2NaOH(aq) \rightarrow Cu(OH)_2(s) + Na_2SO_4(aq)$$

All sodium salts are soluble.

Solubility of ionic solids

Knowledge of the solubility and insolubility of ionic compounds makes writing ionic equations much more straightforward (pages 317–18).

Many ionic solids are soluble in water:
- All group 1 metal compounds are soluble.
- All ammonium compounds are soluble.
- All nitrates are soluble.
- All chlorides are soluble, apart from silver chloride and lead(II) chloride.
- All sulfates are soluble in water, apart from strontium sulfate, barium sulfate and lead(II) sulfate. Calcium sulfate is only very slightly soluble.

Some types of ionic compound are mostly insoluble in water:
- All carbonates are insoluble, apart from group 1 carbonates and ammonium carbonate.
- All hydroxides are insoluble, apart from group 1 hydroxides, ammonium hydroxide and barium hydroxide. Calcium and strontium hydroxides are slightly soluble.

Ionic equations

The essential character of a precipitation reaction, such as that between copper(II) sulfate and sodium hydroxide to form a precipitate of copper(II) hydroxide in a solution of sodium sulfate, becomes clear when a **net ionic equation** is written.

The rules for writing ionic equations are as follows:

Rule 1: write the full equation and balance it. Then write another equation using rules 2, 3 and 4.

Rule 2: for dissolved ionic substances, write the ions separately.

Rule 3: for all solids (whether ionic or not), liquids and gases, write the full formula.

Rule 4: cross out all the 'spectator' ions, i.e. those that appear separately on both sides of the equation.

Using the example of the reaction between copper(II) sulfate and sodium hydroxide, rule 1 gives:

$$CuSO_4(aq) + NaOH(aq) \rightarrow Cu(OH)_2(s) + Na_2SO_4(aq)$$

On balancing, the equation is:

$$CuSO_4(aq) + 2NaOH(aq) \rightarrow Cu(OH)_2(s) + Na_2SO_4(aq)$$

From this full equation, the application of rules 2 and 3 results in:

$$Cu^{2+}(aq) + \cancel{SO_4^{2-}}(aq) + \cancel{2Na^+}(aq) + 2OH^-(aq) \rightarrow$$
$$Cu(OH)_2(s) + \cancel{2Na^+}(aq) + \cancel{SO_4^{2-}}(aq)$$

The spectator ions are in bold type. Apply rule 4 by crossing out the SO_4^{2-} ion on the left with the SO_4^{2-} ion on the right and the $2Na^+$ ions on the left with the $2Na^+$ ions on the right. The net ionic equation is:

$$Cu^{2+}(aq) + 2OH^-(aq) \rightarrow Cu(OH)_2(s)$$

Thinking about what happens in a reaction can speed up the process of writing ionic equations. Copper(II) hydroxide is precipitated, so copper ions must have reacted with hydroxide ions. Therefore, the ionic equation must be:

$$Cu^{2+}(aq) + 2OH^-(aq) \rightarrow Cu(OH)_2(s)$$

Similarly, when solutions of silver nitrate and sodium chloride are mixed, a precipitate of silver chloride is formed. This means that silver ions must have joined with chloride ions to form insoluble silver chloride. Therefore, the ionic equation is:

$$Ag^+(aq) + Cl^-(aq) \rightarrow AgCl(s)$$

The nitrate and sodium ions are the spectator ions.

Strong acids, such as hydrochloric (HCl), sulfuric (H_2SO_4), and nitric (HNO_3), are fully ionised in solution. They are acidic because of the $H^+(aq)$ ions in the solution. When an acid reacts with a base, it is the $H^+(aq)$ ions that react.

Worked example

Write the ionic equation for the reaction of magnesium with dilute hydrochloric acid. The general reaction is:

acid + metal → salt + hydrogen

Answer

The 'long way' of doing this is to follow the rules above.

Rule 1: $Mg(s) + HCl(aq) \rightarrow MgCl_2(aq) + H_2(g)$

Balance the equation:

$$Mg(s) + 2HCl(aq) \rightarrow MgCl_2(aq) + H_2(g)$$

Rules 2 and 3: split dissolved ionic substances into separate ions.

$$Mg(s) + 2H^+(aq) + \cancel{2Cl^-}(aq) \rightarrow Mg^{2+}(aq) + \cancel{2Cl^-}(aq) + H_2(g)$$

e You should practise ionic equations. They are often asked for in AS exams.

> **Rule 4:** cancel the spectator ions (in bold), which are the $2Cl^-$ ions on each side, to give:
>
> $$Mg(s) + 2H^+(aq) \rightarrow Mg^{2+}(aq) + H_2(g)$$
>
> The 'short' way is to understand that the reactions of acids involve H^+ ions. Therefore, the reaction is between magnesium atoms and hydrogen ions. As salts are ionic compounds containing metal ions, the ionic equation is:
>
> $$Mg(s) + 2H^+(aq) \rightarrow Mg^{2+}(aq) + H_2(g)$$

The type of reaction between solid aluminium oxide, which is a base, and dilute sulfuric acid is:

acid + base → salt + water

The reaction is essentially between Al_2O_3 and H^+ ions. Therefore, the ionic equation is:

$$Al_2O_3(s) + 6H^+(aq) \rightarrow 2Al^{3+}(aq) + 3H_2O(l)$$

This is a much simpler equation than the full equation:

$$Al_2O_3(s) + 3H_2SO_4(aq) \rightarrow Al_2(SO_4)_3(aq) + 3H_2O(l)$$

 You must check that ionic equations balance for charge as well as for atoms.

Mole calculations

The mole

In chemistry, the **amount of substance** is always measured in moles (page 9).

> A mole is the amount of substance containing the Avogadro constant number of atoms, molecules or groups of ions.

 The symbol for the mole is mol.

The Avogadro constant, L or N_A, is equal to the number of carbon atoms in exactly 12 g of the carbon-12 isotope.

The value of the Avogadro constant, to three significant figures, is 6.02×10^{23} mol^{-1}.

- 1 mol of sodium contains 6.02×10^{23} sodium atoms.
- 1 mol of water contains 6.02×10^{23} water molecules.
- 1 mol of sodium chloride, NaCl, contains 6.02×10^{23} pairs of Na^+ and Cl^- ions, which is 6.02×10^{23} sodium ions and 6.02×10^{23} chloride ions.
- 1 mol of calcium chloride, $CaCl_2$, contains 6.02×10^{23} calcium ions and $2 \times 6.02 \times 10^{23} = 1.204 \times 10^{24}$ chloride ions.

In most calculations it is simpler to use the following relationships:

> 1 mol of any substance equals the relative molecular mass measured in grams.

> The molar mass of a substance is the mass of 1 mol of that substance.

Count Amedeo Avogadro

Therefore, the molar mass can be calculated as the relative molecular mass in grams. The units are $g\,mol^{-1}$.

- 1 mol of sodium is 23.0 g of sodium, as its molar mass is $23.0\,g\,mol^{-1}$.
- 1 mol of water is 18.0 g of water, as its molar mass is $(1.0 + 1.0 + 16.0) = 18.0\,g\,mol^{-1}$.
- 1 mol of sodium chloride is 58.5 g of sodium chloride, as its molar mass is $23.0 + 35.5 = 58.5\,g\,mol^{-1}$.

Conversions

Conversions involving moles and mass are crucial to many calculations in AS chemistry.

$$\text{amount of substance (moles)} = \frac{\text{mass of substance}}{\text{molar mass of substance}}$$

This is commonly expressed as

$$\text{moles} = \frac{\text{mass}}{\text{molar mass}}$$

Some students like to use the 'mole triangle' to help with this sort of calculation.

- The top (mass) = the two factors at the base (moles and molar mass) multiplied together.
- Either factor at the base (moles or molar mass) = the top divided by the other factor at the base.

Therefore:

$$\text{mass} = \text{moles} \times \text{molar mass}$$

$$\text{moles} = \frac{\text{mass}}{\text{molar mass}}$$

$$\text{molar mass} = \frac{\text{mass}}{\text{moles}}$$

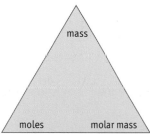

The mole triangle

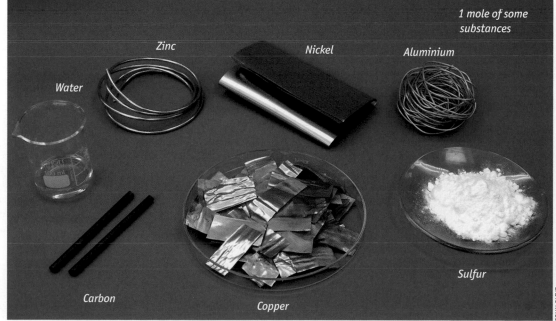

1 mole of some substances

Calculations become easier and more reliable if the units for all quantities are put into the calculation. Some people find this a burden at first, but with practice it becomes second nature.

Mass to mole

ℯ Practise putting units into all your calculations, both in your homework and in examinations. It is an important check to see if you are doing the calculation correctly.

Worked example 1

Calculate the amount (in moles) of iron in 1.23 g.

Answer

$$\text{amount of ion} = \frac{\text{mass}}{\text{molar mass}} = \frac{1.23\,g}{55.8\,g\,mol^{-1}} = 0.0220\,mol$$

ℯ Note that the unit g cancels leaving mol^{-1} on the bottom, which is the same as mol on the top. This is a check that you have the calculation the right way up. Had you thought that amount (in moles) = molar mass/mass, the calculation would have been:

$$\text{amount} = \frac{55.8\,g\,mol^{-1}}{1.23\,g} = 45.4\,mol^{-1}$$

This answer has the wrong unit for amount of substance, so your 'formula' must have been wrong.

Worked example 2

Calculate the amount (in moles) of water in 4.56 g.

Answer

$$\text{amount of water} = \frac{\text{mass}}{\text{molar mass}} = \frac{4.56\,g}{18.0\,g\,mol^{-1}} = 0.253\,mol$$

Worked example 3

Calculate the amount (in moles) of calcium chloride, $CaCl_2$ in 7.89 g.

Answer

molar mass of calcium chloride = $40.1 + (2 \times 35.5) = 111.1$ g mol^{-1}

$$\text{amount of } CaCl_2 = \frac{\text{mass}}{\text{molar mass}} = \frac{7.89\,g}{111.1\,g\,mol^{-1}} = 0.0710\,mol$$

ℯ Note that the identity of the chemical (iron, water, calcium chloride or sucrose) is stated in each calculation. In all but the simplest calculations, phrases such as 'moles = 0.00939' can be misleading.

Worked example 4

Calculate the amount (in moles) of sucrose, $C_{12}H_{22}O_{11}$ in 3.21 g. The molar mass of sucrose is 342.0 g mol^{-1}.

Answer

$$\text{amount of sucrose} = \frac{\text{mass}}{\text{molar mass}} = \frac{3.21\,g}{342.0\,g\,mol^{-1}} = 0.00939\,mol$$

Mole to mass

Worked example 1

Calculate the mass of 0.0222 mol of sodium hydroxide, NaOH (molar mass 40.0 g mol^{-1}).

Answer

mass = molar mass × amount of substance (moles)

mass of sodium hydroxide = 40.0 g mol^{-1} × 0.0222 mol = 0.888 g

---

ⓔ Note that the mol^{-1} cancels out the mol, leaving the unit g, which is the correct unit for mass. Had you thought that mass = molar mass/moles, the calculation would have been:

$$amount = \frac{40.0 \, g \, mol^{-1}}{0.0222 \, mol} = 1802 \, mol^{-2}$$

This answer has an absurd unit and an unlikely large mass. Do think about the magnitude of your answer — examiners see many examples of candidates leaving ridiculous answers.

Worked example 2

Calculate the mass of 0.0333 mol of sulfuric acid, H_2SO_4.

Answer

molar mass of sulfuric acid = (2 × 1.0) + 32.1 + (4 × 16.0) = 98.1 g mol^{-1}

mass of sulfuric acid = molar mass × amount (moles)

= 98.1 g mol^{-1} × 0.0333 mol = 3.27 g

Mass and moles to molar mass

Worked example 1

0.0222 mol of an oxide of sulfur has a mass of 1.42 g. Calculate its molar mass.

Answer

$$molar \, mass \, of \, the \, sulfur \, oxide = \frac{mass}{moles} = \frac{1.42 \, g}{0.0222 \, mol} = 64.0 \, g \, mol^{-1}$$

◀ Analysis of the units in the calculation gives the correct unit of g mol^{-1} in the answer.

Worked example 2

0.0500 mol of an organic acid had a mass of 3.00 g. Calculate the molar mass of the acid.

Answer

$$molar \, mass \, of \, the \, organic \, acid = \frac{mass}{moles} = \frac{3.00 \, g}{0.0500 \, mol} = 60.0 \, g \, mol^{-1}$$

Moles and mass to identity

The identity of a substance or one of its ions can be worked out from mass and amount of substance data.

Worked example 1

An alkene has the empirical formula CH_2. 0.075 mol of the alkene has a mass of 2.1 g. Calculate its molar mass and hence its molecular formula.

Answer

$$\text{molar mass} = \frac{\text{mass}}{\text{moles}} = \frac{2.1\,g}{0.075\,\text{mol}} = 28\ g\ mol^{-1}$$

This is twice the empirical mass of 14.0, so the molecular formula is twice the empirical formula, or C_2H_4.

Worked example 2

0.100 mol of hydrated sodium carbonate, $Na_2CO_3.xH_2O$, has a mass of 28.6 g. Calculate its molar mass and hence the number of molecules of water of crystallisation.

Answer

$$\text{molar mass of hydrated sodium carbonate} = \frac{\text{mass}}{\text{moles}}$$

$$= \frac{28.6\,g}{0.100\,\text{mol}} = 286\ g\ mol^{-1}$$

mass of $Na_2CO_3 = (2 \times 23.0) + 12.0 + (3 \times 16.0) = 106.0$ g

mass of water $= 286$ g $- 106$ g $= 180$ g

There are 180 g of water in 1 mol of the hydrated solid.

$$\text{amount (moles) of water} = \frac{180\,g}{18.0\,g\,mol^{-1}}$$

$$= 10.0 \text{ mol in 1 mol of solid}$$

The number of molecules of water of crystallisation is 10.

> This number is the sum of the mass of Na_2CO_3 and the mass of the water in 1 mol.

e Use the mass of the hydrated salt and the number of moles given in the question to work out the molar mass of the hydrated salt.

If the data are produced by experiment, the amount of water might be calculated as a number very close to 10, say 9.98. You cannot have 9.98 molecules of water. The discrepancy is due to experimental error. In this type of calculation, you must round your answer to the nearest whole number.

Worked example 3

0.0250 mol of a group 2 sulfate has a mass of 4.60 g. Calculate the molar mass of the sulfate and hence identify the group 2 metal ion in the compound.

Answer

$$\text{molar mass} = \frac{\text{mass}}{\text{moles}} = \frac{4.60\,\text{g}}{0.0250\,\text{mol}} = 184\,\text{g mol}^{-1}$$

The formula of group 2 sulfates is of the form MSO_4, where M represents the group 2 metal. Of the $184\,\text{g mol}^{-1}$, $32.1 + (4 \times 16.0) = 96.1\,\text{g}$ comes from the SO_4 group. Therefore, the molar mass of the group 2 metal $= 184\,\text{g mol}^{-1} - 96.1\,\text{g mol}^{-1}$

$$= 87.9\,\text{g mol}^{-1}$$

From the periodic table, the group 2 metal that has a molar mass nearest to $87.9\,\text{g mol}^{-1}$ is strontium, Sr.

◀ The difference between the calculated value of 87.9 and the correct value of 87.6 is due to experimental errors.

Mole to number of particles

This type of calculation often causes confusion. The answer is always a fantastically large number of many billions.

- The calculation is based on the definitions of the mole and of the Avogadro constant (page 54).
- The unit of the Avogadro constant is mol^{-1}, so to get a number, which is dimensionless and so has no unit, this must be multiplied by mol.

$$\text{number of molecules} = \text{amount of substance (moles)} \times \text{Avogadro constant}$$

Therefore,

$$\text{number of water molecules in 1.44 mol water} = 1.44\,\text{mol} \times 6.02 \times 10^{23}\,\text{mol}^{-1}$$
$$= 8.67 \times 10^{23}$$

Usually you will have to work out the amount (moles) first.

Worked example 1

Calculate the number of molecules in 3.33 g of methane, CH_4 (molar mass $= 16.0\,\text{g mol}^{-1}$).

Answer

$$\text{amount of methane} = \frac{\text{mass}}{\text{molar mass}} = \frac{3.33\,\text{g}}{16.0\,\text{g mol}^{-1}} = 0.208\,\text{mol}$$

$$\text{number of molecules} = 0.208\,\text{mol} \times 6.02 \times 10^{23}\,\text{mol}^{-1} = 1.25 \times 10^{23}$$

Calculations of the number of ions require an extra step.

Worked example 2

Calculate the number of sodium ions in 2.22 g of sodium sulfate, Na_2SO_4 (molar mass $= 142.1\,\text{g mol}^{-1}$).

Answer

$$\text{amount of sodium sulfate} = \frac{\text{mass}}{\text{molar mass}} = \frac{2.22\,\text{g}}{142.1\,\text{g mol}^{-1}} = 0.0156\,\text{mol}$$

$$\text{number of ion groups} = 0.0156\,\text{mol} \times 6.02 \times 10^{23}\,\text{mol}^{-1}$$
$$= 9.39 \times 10^{21}$$
$$\text{number of Na}^+ \text{ ions} = 2 \times 9.39 \times 10^{21} = 1.88 \times 10^{22}$$

Each Na_2SO_4 ion group contains two Na^+ ions and one SO_4^{2-} ion. ◀

Questions

1 A compound of rubidium and oxygen contains 72.6% rubidium by mass. Calculate its empirical formula.

2 An organic compound contains the following by mass: carbon 17.8%; hydrogen 3.0%; bromine 79.2%. Calculate its empirical formula.

3 An organic compound contains the following by mass: carbon 36.4%; hydrogen 6.1%; fluorine 57.5%.
   a Calculate its empirical formula.
   b The molar mass of the compound is 66 g mol^{-1}. Deduce its molecular formula.

4 Write the formulae of the following ionic compounds:
   a calcium chloride
   b silver nitrate
   c copper(II) phosphate
   d aluminium oxide

5 Deduce the charge on the cations in the following compounds:
   a $MnCO_3$
   b $V_2(SO_4)_3$

6 Calculate the molar masses of the following compounds:
   a $Ca(OH)_2$
   b $Al_2(SO_4)_3$
   c $FeSO_4.7H_2O$

7 Balance the following equations:
   a $P_4 + O_2 \rightarrow P_2O_5$
   b $Al + O_2 \rightarrow Al_2O_3$
   c $Mg(NO_3)_2 \rightarrow MgO + NO_2 + O_2$
   d $LiOH + H_2SO_4 \rightarrow Li_2SO_4 + H_2O$
   e $Fe^{3+} + I^- \rightarrow Fe^{2+} + I_2$
   f $C_8H_{18} + O_2 \rightarrow CO_2 + H_2O$

8 Write ionic equations for:
   a The precipitation of copper(II) hydroxide, produced on mixing solutions of copper(II) sulfate and sodium hydroxide.
   b The precipitation of barium sulfate, produced on mixing solutions of barium chloride and potassium sulfate.
   c The precipitation of calcium phosphate, produced on mixing solutions of ammonium phosphate and calcium chloride.
   d The neutralisation reaction between solid calcium carbonate and dilute hydrochloric acid.
   e The neutralisation reaction between dilute sulfuric acid and sodium hydroxide solution.

9 Calculate the amount (in moles) of:
   a 1.11 g of calcium carbonate, $CaCO_3$
   b 2.22 g of barium hydroxide, $Ba(OH)_2$

10 Calculate the masses of:
   a 0.0100 mol of sulfuric acid, H_2SO_4
   b 100 mol of sodium metal

11 0.0185 mol of hydrated magnesium sulfate, $MgSO_4.xH_2O$, has a mass of 4.56 g. Calculate the number of molecules of water of crystallisation in the hydrated salt.

12 Calculate the number of molecules in:
   a 1.2 g of water
   b 1.2 mol of water

13 Calculate the number of oxygen atoms in 0.0100 mol of carbon dioxide, CO_2.

14 Calculate the number of hydroxide ions in 10.0 g of barium hydroxide.

Chapter 4

Calculations from chemical equations

Beryllium: emeralds, alloys and nuclear reactors

Atomic number: 4

Electron configuration: $1s^2\ 2s^2$

Symbol: Be

Beryllium is a low-density metal in group 2 of the periodic table. It burns brightly when heated in air, but at room temperature is protected by a layer of oxide. The metal is amphoteric, reacting slowly with acid and rapidly with alkali.

Beryllium occurs in the mineral beryl, which is a beryllium aluminosilicate, $Be_3Al_2Si_6O_{18}$. This mineral often occurs as huge crystals. The largest, found in South Dakota, was over 8 m long and yielded 61 tonnes of beryl.

A huge emerald — this specimen measures 10 cm × 8 cm

ROBERTO DE GUGLIEMO / SCIENCE PHOTO LIBRARY

Emeralds, aquamarine and other gemstones

Beryl is colourless, but impurities cause it to be coloured. Emeralds are crystals of beryl that contain small quantities of green chromium oxide. The best emeralds, found in Brazil, are clear and deep green. Poor quality emeralds are slightly cloudy and appear dull. Artificial emeralds can be made by melting beryl, adding traces of chromium(III) oxide and then allowing the liquid to crystallise slowly.

Aquamarine is a sea-green gemstone which is beryl with iron(III) oxide as an impurity. Rose-coloured beryl (morganite) owes its beautiful pale colour to traces of manganese(IV) oxide.

Beryllium in metallurgy

Beryllium is transparent to X-rays and is used as a window in X-ray tubes. Because of its low density, it is alloyed with other metals to make lightweight components in the space and aero industries. Beryllium and copper form an alloy with high tensile strength, which is used in watch springs and springs that carry electric current.

Inhalation of beryllium compounds is dangerous. The high-charge density of the small +2 ion causes it to bind firmly to the active sites of some enzymes, inhibiting their activity.

Beryllium as a moderator

In nuclear fission, fissile nuclei (e.g. ^{235}U) are bombarded with slow-moving (thermal) neutrons. When the nucleus shatters, it emits several high-speed neutrons. To increase the chance of further fission and maintain the chain reaction, these neutrons have to be slowed down. When beryllium rods are inserted into a reactor, the neutrons are slowed down and the fission reaction speeds up; as they are withdrawn, the reaction slows down. Elements that slow down neutrons in a nuclear reactor are called moderators. Graphite and heavy water (deuterium oxide, D_2O) are also moderators.

Calculations from chemical equations

Introduction

Calculations from chemical equations depend upon the ability to convert masses and volumes into amount of substance (moles) and to use the stoichiometry of an equation.

Calculations must be well laid out, with words explaining what is being calculated at each step. A series of numbers is not sufficient.

It is advisable to include units in calculations, rather than just adding them at the end. This makes it clear that you have used 'formulae' such as 'moles = concentration × volume' correctly and that you have used comparable units, such as dm^3 and not cm^3 for the volume, when the concentration is in $mol \ dm^{-3}$. This method of adding units to the calculation is slower at first, but with practice becomes second nature and is more likely to avoid careless mistakes.

Significant figures

Answers to calculations should be given to an appropriate number of significant figures. The way to do this is to analyse the data:

- Count the number of significant figures in each quantity given in the question.
- Give the answer to the same number of significant figures as the data.
- If there are two pieces of data with different numbers of significant figures (e.g. one given to three significant figures and the other to four), give the answer to the lower number of significant figures.

> ⓔ It is a good idea to check all your calculations.
> - First check that you have worked out the molar masses correctly.
> - Then redo the calculations to make sure that you have not entered any wrong figures into your calculator.
> - Check your units and finally make sure that you have given the answers to the correct number of significant figures.
> - Check that your answer makes sense by doing a mental 'ball park' calculation as well.

The way to work out the correct number of significant figures is based on these rules:

- For whole numbers and decimals of numbers greater than 1, all the figures, *including any zeros at the end*, are counted. For example:

ⓔ You *must* show all your working or you will not score full marks. Examiners mark consequentially, so if you make a mistake early on, you can still score the remaining marks, provided that you have carried out the subsequent steps correctly and your working is clear.

ⓔ If the question specifies a number of significant figures, that number should be given in your answer. Otherwise, give your answer to three significant figures.

ⓔ It is a good idea to keep all the figures on your calculator throughout the calculation or to write down all intermediate answers to four significant figures.

- the number 102 is written to three significant figures (3 s.f.)
- the number 1.02 is also written to 3 s.f.
- the number 1.020 is written to 4 s.f.
- the number 10.20 is also written to 4 s.f.
- the measurement 100 cm^3 is expressed to 3 s.f.
- For decimals in numbers less than 1, count all the numbers, except the zeros *before* and *immediately after* the decimal point. For example:
  - the number 0.102 is written to 3 s.f.
  - the number 0.1020 is written to 4 s.f.
  - the number 0.00102 is written to 3 s.f, not 5 s.f.
  - the number 0.01200 is written to 4 s.f.
- For values written in scientific notation, all the numbers before and after the decimal place are counted as significant figures. The power of 10 is not counted. For example:
  - the number 1.02×10^{-3} is written to 3 s.f.

Mass and volume calculations

Mass-to-mass calculations

A typical question involving mass-to-mass calculation is to ask for the mass of a product that could be obtained from a given mass of one of the reactants. For example, calculate the mass of barium sulfate that is precipitated when a solution containing 5.55 g of barium chloride, $BaCl_2$, is reacted with excess magnesium sulfate. (Molar mass of barium chloride = 137.3 + (2 × 35.5) = 208.3 g mol^{-1})

After the equation has been written, the calculation should be done in three steps:

Step 1: calculate the amount (moles) of reactant. In this example, the reactant is barium chloride.

Step 2: use the stoichiometry of the equation to calculate the amount (moles) of product.

Step 3: convert the moles of product to mass.

The equation is:
$$BaCl_2(aq) + MgSO_4(aq) \rightarrow BaSO_4(s) + MgCl_2(aq)$$

Step 1: amount of barium chloride $= \dfrac{\text{mass}}{\text{molar mass}} = \dfrac{5.55\,\text{g}}{208.3\,\text{g mol}^{-1}} = 0.02664\,\text{mol}$

Step 2: ratio $BaSO_4:BaCl_2$ is 1:1
amount of $BaSO_4$ produced = amount of $BaCl_2$ = 0.02664 mol

Step 3: molar mass of $BaSO_4$ = 137.3 + 32.1 + (4 × 16.0) = 233.4 g mol^{-1}
mass of barium sulfate produced = mol × molar mass
= 0.02664 mol × 233.4 g mol^{-1} = 6.22 g

e The answer must be given to three significant figures because the mass of barium chloride was given to three significant figures. Never round down to one or two significant figures in the middle of a calculation.

e The stoichiometry of an equation is the ratio of the numbers of each reactant and product in the balanced equation.

e When the stoichiometric ratio is 1:1, many candidates fail to make the important point that the amounts (in moles) of the two substances are the same. This *must* be stated clearly, as in the examples below.

The method can be illustrated by a flow diagram. The calculation is to find the mass of substance B produced from (or reacting with) substance A:

$$\text{Mass of A} \xrightarrow{\text{step 1}} \text{Moles of A} \xrightarrow{\text{step 2}} \text{Moles of B} \xrightarrow{\text{step 3}} \text{Mass of B}$$

Steps 1 and 3 involve mass to moles and moles to mass conversions (pages 56–57). Step 2 uses the stoichiometry of the equation.

When the ratio in the equation is not 1:1, care must be taken in step 2.

Worked example

Calculate the mass of sodium sulfate produced when 3.45 g of sodium hydroxide is neutralised by dilute sulfuric acid. The equation is:

$$2NaOH + H_2SO_4 \rightarrow Na_2SO_4 + 2H_2O$$

Answer

Step 1: molar mass of NaOH = 23.0 + 16.0 + 1.0 = 40.0 g mol^{-1}

$$\text{amount (moles) of sodium hydroxide} = \frac{\text{mass}}{\text{molar mass}} = \frac{3.45 \text{ g}}{40 \text{ g mol}^{-1}} = 0.08625 \text{ mol}$$

Step 2: ratio of Na_2SO_4:NaOH is 1:2, so *fewer* moles of sodium sulfate produced

amount of sodium sulfate = $\frac{1}{2} \times 0.08625$ mol = 0.04313 mol

Step 3: molar mass of Na_2SO_4 = (2 × 23.0) + 32.1 + (4 × 16.0) = 142.1 g mol^{-1}

mass of sodium sulfate produced = moles × molar mass
$$= 0.04313 \text{ mol} \times 142.1 \text{ g mol}^{-1} = 6.13 \text{ g}$$

(e) In calculations at AS, the equation will probably be given; you have to calculate the molar masses of two substances.

Volume of gas calculations

These are based on the fact that the volume occupied by 1 mol of all gases at a given temperature and pressure is the same. This volume is called the molar volume.

The molar volume of a gas is the volume occupied by 1 mol of the gas under specified conditions of temperature and pressure.

- Under typical laboratory conditions, molar volume is usually given as 24 dm^3 mol^{-1} = 24 000 cm^3 mol^{-1}.
- At 0°C and 1 atm pressure it is 22.4 dm^3 mol^{-1} = 22 400 cm^3 mol^{-1}.

The relationship between amount of gas (moles) and volume is given by:

$$\text{amount (moles) of gas} = \frac{\text{volume of gas}}{\text{molar volume}}$$

(e) The value of the molar volume will always be given in the question.

Make sure that the units of the two volumes are the same; both must be either dm^3 or cm^3.

Again a 'mole triangle' can be used:

volume of gas = moles × molar volume

$$\text{moles} = \frac{\text{volume of gas}}{\text{molar volume}}$$

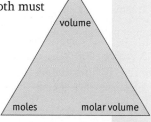

Mass to volume of gas calculations

The route for this type of calculation is similar to that for mass-to-mass calculations.

$$\text{Mass of A} \xrightarrow{\text{step 1}} \text{Moles of A} \xrightarrow{\text{step 2}} \text{Moles of gas B} \xrightarrow{\text{step 3}} \text{Volume of gas B}$$

Steps 1 and 3 use the conversions mass to moles and moles to volume of gas. Step 2 uses the stoichiometry of the equation.

Worked example 1

Calculate the volume of carbon dioxide produced when 2.68 g of calcium carbonate is heated and decomposes according to the equation:

$$CaCO_3(s) \rightarrow CaO(s) + CO_2(g)$$

Under the conditions of the experiment, 1 mol of gas occupies 24 000 cm^3.

Answer

Step 1: molar mass of $CaCO_3 = 40.1 + 12.0 + (3 \times 16.0) = 100.1$ g mol^{-1}

$$\text{amount (moles) of calcium carbonate} = \frac{\text{mass}}{\text{molar mass}}$$

$$= \frac{2.68\,g}{100.1\,g\,mol^{-1}} = 0.0268\ \text{mol}$$

Step 2: ratio of CO_2 to $CaCO_3$ is 1:1

amount (moles) of carbon dioxide = 0.0268 mol

Step 3: volume of carbon dioxide gas = moles × molar volume

$$= 0.0268\ \text{mol} \times 24\,000\ cm^3\ mol^{-1} = 643\ cm^3$$

Worked example 2

The average family motorist uses about 1000 kg of petrol each year. Assume that the molecular formula of petrol is C_8H_{18} and that the equation for the reaction is:

$$C_8H_{18} + 12\tfrac{1}{2}O_2 \rightarrow 8CO_2 + 9H_2O$$

Calculate:

a the yearly volume of carbon dioxide produced at room temperature and pressure

(The molar volume of gas at room temperature and pressure is 24 dm^3 mol^{-1}.)

b the mass of carbon dioxide (the carbon footprint) produced by the car in a year

Answer

a Step 1: molar mass of C_8H_{18} is $(8 \times 12.0) + (18.0 \times 1.0) = 114.0$ g mol^{-1}

$$\text{amount of petrol} = \frac{\text{mass}}{\text{molar mass}} = \frac{1\,000\,000\,g}{114.0\,g\,mol^{-1}} = 8772\ \text{mol}$$

Step 2: ratio of CO_2:$C_8H_{18} = 8$:1 (so more moles of CO_2 produced)

amount of CO_2 produced = 8 × 8772 mol = 7.02×10^4 mol

Step 3: volume of carbon dioxide produced = 7.02×10^4 mol × 24 dm^3 mol^{-1}

$$= 1.68 \times 10^6\ dm^3$$

b 1 mol of CO_2 has a mass of 44.0 g

7.02×10^4 mol of CO_2 has a mass of $7.02 \times 10^4 \times 44.0$ g $= 3.09 \times 10^6$ g

$= 3.09$ tonne

The family car has, on average, a carbon footprint of just over 3 tonnes.

Volume of gas to volume of another gas calculations

There are two alternative methods for performing this type of calculation.

It can be done in three steps, similarly to mass-to-volume calculations.

$$\text{Volume of gas A} \xrightarrow{\text{step 1}} \text{Moles of gas A} \xrightarrow{\text{step 2}} \text{Moles of gas B} \xrightarrow{\text{step 3}} \text{Volume of gas B}$$

Worked example

Calculate the volume of oxygen needed to react with 123 cm^3 of gaseous methane, CH_4.

$$CH_4(g) + 2O_2(g) \rightarrow CO_2(g) + 2H_2O(l)$$

Answer

Step 1: amount of methane $= \dfrac{\text{volume}}{\text{molar volume}} = \dfrac{123 \text{ cm}^3}{24\,000 \text{ cm}^3 \text{ mol}^{-1}} = 0.005125$ mol

Step 2: the ratio of oxygen to methane is 2:1

amount of oxygen $= 2 \times 0.005125$ mol $= 0.01025$ mol

Step 3: volume of oxygen $=$ moles $\times$ molar volume

$= 0.01025$ mol $\times 24\,000 \text{ cm}^3 \text{ mol}^{-1} = 246 \text{ cm}^3$

e You may be given the molar volume for the gases. If not, assume that it is 24 000 cm^3 mol^{-1} or give it a value of V cm^3. The value, V, will cancel.

The alternative, and simpler method, is to use Avogadro's hypothesis.

Equal volumes of gases, measured at the same temperature and pressure, contain the same number of molecules.

This means that if there is twice the number of moles, the volume occupied is doubled — as long as both substances are gases.

Worked example

Calculate the volume of oxygen gas needed to burn completely 200 cm^3 of gaseous butane.

$$2C_4H_{10}(g) + 13O_2(g) \rightarrow 8CO_2(g) + 10H_2O(l)$$

Answer

Both butane and oxygen are gases, therefore:

$$\frac{\text{volume of oxygen}}{\text{volume of butane}} = \frac{\text{moles of oxygen}}{\text{moles of butane}} = \frac{13}{2}$$

$$\text{volume of oxygen} = \frac{13}{2} \times \text{volume of butane}$$

$$= \frac{13}{2} \times 200 \text{ cm}^3 = 1300 \text{ cm}^3$$

e This method can only be used if, under the conditions of the experiment, both substances are gases. You do not need to know the value of the molar volume.

Percentage yield calculations

Many reactions do not produce the same amount of product as that calculated from the chemical equation. This is caused by the reaction being reversible (so equilibrium is reached) or because of competing reactions. This is especially true in organic chemistry.

Percentage yield is defined as:

$$\% \text{ yield} = \frac{\text{actual yield}}{\text{theoretical yield}} \times 100$$

The actual yield is the measured mass of the product obtained in the experiment.

The theoretical yield is the mass that is calculated from the equation for the reaction, assuming that all the reactant is converted into the product.

ⓔ The actual and theoretical yields are usually masses (grams), but *both* could be in moles.

> ### Worked example
> When 1000 g of sulfur dioxide is reacted with excess oxygen, 1225 g of sulfur trioxide is produced:
>
> $$2SO_2 + O_2 \rightarrow 2SO_3$$
>
> Calculate the percentage yield.
>
> **Answer**
> First, calculate the theoretical yield, using the method on page 280.
> **Step 1:** molar mass of $SO_2 = 32.1 + (2 \times 16.0)$
> $$= 64.1 \text{ g mol}^{-1}$$
>
> $$\text{amount of sulfur dioxide} = \frac{\text{mass}}{\text{molar mass}}$$
> $$= \frac{1000 \text{ g}}{64.1 \text{ g mol}^{-1}} = 15.60 \text{ mol}$$
>
> **Step 2:** ratio of $SO_2:SO_3$ is $2:2 = 1:1$
> theoretical amount of sulfur trioxide produced = amount of SO_2 reacted
> $$= 15.60 \text{ mol}$$
> **Step 3:** molar mass of $SO_3 = 32.1 + (3 \times 16.0) = 80.1 \text{ g mol}^{-1}$
> theoretical yield $= \text{moles} \times \text{molar mass}$
> $$= 15.60 \text{ mol} \times 80.1 \text{ g mol}^{-1}$$
> $$= 1250 \text{ g}$$
> Second, use the theoretical yield to calculate the percentage yield:
> $$\frac{\text{actual yield}}{\text{theoretical yield}} \times 100 = \frac{1225}{1250} \times 100$$
> $$= 98.0\%$$

ⓔ Do *not* calculate the percentage yield as:
$$\frac{\text{mass of product} \times 100}{\text{mass of reactant}}$$

Calculation of reaction stoichiometry

If the masses of both reactants (or of one reactant and one product) are known, the stoichiometry of the equation can be worked out. This is done by converting the masses to amounts and examining the ratio of these amounts.

Worked example 1

3.48 g of pure iron was placed in excess copper(II) sulfate solution and stirred until all reaction had ceased. The residue of copper was filtered off, washed and dried. It had a mass of 3.95 g. Use these data to work out which of the two reactions below took place.

$$Fe(s) + CuSO_4(aq) \rightarrow FeSO_4(aq) + Cu(s) \qquad \text{equation 1}$$
$$2Fe(s) + 3CuSO_4(aq) \rightarrow Fe_2(SO_4)_3(aq) + 3Cu(s) \qquad \text{equation 2}$$

Answer

$$\text{amount of iron reacted} = \frac{\text{mass}}{\text{molar mass}}$$

$$= \frac{3.48\,g}{55.8\,g\,mol^{-1}} = 0.0624\,mol$$

$$\text{amount of copper produced} = \frac{\text{mass}}{\text{molar mass}}$$

$$= \frac{3.95\,g}{63.5\,g\,mol^{-1}} = 0.0622\,mol$$

Within experimental error, these numbers are in the ratio of 1:1. Therefore, equation 1 is correct. (Equation 2 has a ratio of moles of iron:moles of copper of 2:3.)

◀ Chemicals react in simple whole number ratios by moles. Therefore, 0.0624:0.0622 is a mole ratio of 1:1.

Worked example 2

When heated, sodium hydrogencarbonate, $NaHCO_3$, decomposes leaving a solid residue. When 4.20 g of sodium hydrogencarbonate was heated until there was no further loss in mass, 2.65 g of solid was left.

Three equations have been suggested for this reaction:

$$2NaHCO_3(s) \rightarrow Na_2O(s) + 2CO_2(g) + H_2O(g) \qquad \text{equation 1}$$
$$2NaHCO_3(s) \rightarrow Na_2CO_3(s) + CO_2(g) + H_2O(g) \qquad \text{equation 2}$$
$$NaHCO_3(s) \rightarrow NaOH(s) + CO_2(g) \qquad \text{equation 3}$$

Which of these equations is consistent with the experimental data?

Answer

molar masses: $NaHCO_3 = 23.0 + 1.0 + 12.0 + (3 \times 16.0) = 84.0\,g\,mol^{-1}$
$Na_2O = (2 \times 23.0) + 16.0 = 62.0\,g\,mol^{-1}$
$Na_2CO_3 = (2 \times 23.0) + 12.0 + (3 \times 16.0) = 106.0\,g\,mol^{-1}$
$NaOH = 23.0 + 16.0 + 1.0 = 40.0\,g\,mol^{-1}$

$$\text{amount of } NaHCO_3 \text{ taken} = \frac{4.20\,g}{84.0\,g\,mol^{-1}} = 0.0500\,mol$$

equation 1: amount of Na_2O produced $= 0.5 \times 0.0500 = 0.0250\,mol$
mass of Na_2O produced $= 0.0250\,mol \times 62.0\,g\,mol^{-1} = 1.55\,g$
This does not equal 2.65 g, so equation 1 is incorrect

equation 2: amount of Na_2CO_3 produced $= 0.5 \times 0.0500 = 0.0250\,mol$
mass of Na_2CO_3 produced $= 0.0250\,mol \times 106.0\,g\,mol^{-1} = 2.65\,g$
This is the same as the experimental mass produced, so equation 2 is in agreement with the experimental data.

◀ The ratio of $NaHCO_3$ to solid product in equations 1 and 2 is 2:1, so the number of moles of solid formed is half the number of moles of reactant.

◀ Equation 3 predicts that 2.00 g of solid would be left, so it cannot be the correct equation.

Limiting reagent

When a reaction is carried out in the laboratory, the reactants are not always present in the exact stoichiometric ratio determined by the equation. As a result, one reactant is used completely; some of the other reactant is left over. The reagent left over is said to be in excess; the one used completely is said to be the **limiting reagent**.

> A limiting reagent is the substance that determines the theoretical yield of product in a reaction.

An analogy is a factory producing sunglasses (S). Every frame (F) needs two lenses (L). The 'equation' for the process is:

$$F + 2L \rightarrow S$$

If the factory owner buys 144 frames and 280 lenses, the maximum number of sunglasses that can be produced is limited by the number of lenses. One hundred and forty-four frames need 288 lenses. There are only 280 lenses, so the lenses are the limiting factor and only 140 sunglasses can be made.

The lenses are the limiting 'reagent' and the frames are the 'reagent' in excess.

To identify the limiting reagent and hence calculate the theoretical yield, use the following method.

Step 1: calculate the amount (in moles) of one reagent and use the reaction stoichiometry to calculate the amount (in moles) of product that could be formed from this reagent.

Step 2: calculate the amount (in moles) of the second reagent and use the reaction stoichiometry to calculate the amount (in moles) of product that could be formed from this second reagent.

Step 3: the reagent that produces the *least* amount (in moles) of product is the limiting reagent.

Step 4: calculate the theoretical yield of the product from the *least* amount (in moles) of product calculated in steps 1 and 2, i.e. from the limiting reagent.

In steps 1 and 2 you must be able to convert mass or volume of reactant into moles and then use the ratio of numbers of moles of reactant to product in the equation (the stoichiometry) to calculate the amount of product. Step 4 is the conversion of amount (moles) of product into mass or volume of product.

> Note the analogy to the mole here. The number of moles of a chemical is a measure of the (very large) number of molecules of that chemical.

> **Worked example 1**
> Solutions containing 12.8 g of sulfuric acid (molar mass 98.1 g mol^{-1}) and 10.0 g of sodium hydroxide (molar mass 40.0 g mol^{-1}) are mixed and produce sodium sulfate and water according to the following equation:
>
> $$2NaOH + H_2SO_4 \rightarrow Na_2SO_4 + 2H_2O$$
>
> Calculate the mass of sodium sulfate (molar mass 142.1 g mol^{-1}) produced.

Answer

Step 1: amount of NaOH $= \dfrac{\text{mass}}{\text{molar mass}} = \dfrac{10.0\,g}{40.0\,g\,mol^{-1}} = 0.250\,mol$

ratio of Na_2SO_4 to NaOH = 1:2

theoretical amount of Na_2SO_4 that would be produced $= \frac{1}{2} \times 0.250$

$= 0.125\,mol$

Step 2: amount of $H_2SO_4 = \dfrac{\text{mass}}{\text{molar mass}} = \dfrac{12.8\,g}{98.1\,g\,mol^{-1}} = 0.130\,mol$

ratio Na_2SO_4 to H_2SO_4 = 1:1

theoretical amount of Na_2SO_4 that would be produced $= 0.130\,mol$

Step 3: the reagent that produces the least product (0.125 mol) is sodium hydroxide, so that is the limiting reagent.

Step 4: mass of Na_2SO_4 produced — moles $\times$ molar mass

$= 0.125\,mol \times 142.1\,g\,mol^{-1}$

$= 17.8\,g$

> Even though there are more moles of sodium hydroxide, it is the limiting reagent. This is because of the stoichiometry of the equation — 2 moles of sodium hydroxide are required to react with 1 mole of sulfuric acid.

Worked example 2

7.95 g of copper(II) oxide is mixed with a solution containing 7.35 g of sulfuric acid. The equation for the reaction is:

$CuO(s) + H_2SO_4(aq) \rightarrow CuSO_4(aq) + H_2O(l)$

a Determine which is the limiting reagent.

b Hence, calculate the mass of $CuSO_4$ produced.

c On evaporation of the solution, 16.3 g of crystals of $CuSO_4.5H_2O$ are formed:

$CuSO_4(aq) \rightarrow CuSO_4.5H_2O(s)$

Calculate the percentage yield.

Answer

a molar mass of CuO $= 63.5 + 16.0 = 79.5\,g\,mol^{-1}$

amount of CuO $= \dfrac{7.95\,g}{79.5\,g\,mol^{-1}} = 0.100\,mol$

The ratio $CuO:CuSO_4$ is 1:1, so 0.100 mol of $CuSO_4$ would be produced.

molar mass of $H_2SO_4 = (2 \times 1.0) + 32.1 + (4 \times 16.0) = 98.1\,g\,mol^{-1}$

amount of $H_2SO_4 = \dfrac{7.35\,g}{98.1\,g\,mol^{-1}} = 0.0750\,mol$

The ratio $H_2SO_4:CuSO_4$ is 1:1, so 0.0750 mol of $CuSO_4$ would be produced. This is the smaller amount, so sulfuric acid is the limiting reagent.

b amount of $CuSO_4$ produced $= 0.0750\,mol$

molar mass of $CuSO_4 = 63.5 + 32.1 + (4 \times 16.0) = 159.5\,g\,mol^{-1}$

mass of $CuSO_4$ produced $= 0.0750\,mol \times 159.6\,g\,mol^{-1} = 12.0\,g$

c 1 mol of $CuSO_4(aq)$ produces 1 mol $CuSO_4.5H_2O(s)$.

molar mass of $CuSO_4.5H_2O = 159.6 + (5 \times 18.0) = 249.6\,g\,mol^{-1}$

theoretical yield $= 0.0750\,mol \times 249.6\,g\,mol^{-1} = 18.7\,g$

% yield $= \dfrac{\text{actual yield}}{\text{theoretical yield}} \times 100 = \dfrac{16.3}{18.7} \times 100 = 87.2\%$

Concentration

A **solution** consists of a substance that is dissolved — the **solute** — and the substance that is doing the dissolving — the **solvent**. For example, when a salt is dissolved in water, the salt is the solute and the water is the solvent. Solutions may be coloured or colourless, but they are always clear and never cloudy.

Units of concentration

Solutions can contain different amounts of solute up to a maximum value, which is called the **solubility** of the solute. Therefore, the same volume of solutions of a solute can contain different amounts of that solute. To enable the amount or the mass of solute in a given volume to be determined, the concentration of the solution must be known.

- The most common unit of concentration is **mol dm^{-3}**.
- Mol dm^{-3} is sometimes called molarity (symbol, M).
- To calculate a concentration in mol dm^{-3}, the amount (moles) of solute is divided by the volume of the solution in dm^3.

Conversions such as amount of solute to concentration can be performed using a version of the 'mole triangle':

$$\text{moles} = \text{concentration} \times \text{volume}$$
$$\text{concentration} = \frac{\text{moles}}{\text{volume}}$$
$$\text{volume} = \frac{\text{moles}}{\text{concentration}}$$

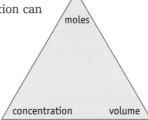

> The volume is the volume of the solution, not of the solvent.

- Another unit of concentration that is sometimes used is **g dm^{-3}**.
- This is the mass of the solute divided by the volume (in dm^3) of the solution.
- The conversion of concentrations in mol dm^{-3} to g dm^{-3} is the same as converting mol to mass, which is to multiply by the molar mass of the solute.
- To convert concentrations in g dm^{-3} to mol dm^{-3}, divide by the molar mass of the solute.

Worked example 1

a Calculate the concentration in mol dm^{-3} of a solution made by dissolving 0.123 mol of sodium hydroxide in water and making up the solution to a total volume of 250 cm^3.

b Calculate the concentration of this solution in g dm^{-3}.

Answer

a volume of solution $= \dfrac{250}{1000} = 0.250 \text{ dm}^3$

concentration $= \dfrac{\text{moles of solute}}{\text{volume of solution}} = \dfrac{0.123 \text{ mol}}{0.250 \text{ dm}^3} = 0.492 \text{ mol dm}^{-3}$

b molar mass of NaOH $= 23.0 + 16.0 + 1.0 = 40.0 \text{ g mol}^{-1}$

concentration of solution $= 0.492 \text{ mol dm}^{-3} \times 40.0 \text{ g mol}^{-1} = 19.7 \text{ g dm}^{-3}$

> Mol cancels with mol^{-1}, leaving the final unit as g dm^{-3}. This is a useful check to make sure that you have carried out the conversion correctly.

Worked example 2

Calculate the mass of hydrated sodium carbonate, $Na_2CO_3.10H_2O$, that is required to make up 250 cm^3 of a 0.100 mol dm^{-3} solution.

Answer

amount of sodium carbonate required = concentration × volume of solution

$$\text{amount} = 0.100 \text{ mol dm}^{-3} \times \frac{250 \text{ dm}^3}{1000} = 0.0250 \text{ mol}$$

molar mass of $Na_2CO_3.10H_2O = (2 \times 23.0) + 12.0 + (3 \times 16.0) + (10 \times 18.0)$
$$= 286.0 \text{ g mol}^{-1}$$

mass required = moles × molar mass = 0.0250 mol × 286.0 g mol^{-1} = 7.15 g

In the question, the volume of the solution was given in cm^3. Therefore, it had to be converted to dm^3.

Worked example 3

Calculate the amount of sulfuric acid in 22.4 cm^3 of 0.0502 mol dm^{-3} solution.

Answer

amount (in moles) of sulfuric acid = concentration × volume

$$\text{amount} = 0.0502 \text{ mol dm}^{-3} \times \frac{22.4 \text{ dm}^3}{1000} = 0.00112 \text{ mol}$$

Worked example 4

What volume of sodium hydroxide solution of concentration 0.100 mol dm^{-3} contains 0.00250 mol of sodium hydroxide?

Answer

$$\text{volume of sodium hydroxide solution} = \frac{\text{moles}}{\text{concentration}}$$

$$\text{volume} = \frac{0.00250 \text{ mol}}{0.100 \text{ mol dm}^{-3}} = 0.0250 \text{ dm}^3 = 25.0 \text{ cm}^3$$

Concentrations of very dilute solutions

There are two units used commonly:
- nanomoles per dm^3 (nmol dm^{-3}); 1 nmol = 10^{-9} mol
- parts per million (ppm)

'Parts per million' is normally the mass of a solute per million grams of solvent. (The assumption is made that the masses of the solvent and the solution are the same because the solution is very dilute.)

Worked example 1

Aluminium sulfate is used as a coagulant in domestic water. Analysis of some water found that the aluminium ion concentration was 6.3 nmol dm^{-3}. Calculate the mass of aluminium ions in 1 dm^3 of the treated water.

Answer

6.3 nmol = 6.3×10^{-9} mol

molar mass of Al^{3+} = 27.0 g mol^{-1}

mass of Al^{3+} ions in the water = 6.3×10^{-9} mol × 27.0 g mol^{-1}
$$= 1.70 \times 10^{-7} \text{ g} (= 1.70 \times 10^{-4} \text{ mg} = 170 \text{ ng})$$

Chapter 4: Calculations from chemical equations   73

Worked example 2

A bottle of mineral water was labelled as containing 2.0 ppm of sodium ions.

a Calculate its concentration in nmol dm^{-3}.

b Calculate the number of sodium ions in 1.0 dm^3 of the mineral water.
(Avogadro constant = 6.02×10^{23} mol^{-1})

Answer

a 1 000 000 g of water contain 2.0 g of sodium ions

1000 g of water has a volume of 1.0 dm^3 and contains

$$\frac{2.0}{1000 \, g} = 0.0020 \, g \text{ of sodium ions}$$

$$\text{amount (moles) of sodium ions} = \frac{0.0020 \, g}{23.0 \, g \, mol^{-1}}$$

$$= 8.7 \times 10^{-5} \, mol = 8.7 \times 10^4 \, nmol$$

The concentration of sodium ions is 8.7×10^4 nmol dm^{-3}.

b number of ions = moles × Avogadro constant

$$= 8.7 \times 10^{-5} \, mol \times 6.02 \times 10^{23} \, mol^{-1} = 5.2 \times 10^{19}$$

◀ 1 mol = 1×10^9 nmol,
so 8.7×10^{-5} mol
= $8.7 \times 10^{-5} \times 1 \times 10^9$
= 8.7×10^4 nmol

Questions

1 Calculate the mass of iron(III) hydroxide precipitated in the reaction between 12.7 g of iron(III) sulfate and excess sodium hydroxide solution.

$$Fe_2(SO_4)_3(aq) + 6NaOH(aq) \rightarrow$$
$$2Fe(OH)_3(s) + 3Na_2SO_4(aq)$$

2 Copper reacts with silver nitrate solution according to the equation:

$$Cu(s) + 2AgNO_3(aq) \rightarrow Cu(NO_3)_2(aq) + 2Ag(s)$$

Calculate the mass of copper needed to react with a solution containing 12.6 g of silver nitrate.

3 Calculate the volume of oxygen produced when 33.3 g of sodium nitrate is heated. Sodium nitrate decomposes according to the equation:

$$2NaNO_3(s) \rightarrow 2NaNO_2(s) + O_2(g)$$

Under the conditions of the experiment, 1 mol of gas occupies a volume of 25.0 dm^3.

4 When 3000 g of hydrogen is reacted with excess nitrogen under high pressure and temperature in the presence of an iron catalyst, 2550 g of ammonia is produced.

$$N_2 + 3H_2 \rightarrow 2NH_3$$

Calculate the theoretical yield and hence the percentage yield in the process.

5 Calculate the amount of sulfuric acid in 23.4 cm^3 of a 0.0545 mol dm^{-3} solution of the acid.

6 Calculate the volume of a 0.106 mol dm^{-3} solution of sodium hydroxide, which contains 0.00164 mol.

7 Calculate the mass of ethanedioic acid, $H_2C_2O_4.2H_2O$, that is needed to make 500 cm^3 of a 0.0500 mol dm^{-3} solution.

8 4.50 g of iron powder was added to 50.0 cm^3 of 2.00 mol dm^{-3} copper(II) sulfate solution. The equation for the reaction is:

$$Fe(s) + CuSO_4(aq) \rightarrow Cu(s) + FeSO_4(aq)$$

a Calculate which reagent is limiting.

b Calculate the mass of copper produced.

c What observation would tell you that you were correct in identifying the limiting reagent?

9 21.5 g of sodium hydroxide was added carefully to 500.0 cm^3 of 1.00 mol dm^{-3} sulfuric acid solution.

a Write the equation for the reaction.

b Calculate which reagent is limiting.

c Calculate the mass of sodium sulfate produced.

d What would be observed if pieces of red and blue litmus paper were added to the solution after the reaction?

Chapter 5
Energetics

Boron: glass, alloys and abrasives

Atomic number: 5

Electron configuration: $1s^2\ 2s^2\ 2p_x{}^1$

Symbol: B

Boron is a metalloid at the top of group 3 of the periodic table. It reacts with non-metals such as chlorine to form covalent compounds in which boron has an incomplete octet. Its oxide is acidic, reacting with alkalis to form borate salts.

The main ore is borax, $Na_2B_4O_5(OH)_4.8H_2O$, which is found in dry areas such as the Mojave Desert in California.

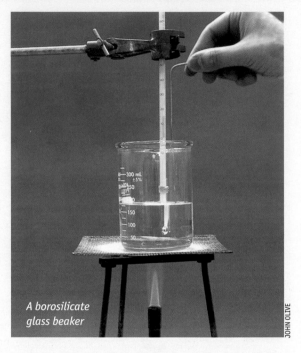

A borosilicate glass beaker

JOHN OLIVE

Glass

Adding boron compounds to soda glass makes the glass heat resistant. Borate ions consist of long chains of boron and oxygen atoms, which fit well with the silicon–oxygen chains of glass. Borates make glass harder, with a higher softening temperature. Borosilicate glass is used for chemical apparatus (e.g. 'hard glass' test tubes) that can be strongly heated with a Bunsen flame. It shatters less easily than ordinary glass, so it is used to make beakers and condensers. Glass ovenproof dishes are also made of borosilicate glass.

A trace of boron implanted in a crystal of silicon causes that part of the crystal to become a semi-conductor. The boron atom fits into the silicon lattice. As a boron atom has one less outer-shell electron than a silicon atom, a positive 'hole' is created in the lattice. This 'hole' can move and acts as a carrier of positive charge.

Alloys

Boron is a very hard element that does not conduct electricity at room temperature. Above 400°C, it starts to conduct. It forms an alloy with aluminium, resulting in a strong metal that is almost as good a conductor as copper, but is much less dense. This alloy is used in overhead electrical transmission cables. The lower weight, compared with wires made of pure copper, means that the pylons supporting the cables can be spaced further apart.

Abrasives

Boron nitride exists in two forms. When the two elements combine, a slippery white solid (formula BN) is formed. This has a structure similar to that of graphite, but with alternate boron and nitrogen atoms in the planar hexagonal rings. If this form is heated to high temperatures under pressure, a diamond-like structure is obtained, which is as hard as diamond and is used as an abrasive. Boron carbide is almost as hard and also used as an abrasive, e.g. for polishing ceramics.

Energetics

Introduction

The **first law of thermodynamics** states that energy can neither be created nor destroyed. However, one form of energy can be converted into another. This law is sometimes called the **law of conservation of energy**. For example, when petrol burns in a car engine, the chemical energy of the petrol–air mixture is converted into the kinetic energy of the car, as well as into heat and sound.

There are various forms of energy:

- **work** — when a force moves an object.
- **kinetic energy** — the energy due to the motion of a body. The value is equal to half the mass of the body multiplied by the square of its speed.
- **gravitational potential energy** — the energy due to the position of the object relative to the centre of the earth. For example, the water in a reservoir is at a higher potential energy than the water below the dam.
- **chemical potential energy** — the energy stored in a molecule. It is due to the relative position of the atoms.
- **radiant energy** — the energy in light and sound. Light of high frequency has more energy than low-frequency light of the same intensity. The energy of a photon of radiation is given by the expression $E = h\nu$, where h is Planck's constant and ν is the frequency of the radiation. A single photon of blue light has energy, $E = 6.63 \times 10^{-34} \times 6.4 \times 10^{14} = 4.2 \times 10^{-19}$ J.
- **electrical energy** — this can be in the form of electric potential energy or the energy due to an electric current.

Albert Einstein suggested that mass is related to energy by the equation $E = mc^2$, where c is the speed of light. The conversion of mass to energy only occurs during nuclear reactions, such as the emission of radioactivity, nuclear fission and nuclear fusion.

Heat is a form of kinetic energy. When an object is heated, the random motion of the atoms and molecules in that body increases. This motion can be in the form of vibration, rotation or the movement of molecules. Heat flows spontaneously from a hotter object to a colder object.

Radiation consists of a stream of packets of electromagnetic energy called photons. Visible light photons have more energy than infrared photons, because they have a higher frequency (for blue light, $\nu = 6.4 \times 10^{14}$ Hz; for infrared radiation, ν is around 3.0×10^{14} Hz).

> ℮ Do not confuse heat with temperature. Temperature is determined by the *average* energy of the random motion of the atoms and molecules. A beaker of water at 50°C is hotter than a bath full of water at 40°C, but the amount of heat energy in the bathwater is considerably greater than the heat in the beaker of water.

The SI unit of energy is the **joule**, symbol J, but the energy changes of chemical reactions are usually expressed in **kilojoules**, symbol kJ.

1 kJ = 1000 J

When a substance is heated, its temperature increases. The amount of heat required and the temperature change are given by the expression:

heat required = mass × specific heat capacity × the rise in temperature

$$\text{heat} = m \times c \times \Delta T$$

where m = the mass, c = the specific heat capacity and ΔT is the temperature change

The two temperatures can be measured in degrees celsius, °C, or kelvin, K. The numerical value of ΔT is the same.

The specific heat capacity of a substance is the heat required to increase the temperature of 1g of the substance by 1°C.

The specific heat capacity of water is 4.18 J g^{-1} °C^{-1}.

The specific heat capacity of iron is 0.45 J g^{-1} °C^{-1}.

> **Worked example**
> Calculate the heat required to increase the temperature of 100 cm^3 of water from 17.6°C to 50.5°C.
>
> **Answer**
> density of water = 1 g cm^{-3}
> mass of 100 cm^3 of water = 100 g
> ΔT = 50.5 − 17.6 = 32.9°C
> heat required = $m \times c \times \Delta T$ = 100 g × 4.18 J g^{-1} °C^{-1} × 32.9°C
> = 13 800 J = 13.8 kJ

Enthalpy changes

When a chemical reaction takes place, chemical energy is interchanged with heat and work.

If chemical energy is changed into heat energy, the chemicals become hot and heat then flows to the surroundings. Reactions that produce heat are said to be **exothermic**. For example, when a mixture of powdered aluminium and iron(III) oxide is heated, a violent reaction takes place. The temperature rises so much that the iron produced melts. This exothermic reaction is used to join railway lines together.

In an exothermic reaction, chemical energy is converted into heat energy and the temperature of the system rises.

The term **system** is used in thermochemistry to represent the reaction mixture, including any solvent and the reaction vessel. The **surroundings** are everything outside the system, which in practice is the air in the room in which the experiment is taking place.

If, during a reaction, the mixture cools, the reaction is said to be **endothermic**. For example, when ethanoic acid is added to solid sodium hydrogencarbonate a rapid reaction takes place and the temperature drops considerably.

In an endothermic reaction, heat energy is converted into chemical energy and the temperature of the system falls.

The release or absorption of heat energy by a system at constant pressure is caused by a change in the **enthalpy** of the chemicals.

The enthalpy, *H*, is the chemical energy in the system at constant pressure that can be converted into heat.

TOPFOTO

Using an exothermic reaction to weld railway lines

The enthalpy of a substance depends upon:
- the physical state of the substance (whether it is a solid, liquid or gas)
- the amount (in moles) of the substance
- the temperature and pressure

In a chemical change, the enthalpy of the products will be different from the enthalpy of the reactants. The change in enthalpy, Δ**H**, is given by:

$$\Delta H = H_{products} - H_{reactants}$$

If enthalpy is converted into heat energy (an *exothermic* reaction), the value of $H_{products}$ is less than $H_{reactants}$, so Δ*H* is *negative*. This is shown in an energy level diagram (Figure 5.1).

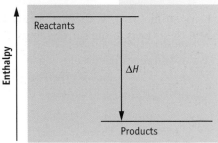

Figure 5.1 An enthalpy level diagram for an exothermic reaction

◀ Δ*H* is **negative** for an **exothermic** reaction.

An example of an endothermic reaction — ethanoic acid and solid sodium hydrogencarbonate (a) at the start of the reaction and (b) after 15 minutes.

If heat energy is converted into enthalpy (an *endothermic* reaction), the value of $H_{products}$ is greater than $H_{reactants}$, so ΔH is *positive*. This is shown in Figure 5.2.

ΔH is **positive** for an **endothermic** reaction.

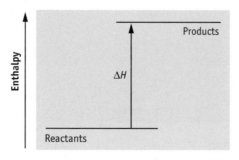

Figure 5.2
An enthalpy level diagram for an endothermic reaction

Absolute enthalpy values cannot be found. The only value that can be measured is the difference in enthalpy when a reaction or physical change takes place. For example, when 1 mol of methane, CH_4, is burnt in excess oxygen, 890 kJ of heat energy is produced. This means that the enthalpy change, ΔH, is -890 kJ mol^{-1}.

> **e** Note that as heat is produced, chemical energy must have been converted into heat. Therefore, ΔH (the change in enthalpy) is negative: $H_{products} < H_{reactants}$.

The thermochemical equation that shows the combustion of 1 mol of methane is:
$$CH_4(g) + 2O_2(g) \rightarrow CO_2(g) + 2H_2O(l) \quad \Delta H = -890 \text{ kJ mol}^{-1}$$

e It is essential to put state symbols into all thermo-chemical equations.

In practice, most enthalpy changes are quoted as **standard enthalpy changes**. This means that the heat produced is measured under **standard conditions**, which are:

- a pressure of 1 atm
- a stated temperature (usually 25°C)
- all solutions at a concentration of 1 mol dm^{-3}

The symbol for standard conditions is a superscript $\ominus$, as in $\Delta H^{\ominus}$.

Hot and cold packs

Self-heating soup tins have two compartments. One compartment contains the soup and the other contains solid calcium oxide and a sachet of water. When the sachet is broken, the water reacts with the calcium oxide in an exothermic reaction and the heat produced warms the soup in the inner container.

$$CaO(s) + H_2O(l) \rightarrow Ca(OH)_2(aq) \qquad \Delta H = -80 \, kJ \, mol^{-1}$$

20 g of calcium oxide mixed with water causes the temperature of 200 cm^3 of soup to rise by about 25°C.

It is important that when an athlete pulls a muscle, a cold pack is applied quickly. One such pack consists of a divided plastic bag, one part of which contains a salt. The other part contains water. When crushed, the two mix and the salt dissolves endothermically. The salt most commonly used is ammonium nitrate:

$$NH_4NO_3(s) + aq \rightarrow NH_4^+(aq) + NO_3^-(aq)$$
$$\Delta H = +26 \, kJ \, mol^{-1}$$

20 g of ammonium nitrate causes the temperature of 100 cm^3 of water to fall by 15°C.

Hess's law

Hess's law states that the enthalpy of a substance at a particular temperature and pressure is independent of how it is formed. It is usually stated as the enthalpy change of a physical or chemical process is the same whatever the path from the reactants to the products. Hess's law is an example of the law of conservation of energy.

> The enthalpy change for any reaction is independent of the route taken from reactants to products.

Consider the cycle:

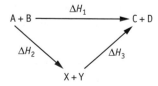

The enthalpy change from A + B *directly* to C + D is ΔH_1.

The enthalpy change from A + B to C + D *via* X and Y is $\Delta H_2 + \Delta H_3$.

Thus, by Hess's law:

$\Delta H_1 = \Delta H_2 + \Delta H_3$ or $\Delta H_3 = \Delta H_1 - \Delta H_2$

It is impossible to measure the enthalpy change directly for the reaction:

$C(s) + \frac{1}{2}O_2(g) \rightarrow CO(g)$

However, the enthalpy changes for the following reactions can be measured:

$C(s) + O_2(g) \rightarrow CO_2(g)$ $\quad \Delta H_1 = -394$ kJ mol^{-1}

$CO(g) + \frac{1}{2}O_2(g) \rightarrow CO_2(g)$ $\quad \Delta H_2 = -283$ kJ mol^{-1}

A Hess's law triangle can be drawn and the enthalpy change for the first equation can be calculated.

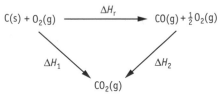

The enthalpy change directly from C to CO_2 is ΔH_1.

The indirect enthalpy change via CO is $\Delta H_r + \Delta H_2$.

These are equal, so:

$\Delta H_1 = \Delta H_r + \Delta H_2$

$\Delta H_r = \Delta H_1 - \Delta H_2 = -394 - (-283) = -111$ kJ mol^{-1}

This can also be shown by an energy level diagram:

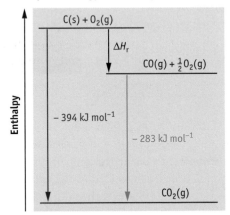

$\Delta H_r + (-283) = -394$

$\Delta H_r = -394 + 283 = -111$ kJ mol^{-1}

e Compatible data must be used. The enthalpy of a substance varies slightly with temperature, so the data must all be under the same conditions. This is why enthalpies in data books are quoted under conditions of 1 atm and 298 K.

Standard enthalpies

Standard enthalpy of formation, $\Delta H_f^{\ominus}$

The standard enthalpy of formation is the enthalpy change when *1 mol* of a substance is formed from its *elements* in their *standard* states at 1 atm pressure and at a stated temperature, usually 298 K (25 °C). The units are kJ mol^{-1}.

Standard state means the form in which the element is normally found under the quoted conditions of pressure and temperature. This means that bromine is a liquid and sulfur is a solid.

The standard enthalpy of formation of carbon monoxide is the heat change for the reaction:

$$C(s) + \tfrac{1}{2}O_2(g) \rightarrow CO(g)$$

The standard enthalpy of formation of ethanol is the heat change for the reaction:

$$2C(s) + 3H_2(g) + \tfrac{1}{2}O_2(g) \rightarrow C_2H_5OH(l)$$

> ℯ Note that state symbols are always included in thermochemical equations. The value for the enthalpy of formation of gaseous ethanol is different from that for liquid ethanol:
> ■ $\Delta H_f^{\ominus}$ ($C_2H_5OH(l)$) $= -277.1$ kJ mol^{-1}
> ■ $\Delta H_f^{\ominus}$ ($C_2H_5OH(g)$) $= -238.5$ kJ mol^{-1}

Standard enthalpy of reaction, $\Delta H_r^{\ominus}$

The standard enthalpy of reaction is the enthalpy change when the number of moles of the substances in the equation *as written* react under standard conditions of 1 atm pressure and a stated temperature, usually 298K (25°C). The units are kJ mol^{-1}.

Consider the reaction:

$$2Na(s) + Cl_2(g) \rightarrow 2NaCl(s)$$

The enthalpy change for the reaction is the enthalpy change when 2 mol of solid sodium reacts with 1 mol of gaseous chlorine to produce 2 mol of solid sodium chloride. The value of $\Delta H_r^{\ominus}$ is -822 kJ.

Compare this with the reaction:

$$Na(s) + \tfrac{1}{2}Cl_2(g) \rightarrow NaCl(s)$$

The value of $\Delta H_r^{\ominus}$ is -411 kJ, which is half the value for the first reaction. This is because half the amounts of sodium and chlorine are reacting.

Calculation of enthalpy of reaction from standard enthalpy of formation data

The enthalpy of reaction can be calculated from the standard enthalpy of formation using a Hess's law cycle:

> ℯ The correct formulae and state symbols, including those of the elements, must be shown in the cycle.

$$\Delta H_f \text{ of } A + 2 \times \Delta H_f \text{ of } B + \Delta H_r = 3 \times \Delta H_f \text{ of } C + 4 \times \Delta H_f \text{ of } D$$
$$\Delta H_r = \{3\Delta H_f(C) + 4\Delta H_f(D)\} - \{(\Delta H_f(A) + 2\Delta H_f(B))\}$$

A formula that enables the enthalpy of a reaction to be calculated from enthalpy of **formation** data is:

$$\Delta H_r = \sum \Delta H_f \text{ (products)} - \sum \Delta H_f \text{ (reactants)}$$

The symbol $\sum$ means 'the sum of'.
This 'formula' can *only* be used when enthalpy of formation data are given in the question.

Worked example

Use the data below to draw a Hess's law diagram and hence calculate the enthalpy change for the reaction:

$$CH_2{=}CH{-}CH{=}CH_2(g) + 2HBr(g) \rightarrow CH_3CHBrCHBrCH_3(g)$$

Substance	Enthalpy of formation/kJ mol^{-1}
Butadiene, $CH_2{=}CH{-}CH{=}CH_2(g)$	+162
Hydrogen bromide, HBr(g)	−36.4
2,3-dibromobutane, $CH_3CHBrCHBrCH_3(g)$	−103

Answer

$$\Delta H_f \text{ (butadiene)} + 2\Delta H_f \text{ (HBr)} + \Delta H_r = \Delta H_f \text{ (2,3-dibromobutane)}$$
$$\Delta H_r = \Delta H_f \text{ (2,3-dibromobutane)} - \{\Delta H_f \text{ (butadiene)} + 2\Delta H_f \text{ (HBr)}\}$$
$$= -103 - \{+162 + (2 \times -36.4)\} = -192.2 \text{ kJ mol}^{-1}$$

Note that, because there are 2 mol of HBr in the equation, the cycle contains $2 \times \Delta H_f$ of HBr.

e If the question had not asked for a Hess's law diagram, the answer could have been obtained by using the formula:

$$\Delta H_r = \sum \Delta H_f \text{ (products)} - \sum \Delta H_f \text{ (reactants)}$$
$$\Delta H_r = \Delta H_f \text{ (2,3-dibromobutane)} - \{\Delta H_f \text{ (butadiene)} + 2\Delta H_f \text{ (hydrogen bromide)}\}$$
$$= -103 - \{+162 + (2 \times -36.4)\} = -192.2 \text{ kJ mol}^{-1}$$

Experimental method to find the enthalpy of a reaction using an insulated container

If a reaction takes place at a reasonable rate at room temperature, the heat change can be measured using an expanded polystyrene cup as a calorimeter. Polystyrene is a thermal insulator. Not only does it minimise heat loss to the surroundings, it absorbs little heat itself. A polystyrene cup can be used when measuring the temperature change that occurs when iron reacts with copper(II) sulfate solution according to the equation:

$$Fe(s) + CuSO_4(aq) \rightarrow FeSO_4(aq) + Cu(s)$$

The enthalpy of the reaction is measured using the following procedure:

- A measured volume of copper(II) sulfate solution of known concentration is pipetted into an expanded polystyrene cup held in a glass beaker.
- Some powdered iron is weighed out, such that the iron is in excess.
- The temperature of the aqueous copper(II) sulfate is measured every 30 seconds for 2 minutes. At 2.5 minutes, the powdered iron is tipped in.

■ The solution is stirred and the temperature recorded every 30 seconds until a maximum temperature is reached, and then for a further 2 minutes.

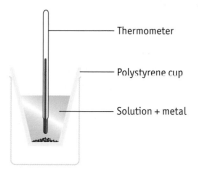

Figure 5.3 Experiment to determine enthalpy of reaction

Three quantities have to be evaluated before the enthalpy of the reaction can be calculated:

■ the **mass of solution** that was heated. The density of the solution is assumed to be the same as pure water (1 g cm^{-3}). Therefore, if 50.0 cm^3 of copper(II) sulfate had been used, the mass would have been 50.0 g.

■ the **temperature rise.** The reaction is slow, so some heat is lost during the experiment. To find a more accurate value of ΔT than simply the difference between the temperature at the start and that at the finish, a temperature–time graph is plotted and the value of ΔT found by extrapolation:
  – Draw a straight line through the first five points and extend (extrapolate) the line to at least 3 minutes.
  – Draw a straight line from the maximum to the final temperature and extrapolate it backwards to before the iron was added.
  – Measure the *difference* in height between these two lines at the time that the iron was added (2.5 minutes in the experiment described above). This height is ΔT (Figure 5.4).

■ the **amount** (moles) of copper(II) sulfate that reacts:
  amount of copper(II) sulfate = concentration × volume in dm^3

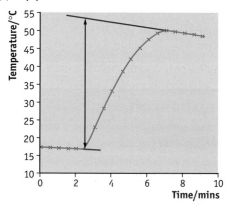

ⓔ An alternative method is to weigh the empty polystyrene cup, add the copper(II) sulfate solution and weigh it again. The difference is the mass of the copper(II) sulfate solution.

Figure 5.4 A temperature–time graph for determining enthalpy of reaction

ⓔ Always extrapolate *forwards* from the last reading before the reaction started, and *backwards* from the highest temperature reached (lowest temperature if the reaction was endothermic).

Worked example

Use the data given to calculate the enthalpy of reaction for:

$$Fe(s) + CuSO_4(aq) \rightarrow FeSO_4(aq) + Cu(s)$$

In this experiment:

- volume of 1.00 mol dm^{-3} copper sulfate solution = 50.0 cm^3
- temperature rise (from Figure 5.4) = 53.3 − 16.9 = 36.4°C
- the specific heat capacity of the solution is 4.18 J g^{-1} °C^{-1}

Answer

mass of solution = volume × density

$\qquad$ = 50.0 cm^3 × 1 g cm^{-3}

$\qquad$ = 50.0 g

heat produced = mass × specific heat capacity × rise in temperature

$\qquad$ = 50.0 g × 4.18 J g^{-1} °C^{-1} × 36.4°C

$\qquad$ = 7610 J = 7.61 kJ

amount of copper sulfate taken = 1.00 mol dm^{-3} × $\dfrac{50.0 \text{ dm}^3}{1000}$

$\qquad\qquad\qquad$ = 0.0500 mol

heat produced per mol = $\dfrac{\text{heat produced}}{\text{number of moles}}$

$\qquad\qquad$ = $\dfrac{7.60 \text{ kJ}}{0.0500 \text{ mol}}$

$\qquad\qquad$ = 152 kJ mol^{-1}

$\Delta H_r = -152$ kJ mol^{-1}

> **e** Remember that it is the mass of the *solution*, not the mass of copper sulfate, that is used in the expression heat = mass × specific heat capacity × ΔT.

> **e** A common error is for the sign to be wrong in the final answer. Remember that, because the temperature rose, heat was produced. This means that the reaction is exothermic and so the value of ΔH is *negative*.

Possible sources of error

- The reaction is slow, so some heat is lost to the surroundings. Extrapolating the graph, rather than calculating ΔT as $T_{max} - T_{start}$, allows for most of this heat loss. The error makes the experimentally determined value of $\Delta H_r^{\ominus}$ less exothermic than the correct value.
- Some heat will be absorbed by the iron metal and by the thermometer.

◀ A lid on the calorimeter would help to reduce heat loss.

Standard enthalpy of combustion, $\Delta H_c^{\ominus}$

The standard enthalpy of combustion is the enthalpy change when *1 mol* of the substance is burnt in *excess* oxygen under *standard conditions* of 1 atm pressure and a stated temperature (usually 25°C). The units are kJ mol^{-1}.

The standard enthalpy of combustion for ethanol is the enthalpy change when 1 mol of ethanol reacts according to the equation:

$$C_2H_5OH(l) + 3O_2(g) \rightarrow 2CO_2(g) + 3H_2O(l)$$

> **e** Note that all combustion reactions are exothermic, and so all ΔH_c values are negative.

For ethane, it is the enthalpy change when 1 mol of ethane reacts according to the equation:

$$C_2H_6(g) + 3\tfrac{1}{2}O_2(g) \rightarrow 2CO_2(g) + 3H_2O(l)$$

> **e** There must only be 1 mol of the substance on the left-hand side in an enthalpy of combustion equation. The normal balanced equation for the combustion of ethane is:
>
> $$2C_2H_6(g) + 7O_2(g) \rightarrow 4CO_2(g) + 6H_2O(l)$$
>
> Therefore, for the enthalpy of combustion of ethane, this equation must be halved, so that there is only 1 mol of ethane on the left-hand side:
>
> $$C_2H_6(g) + 3\tfrac{1}{2}O_2(g) \rightarrow 2CO_2(g) + 3H_2O(l)$$

Combusting oil wells in southern Iraq

Experimental method to find the enthalpy of combustion of a liquid

To find the enthalpy of combustion of a liquid, a known mass of the liquid is burnt and the heat produced used to warm up a known volume of water. An appropriate laboratory apparatus is shown in Figure 5.5.

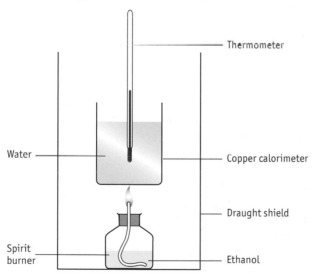

Figure 5.5
Apparatus used to determine the enthalpy of combustion of a liquid

To determine the enthalpy of combustion of a liquid, the following procedure is used:
- The spirit burner containing the liquid is weighed.
- A known volume of water is added to the copper calorimeter.
- The temperature of the water in the calorimeter is measured every minute for four minutes.
- The burner is lit after 4.5 minutes.

- The temperature of the water is measured every minute.
- When the temperature has reached about 20°C above room temperature, the flame in the spirit burner is extinguished and the burner immediately reweighed.
- The temperature readings are stopped 5 minutes after the temperature has reached a maximum value.

Three quantities have to be evaluated before the enthalpy of combustion can be calculated:

- the **mass of water** that was heated. The density of water is $1\,g\,cm^{-3}$. Therefore, if $100\,cm^3$ of water had been measured out, the mass would have been 100 g.
- the **temperature rise**. A significant amount of heat is lost during the experiment, so to find a more accurate value of ΔT, a temperature–time graph is plotted and the value of ΔT found by extrapolation:
  - Draw a straight line through the first five points and extend (extrapolate) it to at least 3 minutes.
  - Draw a straight line from the maximum temperature to the final temperature and extrapolate it backwards to before the time when the burner was lit.
  - Measure the *difference* in heights between these two lines at the time when the heating started. This difference is ΔT (Figure 5.6).
- the **amount in moles** of ethanol burnt. The mass of ethanol used is the mass of the burner and ethanol at the start minus the mass after burning:

$$\text{amount in moles} = \frac{\text{mass}}{\text{molar mass}}$$

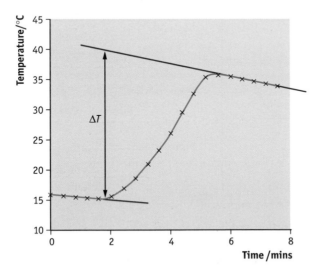

Figure 5.6
A temperature–time graph for determining enthalpy of combustion

Worked example

Use the data given to calculate the enthalpy of combustion of ethanol. The equation is:

$$C_2H_5OH(l) + 3O_2(g) \rightarrow 2CO_2(g) + 3H_2O(l)$$

In this experiment:

- mass of water = 100 g
- temperature rise (from Figure 5.5) = 24.6°C
- mass of ethanol burnt = 0.388 g

The specific heat capacity of water is 4.18 J g^{-1} °C^{-1}

Answer

heat produced = mass of water heated × specific heat capacity of water × ΔT

$$= 100 \text{ g} \times 4.18 \text{ J g}^{-1} \text{ °C}^{-1} \times 24.6\text{°C} = 10\ 282 \text{ J} = 10.28 \text{ kJ}$$

$$\text{amount of ethanol burnt} = \frac{\text{mass}}{\text{molar mass}} = \frac{0.388 \text{ g}}{46 \text{ g mol}^{-1}} = 0.00843 \text{ mol}$$

$$\text{heat produced per mol} = \frac{\text{heat produced}}{\text{number of moles}} = \frac{10.28 \text{ kJ}}{0.00843 \text{ mol}}$$

$$= 1219 \text{ kJ mol}^{-1}$$

ΔH_c of ethanol = −1219 = −1.22 × 10^3 kJ mol^{-1} (3 significant figures)

Possible sources of error

The experimentally determined value in the example (−1220 kJ mol^{-1}) is less exothermic than the accepted data book value. This is caused by a number of factors:

- As the experiment takes a long time, not all the heat lost to the surroundings is compensated for by the extrapolation of the graph.
- Some of the heat released in burning heats up the air and not the water.
- The beaker absorbs some of the heat produced.
- Some of the ethanol may not burn completely to carbon dioxide and water. (Incomplete combustion would cause black soot to be deposited on the bottom of the beaker.)
- The conditions are not standard. Water vapour, and not liquid water, is produced.

These errors result in less heat being absorbed by the water, so the value for ΔT is too low. Hence, the value calculated for ΔH_c is also low.

The calorie is an old unit of energy. It is related to the joule by the expression 1 cal = 4.18 J.

When food is metabolised, it is converted into carbon dioxide and water. So the 'Calorific value' of a food is the same as the heat produced when that substance is burned.

The specific enthalpy of combustion of bread is −11 kJ g^{-1}. This means that when 1 g of bread is metabolised, 11 kJ or 2.6 kcal of energy are released (see Table 5.1).

Calculation of standard enthalpy of formation from combustion enthalpy data

Many enthalpies of formation cannot be measured directly. However, their values can be calculated from enthalpy of combustion data, using Hess's law:

(e) It is the mass of the *water* and not the mass of ethanol that is used in the expression: heat = mass × specific heat capacity × ΔT.

(e) Remember that, because the temperature rose, heat was produced. This means that the reaction is exothermic, and so the value of ΔH is *negative*.

(e) All combustion reactions are exothermic.

Table 5.1 Calorific values of some foods

Food	Calorific value/ kcal 100 g^{-1}
Apples	60
Beef	240
Beer	34
Biscuits	110
Bread	260
Butter	810
Eggs	140
Glucose	370
Milk	70
Potatoes	83
Sugar	390

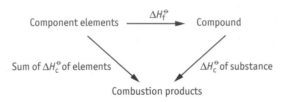

$$\Delta H_f^\ominus + \Delta H_c^\ominus \text{ of compound} = \text{sum of } \Delta H_c^\ominus \text{ of elements}$$
$$\Delta H_f^\ominus = \text{sum of } \Delta H_c^\ominus \text{ of elements} - \Delta H_c^\ominus \text{ of compound}$$

Worked example

Calculate the enthalpy of formation of ethanol, C_2H_5OH, given the following enthalpy of combustion data:

Substance	$\Delta H_c^\ominus$ /kJ mol^{-1}
Ethanol, $C_2H_5OH(l)$	−1371
Carbon, C(s)	−394
Hydrogen, $H_2(g)$	−286

Answer

The Hess's law cycle is:

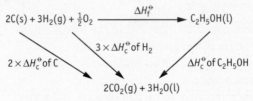

$$2 \times \Delta H_c^\ominus \text{ of C} + 3 \times \Delta H_c^\ominus \text{ of } H_2 = \Delta H_f^\ominus \text{ of ethanol} + \Delta H_c^\ominus \text{ of ethanol}$$
$$\Delta H_f^\ominus \text{ of ethanol} = 2 \times (-394) + 3 \times (-286) - (-1371)$$
$$= -275 \text{ kJ mol}^{-1}$$

The sum of the enthalpies of combustion of the elements equals the sum of the enthalpy changes via the formation of ethanol, and its combustion.

Standard enthalpy of neutralisation, $\Delta H_{neut}^\ominus$

The standard enthalpy of neutralisation is the enthalpy change when *1 mol of water* is produced by the neutralisation of a solution of an acid by *excess* base under standard conditions, with all solutions of *concentration 1 mol dm^{-3}*. The units are kJ mol^{-1}.

For hydrochloric acid, the standard enthalpy of neutralisation is the enthalpy change represented by:

$$HCl(aq) + NaOH(aq) \rightarrow NaCl(aq) + H_2O(l)$$

For sulfuric acid, it is the enthalpy change for:

$$\tfrac{1}{2}H_2SO_4(aq) + NaOH(aq) \rightarrow \tfrac{1}{2}Na_2SO_4(aq) + H_2O(l)$$

All strong acids are completely ionised in solution. Therefore, the ionic equation representing the enthalpy of the neutralisation of any strong acid by a strong base is:

$$H^+(aq) + OH^-(aq) \rightarrow H_2O(l) \quad \Delta H^\ominus = -57.9 \text{ kJ mol}^{-1}$$

Experimental method to find the enthalpy of neutralisation of an acid

The acid should have a concentration of 1.00 mol dm^{-3}. The alkali used to neutralise it must be in slight excess, so that all the acid reacts. The following procedure is used:

- Using a pipette, measure out 25.0 cm^3 of the 1.00 mol dm^{-3} acid solution into an expanded polystyrene cup.
- Measure the temperature of the acid.
- Measure the temperature of the alkali (usually sodium hydroxide solution of concentration 1.1 mol dm^{-3}).
- Calculate the mean of the two temperatures.
- Measure out 25.0 cm^3 of the alkali using a pipette, and add it to the acid solution.
- Stir the mixture with the thermometer and measure the maximum temperature reached.

e An alternative method is to use measuring cylinders to measure out the acid and the alkali, and to weigh the polystyrene cup before and after addition of the solutions.

> *Worked example*
>
> The results of an experiment to find the enthalpy of neutralisation of hydrochloric acid were:
> - volume of 1.00 mol dm^{-3} hydrochloric acid = 25.0 cm^3
> - volume of 1.1 mol dm^{-3} sodium hydroxide = 25.0 cm^3
> - initial temperature of acid = 17.4°C
> - initial temperature of alkali = 17.2°C
> - maximum temperature after mixing = 24.1°C
>
> Specific heat capacity of the final solution = 4.18 J g^{-1} °C^{-1}
>
> Calculate the enthalpy of neutralisation.
>
> **Answer**
>
> average starting temperature = $\frac{1}{2}$(17.4 + 17.2) = 17.3°C
>
> ΔT = 24.1 – 17.3 = 6.8°C
>
> volume of solution that absorbed heat = 25.0 + 25.0 = 50.0 cm^3
>
> mass of solution = volume × density
>
> $\qquad$ = 50.0 cm^3 × 1.00 g cm^{-3} = 50.0 g
>
> heat produced = mass × specific heat capacity × rise in temperature
>
> $\qquad$ = 50.0 g × 4.18 J g^{-1} °C^{-1} × 6.8°C
>
> $\qquad$ = 1421 J = 1.421 kJ
>
> amount of acid = concentration × volume in dm^3
>
> $\qquad$ = 1.00 mol dm^{-3} × $\dfrac{25.0 \text{ dm}^3}{1000}$ = 0.0250 mol
>
> heat produced = $\dfrac{\text{heat produced}}{\text{number of moles}}$ = $\dfrac{1.421 \text{ kJ}}{0.0250 \text{ mol}}$ = 56.8 kJ mol^{-1}
>
> $\Delta H^{\ominus}_{\text{neut}}$ = −57 kJ mol^{-1}

◀ The temperature rose, so the reaction is exothermic. Therefore ΔH is *negative*.

The mass in the 'heat produced' expression is the total mass of the solution, not the mass of the hydrochloric acid.

There is no need to plot a temperature–time curve, because the reaction is so rapid that an insignificant amount of heat is lost to the surroundings.

e The answer should only be given to 2 significant figures because ΔT is only measured to 2 significant figures.

Sources of error
The experimentally derived value for the standard enthalpy of neutralisation is slightly low because of the heat absorbed by the polystyrene cup and by the thermometer.

Application of this method
This method can be used for any rapid reaction that takes place in solution and that is not very exothermic — for example, the precipitation of insoluble salts such as barium sulfate.

Standard enthalpy of atomisation, $\Delta H_a^{\ominus}$

The standard *enthalpy of atomisation* of an element is the enthalpy change when *1 mol* of *gaseous* atoms is formed from the element in its standard state at 1 atm pressure and a stated temperature (usually 298 K). The units are $kJ\,mol^{-1}$.

For bromine, it is the enthalpy change involved in the reaction:
$$\tfrac{1}{2}Br_2(l) \rightarrow Br(g)$$

For sodium, it is the enthalpy change involved in the reaction:
$$Na(s) \rightarrow Na(g)$$

Bond enthalpy, ΔH_B

Bond enthalpy is the enthalpy change when a bond in a gaseous molecule is broken. The units are $kJ\,mol^{-1}$.

The H–H bond enthalpy is the enthalpy change for:
$$H_2(g) \rightarrow 2H(g) \qquad \Delta H_B = +436 \text{ kJ mol}^{-1}$$

Some bond enthalpy values are given in Table 5.2. These are *average* values. For example, the C–H bond enthalpy is the average of the C–H bonds in methane, ethane, ethanol and so on.

The C–H bond enthalpy in methane is one-quarter of the total enthalpy change for:
$$CH_4(g) \rightarrow C(g) + 4H(g) \qquad \Delta H = +1740 \text{ kJ mol}^{-1}$$
$$\Delta H_B(\text{C–H}) = \frac{1740}{4} = +435 \text{ kJ mol}^{-1}$$
$$\Delta H_B(\text{C–H}) \text{ in ethane} = +420 \text{ kJ mol}^{-1}$$

The *average* C–H bond enthalpy in a large number of organic compounds is +413 kJ mol^{-1}. If a bond enthalpy is worked out for a particular compound, the value obtained is slightly different from the average value.

> **e** Bond breaking is always endothermic; bond making is always exothermic.

> **e** Biologists sometimes refer to 'high-energy bonds' in ATP and the release of energy when a phosphate residue is removed from ATP and ADP is formed. This is an over-simplification. The breaking of the P–O bond as ATP is converted to ADP is endothermic. However, energy is then released in the hydration of the phosphate and hydrogen ions, making the overall process exothermic.

Bond	Average bond enthalpy/kJ mol^{-1}	Bond	Average bond enthalpy/kJ mol^{-1}
C–C	+348	H–H	+436
C=C	+612	H–Cl	+431
C–H	+413	H–Br	+366
C–O	+360	H–I	+299
C=O	+743	O–H	+463
C–Cl	+338	Cl–Cl	+242
C–Br	+276	Br–Br	+193
C–I	+238	I–I	+151

Table 5.2 Some average bond enthalpies

Calculation of enthalpy of reaction from average bond enthalpies

The simplest way to do this type of calculation is in three steps.

Step1: list all the bonds broken. Write down the energy required (a **positive** number) to break each bond. Add, to find the total energy required.

Step 2: list the bonds made. Write down the energy released (a **negative** number) to make each bond. Add, to find the total energy released.

Step 3: add the two totals to give $\Delta H_r^{\ominus}$.

ⓔ Bond breaking is endothermic; bond making is exothermic.

> *Worked example 1*
> Use the data in Table 5.2 to calculate the enthalpy of the reaction:
>
>
> **Answer**
>
> Step 1:
>
> | Bonds broken | Bond enthalpy |
> |------|------|
> | C=C | +612 |
> | Cl–Cl | +242 |
> | **Total** | **+854** |
>
> Step 2:
>
> | Bonds made | Bond enthalpy |
> |------|------|
> | C–C | –348 |
> | 2 × C–Cl | 2 × (–338) |
> | **Total** | **–1024** |
>
> Step 3:
> $$\Delta H_r^{\ominus} = +854 + (-1024) = -170 \text{ kJ mol}^{-1}$$

ⓔ If you have any difficulty deciding which bonds break in the reaction, break *all* the bonds in the reactants and then make *all* the bonds in the products. This method is shown in worked example 2.

Worked example 2

a Use the data in Table 5.2 to calculate the enthalpy of the reaction:

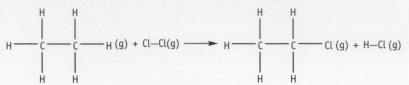

b The enthalpy change for this reaction is $-144\,\text{kJ mol}^{-1}$. Explain why this is different from the answer in **a**.

Answer

a Step 1:

Bonds broken	Bond enthalpy
C–C	+348
$6 \times$ C–H	$6 \times (+413)$
Cl–Cl	+242
Total	**+3068**

Step 2:

Bonds made	Bond enthalpy
$5 \times$ C–H	$5 \times (-413)$
C–C	−348
C–Cl	−338
H–Cl	−431
Total	**−3182**

Step 3:

$$\Delta H_r^{\ominus} = +3068 - 3182 = -114 \text{ kJ mol}^{-1}$$

b The calculation uses average bond enthalpies. The bond enthalpies of the C–H bond in ethane and the C–Cl bond in chloroethane are slightly different from the average C–H and C–Cl bond enthalpies.

Worked example 3

Given the data in Table 5.2 and the enthalpy change of the reaction between buta-1,3-diene and chlorine, calculate the C=C bond enthalpy in buta-1,3-diene.

Given the enthalpy change for one of its reactions and all other relevant bond enthalpies, the bond enthalpy in a compound can be calculated by assigning it to an unknown, z, and proceeding as in worked example 3.

Answer

Let the C=C bond enthalpy in buta-1,3-diene = z

Step 1:

Bonds broken	Bond enthalpy
$2 \times$ C=C	$2z$
$2 \times$ Cl–Cl	$2 \times (+242) = 484$
Total	**$(2z + 484)$**

Step 2:

Bonds made	Bond enthalpy
$2 \times$ C–C	$2 \times (-348) = -696$
$4 \times$ C–Cl	$4 \times (-338) = -1352$
Total	**-2048**

Step 3:

$\Delta H_r^\circ = -328 = (2z + 484) - 2048$

$2z = -328 - 484 + 2048 = +1236$

C=C bond enthalpy in buta-1,3-diene = z

$$= \frac{+1236}{2}$$
$$= +618 \text{ kJ mol}^{-1}$$

> There is no need to break the six C–H bonds and then remake them in the product. Remember that there are two C=C bonds in buta-1,3-diene.

The bond that is likely to be broken in a chemical reaction can be indicated by bond enthalpies. For example, when hydroxide ions are added to iodoethane, either a C–H or the C–I bond could be broken and a new C–OH bond formed. The bond enthalpies of the two bonds are:

$+413 \text{ kJ mol}^{-1}$ for C–H; $+238 \text{ kJ mol}^{-1}$ for C–I

The weaker C–I bond is the one predicted to break. This is borne out by the fact that ethanol, C_2H_5OH, is formed rather than 2-iodoethanol, CH_2ICH_2OH.

Bond enthalpies can also be used to predict relative rates of reaction.

Worked example

Use the data in Table 5.2 to predict the relative rates of the reaction of hydroxide ions with 1-chloropropane, 1-bromopropane and 1-iodopropane:

$CH_3CH_2CH_2X + OH^- \rightarrow CH_3CH_2CH_2OH + X^-$

where X is a halogen atom.

Answer

The carbon–halogen bond is broken in this reaction.

The C–I bond has the lowest bond enthalpy and is, therefore, the weakest. So, 1-iodopropane reacts the fastest. The C–Cl bond is the strongest. Therefore, 1-chloropropane reacts most slowly.

If the bond broken is weaker, the activation energy is lower. Therefore, a greater proportion of collisions will result in a reaction (see page 236).

Enthalpy titrations

The concentration of a solution of an acid can be determined by measuring the temperature change when an alkaline solution of known concentration is added gradually, until it is in excess.

The method is as follows:
- Pipette 25.0 cm³ of the acid solution into an expanded polystyrene cup.
- Measure the temperature of the acid.
- Add a standard 1.00 mol dm⁻³ solution of sodium hydroxide from a burette in 2 cm³ portions, stirring and measuring the temperature after each addition.
- Continue to add a further 8 cm³ of the sodium hydroxide after the maximum temperature has been reached.
- Plot a graph of temperature against volume of sodium hydroxide.

To find the volume of alkali required for neutralisation, a line is drawn through the points up to one reading before the maximum temperature and then extrapolated. Another line is drawn from the maximum temperature to the last reading and extrapolated back. The point at which the two extrapolated lines intersect is the volume needed for neutralisation. For example, using the data in Figure 5.7, the volume needed for neutralisation is 20.6 cm³.

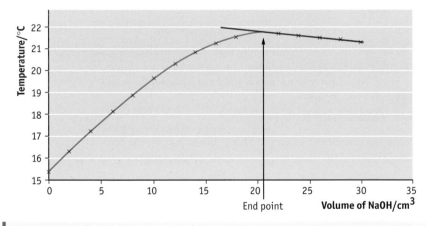

Figure 5.7 An enthalpy titration graph

Worked example
In an enthalpy titration, the volume of alkali needed to neutralise 25 cm³ of dilute hydrochloric acid was 20.6 cm³. The equation for the reaction is:

$$HCl + NaOH \rightarrow NaCl + H_2O$$

Calculate the concentration of the acid.

Answer
amount of sodium hydroxide = concentration × volume in dm³

$$= 1.00 \text{ mol dm}^{-3} \times (20.6/1000) \text{ dm}^3 = 0.0206 \text{ mol}$$

amount of HCl = amount of NaOH = 0.0206 mol

$$\text{concentration of HCl} = \frac{\text{mol}}{\text{volume}} = \frac{0.0206 \text{ mol}}{(25.0/1000) \text{ dm}^3} = 0.824 \text{ mol dm}^{-3}$$

Questions

1 Calculate the amount of heat needed to raise the temperature of 50.0 g of iron from 21.0°C to 100°C. The specific heat capacity of iron is 0.450 J g^{-1} °C^{-1}.

2 Calculate the temperature reached when 100 g of water at 18.2°C absorbs 9.45 kJ of heat energy. The specific heat capacity of water is 4.18 J g^{-1}°C^{-1}.

3 Draw labelled energy level diagrams for the following reactions:
   a $S(s) + O_2(g) \rightarrow SO_2(g)$ $\Delta H^\ominus = -297$ kJ mol^{-1}
   b $6CO_2(aq) + 6H_2O(l) \rightarrow C_6H_{12}O_6(aq) + 6O_2(g)$
   $\Delta H^\ominus = +2800$ kJ mol^{-1} $C_6H_{12}O_6$

4 Write the thermochemical equation that represents:
   a the combustion of butane, $C_4H_{10}(g)$
   b the formation of butane

5 Draw a Hess's law cycle and use it, together with the data given, to calculate the enthalpy change for the reaction:

   $P(s) + 1\frac{1}{2}Cl_2(g) \rightarrow PCl_3(l)$

   Data: $P(s) + 2\frac{1}{2}Cl_2(g) \rightarrow PCl_5(s)$
   $\Delta H^\ominus = -463$ kJ mol^{-1}

   $PCl_3(l) + Cl_2(g) \rightarrow PCl_5(s)$
   $\Delta H^\ominus = -124$ kJ mol^{-1}

6 Calculate the enthalpy change for the reaction:
   $C_2H_6(g) + 2\frac{1}{2}O_2(g) \rightarrow 2CO(g) + 3H_2O(l)$

   Standard enthalpies of formation are given in the table below.

Substance	$\Delta H_f^\ominus$/kJ mol^{-1}
Ethane, $C_2H_6(g)$	−84.7
Carbon monoxide, $CO(g)$	−111
Water, $H_2O(l)$	−286

7 Draw a Hess's law diagram and calculate the enthalpy change for the reaction:
   $C_2H_4(g) + H_2O(l) \rightarrow C_2H_5OH(l)$

   Standard enthalpies of formation are given in the table below.

Substance	$\Delta H_f^\ominus$/kJ mol^{-1}
Ethene, $C_2H_4(g)$	+52.3
Water, $H_2O(l)$	−286
Ethanol, $C_2H_5OH(l)$	−278

8 Write the equation that represents the standard enthalpy of formation of glucose, $C_6H_{12}O_6(s)$, and calculate its value. Standard enthalpies of combustion are given in the table below.

Substance	$\Delta H_c^\ominus$/kJ mol^{-1}
Carbon, $C(s)$	−393.5
Hydrogen, $H_2(g)$	−285.8
Glucose, $C_6H_{12}O_6(s)$	−2816

9 Write the thermochemical equation that represents the standard enthalpy of neutralisation of:
   a nitric acid, $HNO_3(aq)$
   b ethanedioic acid, $H_2C_2O_4(aq)$

10 Explain why the standard enthalpy of neutralisation of strong acids by aqueous sodium hydroxide is always close to −57.6 kJ mol^{-1}.

11 Using the average bond enthalpies given in the table, calculate the enthalpy change for the reaction:
   $CH_3CH=CH_2(g) + HBr(g) \rightarrow CH_3CHBrCH_3(g)$

Bond	ΔH_B/kJ mol^{-1}	Bond	ΔH_B/kJ mol^{-1}
C−C	+348	C−Br	+276
C=C	+612	H−Br	+366
C−H	+413		

12 Calculate the C−Cl bond enthalpy in CH_2ClCH_2Cl, given the bond enthalpies in the table and the standard enthalpy of the reaction:
   $CH_2=CH_2(g) + Cl_2(g) \rightarrow CH_2ClCH_2Cl(g)$
   $\Delta H_r = -217.4$ kJ mol^{-1}

Bond	ΔH_B/kJ mol^{-1}	Bond	ΔH_B/kJ mol^{-1}
C−C	+348	C−H	+413
C=C	+612	Cl−Cl	+242

The following questions are of the type that will only be asked in Unit Test 3B.

13 Plan an experiment that would enable you to calculate the exothermic enthalpy of dissolving anhydrous copper(II) sulfate, $CuSO_4(s)$, in water. The equation for the process, which is slow, can be represented by:
   $CuSO_4(s) + aq \rightarrow CuSO_4(aq)$

Your answer should include what you would do, what measurements you would take and how you would use those measurements to calculate the enthalpy change.

14 Plan an experiment that would enable you to calculate the enthalpy change for the decomposition of calcium carbonate:

$$CaCO_3(s) \rightarrow CaO(s) + CO_2(g)$$

You are supplied with samples of calcium oxide, calcium carbonate and dilute hydrochloric acid, plus any laboratory equipment that you might require.

15 Ethanoic acid, CH_3COOH, is a weak acid. Therefore, its standard enthalpy of neutralisation is less exothermic than that of a strong acid.

  a Use the following experimental data to calculate the standard enthalpy of neutralisation of a 1.00 mol dm^{-3} solution of ethanoic acid:
  - volume of 1.00 mol dm^{-3} ethanoic acid solution = 40.0 cm^3
  - volume of 1.10 mol dm^{-3} sodium hydroxide solution = 40.0 cm^3
  - mass of mixed solution after the experiment = 80.0 g
  - average temperature of reactant solution at start = 18.5 °C
  - maximum temperature after mixing = 24.9 °C

  The specific heat capacity of the solution is 4.18 J g^{-1} °C^{-1}.

  b Why was the concentration of the sodium hydroxide solution higher than that of the ethanoic acid solution?

16 One type of hot pack, sometimes carried by climbers and hikers, consists of a double plastic bag. One section contains magnesium sulfate and the other section contains water. When mixed, the magnesium sulfate dissolves exothermically.

$$MgSO_4(s) + aq \rightarrow Mg^{2+}(aq) + SO_4^{2-}(aq)$$
$$\Delta H = -91\ kJ\ mol^{-1}$$

Calculate the temperature rise when 50 cm^3 of water and 6.0 g of magnesium sulfate are mixed, forming a solution of mass 56 g. Assume that the specific heat capacity of the solution is 4.08 J g^{-1} °C^{-1}.

17 A spirit burner was filled with methanol and weighed. It was then lit and placed underneath a beaker containing 175 cm^3 of water at 18.5 °C. After a few minutes, the spirit lamp was blown out and the temperature of the water was found to be 40.2 °C. The burner was reweighed and was found to be 0.85 g lighter than at the start.

The density of water is 1.00 g cm^{-3} and its specific heat capacity is 4.18 J g^{-1} °C^{-1}.

The equation for this combustion reaction is:

$$CH_3OH(l) + 1\tfrac{1}{2}O_2(g) \rightarrow CO_2(g) + 2H_2O(g)$$

  a Calculate the enthalpy of combustion of methanol, CH_3OH.

  b The data book value for the standard enthalpy of combustion of methanol is -726 kJ mol^{-1}. Give three reasons why this is significantly more exothermic than the value calculated from the experiment.

18 Propane gas, $C_3H_6(g)$, was burnt in a modified Bunsen burner that contained a flow meter to measure the volume of gas burnt. This was set up under a beaker containing 100 g of water, initially at 19.4 °C. After 208 cm^3 of gas had been burnt, the temperature of the water had risen to 61.5 °C.

Calculate the enthalpy of combustion of propane gas.

Data: Under the conditions of the experiment, 1 mol of gas has a volume of 24.0 dm^3.

The specific heat capacity of water is 4.18 J g^{-1} °C^{-1}.

Chapter 6

Bonding

Carbon: gems, pencils and buckyballs

Atomic number: 6

Electron configuration: $1s^2\ 2s^2\ 2p_x^1\ 2p_y^1\ 2p_z^0$

Symbol: C

Carbon is in group 4 of the periodic table. It is a unique element in that carbon atoms form strong bonds with other carbon atoms. This enables millions of compounds with rings and chains of carbon atoms to be made. All living organisms are based on carbon.

This glass geodesic dome, designed by the architect Richard Buckminster Fuller, served as the USA pavilion at EXPO '67 in Montreal, Canada

Diamonds

Diamonds are pure carbon and burn in air to form carbon dioxide. They are formed under high pressure and at high temperature, deep in the Earth. Diamonds are the most expensive and sought-after gemstones. Apart from cubic boron nitride, nothing is hard enough to scratch a diamond.

Each carbon atom is covalently bonded to *four* neighbouring carbon atoms arranged tetrahedrally around it. This structure repeats throughout the crystal. Cutting or melting a diamond requires many covalent bonds to be broken, which is why diamonds are so hard and have such a high melting temperature.

Pencils

Pure graphite is slippery — layers slide off under pressure. The 'lead' in a pencil is graphite that has been mixed with clay and heated. The addition of baked clay makes the lead hard, so a 4H pencil lead contains much more clay than it does graphite.

Graphite is also pure carbon. Each atom is covalently bonded to three others in a layer of interlocking hexagonal rings. There are weak van der Waals forces between the layers, which can be peeled off under a shearing force. This makes graphite a good lubricant and also suitable for pencils. The fourth bonding electron is delocalised, so graphite can conduct electricity. It is used as the cathode in dry batteries and as the anode in aluminium extraction.

Buckminsterfullerene

In 1985, a new form of carbon was discovered. It has the formula C_{60} and its structure is similar to that of a geodesic dome designed by Buckminster Fuller. The geometry of the molecule resembles that of a football. The panels of a football are made up of hexagons and pentagons; the C_{60} molecule has interlocking rings containing six or five carbon atoms, each with three σ-bonds and one π-bond. In 1991, another form of carbon was discovered in which the carbon atoms form long tubular molecules with an internal cavity of about 15 nanometres. These are called carbon nanotubes.

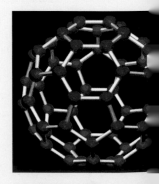

A buckyball

Bonding

Introduction

There are three ways in which atoms can be chemically bonded to other atoms:
- **Metallic** bonding occurs between the atoms of metallic elements.
- **Ionic** bonding in compounds is usually between a metal and a non-metal.
- **Covalent** bonding occurs between two atoms of a non-metallic element (e.g. between two oxygen atoms in O_2) or between two atoms of different elements (e.g. between a hydrogen and a chlorine atom in HCl). Most covalent bonds exist between two non-metallic elements. However, some do occur between a metal and a non-metal, for example in beryllium chloride, $BeCl_2$.

Metallic bonding

The simplest theory for metallic bonding is that the metal atoms lose their valence (outer shell) electrons and form cations. These cations are arranged in a regular lattice, with one layer of ions above another layer, and are surrounded by a 'sea' of electrons that can move through the lattice (Figure 6.1). The electrons are not localised in-between the metal ions; they are **delocalised** through the structure. The bonding is the attraction between the positive ions, which are fixed in position, and the negative electrons, which are constantly moving between the ions. There are slight spaces between the metal ions, which are occupied by the delocalised electrons. Therefore, the metallic radius is slightly larger than the ionic radius.

> The structure of a metal is a regular arrangement of cations held together by a sea of electrons.

The strength of the metallic bond depends upon:
- the charge on the metal ion (which is equal to the number of delocalised electrons)
- the metallic radius (Table 6.1)
- the structure of the metallic lattice, for example body-centred cubic for the group 1 metals and face-centred cubic for calcium, magnesium and aluminium

Figure 6.1
Structure of a layer in a group 2 metal

Physical properties of metals

Density and melting temperature

The group 1 metals have low densities and low melting temperatures (Table 6.1). The electron configuration of their atoms is ns^1, where n is the orbit number of the outer s orbital, and they lose the outer s-electron.

The group 2 metals lose both the ns^2 electrons. This causes the ion to be smaller than the group 1 ion in the same period. Therefore, because of the smaller radius

> **e** You do not need to know how the ions are arranged in metals.

and the higher charge, the group 2 metals are denser and have higher melting temperatures.

The d-block metals use their $(n-1)d$ electrons as well as their ns^2 electrons in bonding. The result is that they are much harder, denser and have significantly higher melting temperatures.

	Electron configuration	Metallic radius/nm	Ionic radius/nm	Melting temperature/°C	Density/ g cm⁻³
Sodium	[Ne] $3s^1$	0.19	0.095	98	0.97
Magnesium	[Ne] $3s^2$	0.16	0.065	649	1.74
Aluminium	[Ne] $3s^2\,3p^1$	0.14	0.050	660	2.70
Potassium	[Ar] $4s^1$	0.23	0.13	63	0.86
Calcium	[Ar] $4s^2$	0.20	0.099	839	1.54
Titanium	[Ar] $3d^2\,4s^2$	0.15	(+2) 0.090	1675	4.54
Iron	[Ar] $3d^6\,4s^2$	0.13	(+2) 0.076	1535	7.86
Copper	[Ar] $3d^{10}\,4s^1$	0.13	(+2) 0.069	1083	8.92

Table 6.1 Physical properties of metals

The metallic radius is half the distance between the centres of two adjacent metal ions in the metallic lattice and is slightly greater than the ionic radius found in ionic compounds.

The melting temperature of metals increases across a period because the metallic radius decreases and more electrons are released for bonding. It decreases down a group because the metallic radius increases and the force of attraction becomes less.

A stronger force of attraction means that more energy has to be supplied to overcome that force and melt the solid. Therefore, a higher temperature is needed than in metals with weaker metallic bonds.

Electrical conductivity
Electricity is a flow of charge. All metals conduct electricity when solid and when molten. The sea of electrons is mobile and moves through the lattice of metal ions. Therefore, the electric current in a metal is a flow of electrons.

Thermal conductivity
Metals are good conductors of heat. This is also because of the free-moving electrons, which pass kinetic energy along a piece of metal.

Malleability
Metals can be hammered or pressed into different shapes. One layer of metal ions can slide over another layer. This is because there are always electrons between the layers, preventing strong forces of repulsion between the positive ions in one layer and the positive ions in another layer. Some metals, such as lead, are extremely soft. The d-block metals are much harder because there are more electrons that bind the layers together.

Metals and graphite conduct electricity by the movement of electrons through the solid. Ionic substances conduct electricity by the movement of ions. This only happens when the ionic compound is molten or dissolved, *not* when it is a solid.

Chemical properties of metals

Ionisation energy
Metals are on the left-hand side of the periodic table. Metal atoms have one, two or three electrons in their outer orbits (apart from lead and tin, which have four). Therefore, the effective nuclear charge is smaller than for the elements that occur

later in the same period. This means that the outer electron is held less firmly than in non-metals and the first ionisation energies are smaller than for non-metals in the same period. The variation of first ionisation energy of the elements from neon to calcium is shown in Figure 6.2.

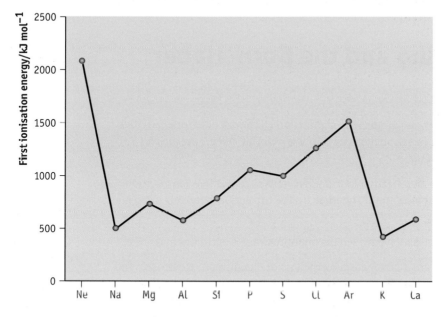

Figure 6.2 First ionisation energies

Metals are shown in red and non-metals in blue.

Electronegativity

Metals have the lowest electronegativity values in the periodic table (page 161). Electronegativity decreases down a group and increases across a period. Therefore, caesium has the lowest electronegativity of all the elements.

Formation of positive ions

The low values of the first ionisation energies and electronegativities make it energetically feasible for metals to lose electrons in bonding and to form positive ions that are then either bonded in an ionic lattice or hydrated.

Reactions

Reaction with acids

The more reactive metals react with dilute acids to form hydrogen gas and a solution of a salt. For example, the ionic equation for the reaction of zinc with a strong acid such as sulfuric acid is:

$$Zn(s) + 2H^+(aq) \rightarrow Zn^{2+}(aq) + H_2(g)$$

Reaction with water

The most reactive metals react with cold water to form hydrogen gas and an alkaline solution of the metal hydroxide. For example, the ionic equation for calcium reacting with water is:

$$Ca(s) + 2H_2O(l) \rightarrow Ca^{2+}(aq) + 2OH^-(aq) + H_2(g)$$

Reaction with ions of a less reactive metal

A more reactive metal will reduce the ions of a less reactive metal. For example, when a piece of iron is placed in a solution of copper sulfate, the blue solution fades as a red-brown deposit of copper metal is formed:

$$Fe(s) + Cu^{2+}(aq) \rightarrow Fe^{2+}(aq) + Cu(s)$$

Ionic bonding and the Born–Haber cycle

Ionic bonds form between atoms with significantly different electronegativities. For example, sodium (electronegativity 0.9) and chlorine (electronegativity 3.0) form an ionic compound.

Ionic compounds are electrically neutral. They consist of positive ions (cations) and negative ions (anions) held together by the attraction between opposite charges.

An ionic bond is the attraction between the opposite charges of cations and anions.

A **cation** is formed from a metal atom by the loss of one or more electrons. The ion has more protons than electrons and, therefore, is positively charged.

An **anion** is formed from a non-metal atom by the gain of one or more electrons. The ion has more electrons than protons and, therefore, is negatively charged.

In simple terms, ion formation can be thought of as one atom (the metal) giving one or more electrons to another atom (a non-metal). For example, sodium (11 protons and 11 electrons) gives one electron to a chlorine atom (17 protons and 17 electrons), forming a Na^+ ion (11 protons and 10 electrons) and a Cl^- ion (17 protons and 18 electrons). However, sodium chloride can be made in a number of ways and it is impossible to tell which electron comes from which atom.

The electron configurations of a sodium ion and a chloride ion, showing all the electrons, are given below:

Sodium ion
2,8

and

Chloride ion
2,8,8

The dot-and-cross diagram for sodium chloride, showing the outer electrons only, indicates where the electrons come from if sodium chloride were made by the reaction of sodium with chlorine:

In the solid, each sodium ion is surrounded by six chloride ions and each chloride ion is surrounded by six sodium ions. The ratio of Na^+ ions to Cl^- ions is 1:1.

e The electron configuration of sodium changes from 2,8,1 to 2,8 — it has eight electrons in its outer orbit. These must be included in any electron configuration diagram, even if you are asked to show the outer electrons only.

The arrangement is shown in Figure 6.3. Each red sphere represents the centre of a sodium ion and each green sphere the centre of a chloride ion.

Strength of ionic bonds

The strength of an ionic bond is mainly determined by:
- the charges on the ions. The force of attraction depends upon the product of the charges. Therefore, the attraction between a 2+ ion and a 1– ion is twice that between a 1+ and a 1– ion.
- the radii of the ions. The force of attraction varies *inversely* with the sum of the ionic radii. This means that smaller ions form stronger ionic bonds.

Thus, magnesium chloride is more strongly ionically bonded than sodium chloride. The magnesium ion is 2+ and has a smaller radius than the 1+ sodium ion.

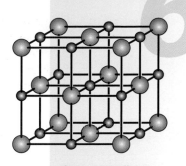

Figure 6.3 Sodium chloride ionic lattice

Ionic radius

The position of the centres of ions in an ionic crystal can be identified by electron diffraction patterns. The distance between a sodium ion and one of the adjacent chloride ions is the sum of the ionic radii, r (of Na^+) + r (of Cl^-). Positive ions are *smaller* than their parent atoms; negative ions are *bigger* than their parent atoms (Table 6.2).

> The melting temperature is determined by the ionic bond strength. Strong bonds give rise to a high melting temperature.

	Ion	Electron configuration of ion	Atomic radius/nm	Ionic radius/nm
Sodium	Na^+	2,8	0.19	0.095
Magnesium	Mg^{2+}	2,8	0.16	0.065
Calcium	Ca^{2+}	2,8,8	0.20	0.10
Barium	Ba^{2+}	2,8,18,8	0.22	0.14
Chlorine	Cl^-	2,8,8	0.10	0.18
Bromine	Br^-	2,8,18,8	0.11	0.20
Iodine	I^-	2,8,18,18,8	0.13	0.22
Oxygen	O^{2-}	2,8	0.073	0.14
Hydroxide	OH^-			≈ 0.12

Table 6.2 Atomic and ionic radii

> Note that the best ionic size match in the group 2 oxides and hydroxides is with the biggest cation, Ba^{2+}.

Trend in ionic radius down a group

As a group is descended, the ions have more electron shells. Therefore, down the group, the ions get larger. For example, the chloride ion has electrons in three shells and is smaller than the bromide ion, which has electrons in four shells. So, the radii of the group 7 ions increase in the order $F^- < Cl^- < Br^- < I^-$.

Trend in ionic radius from group 5 in one period to group 3 in the next period

The ions N^{3-}, O^{2-}, F^-, Na^+, Mg^{2+} and Al^{3+} all have the electron configuration 2,8. These six ions are said to be **isoelectronic**, as they have identical electron configurations. Their radii decrease from N^{3-}, which is the largest, to Al^{3+}, which is the smallest. This is because the charge on the nucleus increases from 7 in N^{3-} to 13 in Al^{3+}. The greater nuclear charge is experienced by the same number of electrons, so the attraction between the nucleus and the orbiting electrons goes up, making the ions progressively smaller.

> An atom and the ion formed from it have the same number of protons in the nucleus. It is only the number of electrons that changes.

Evidence for the existence of ions

Conduction of electricity

An electric current is a flow of charged particles. When a metal, either a solid (e.g. copper) or a liquid (mercury), conducts electricity, the delocalised electrons move across the applied electric potential. The current is a flow of electrons.

Ionic solids, such as sodium chloride, do not conduct electricity because:
- they do not have delocalised electrons
- the ions are fixed in the lattice

However, when the solid is melted or dissolved, the lattice breaks down and the ions are free to move across an applied potential. The current is a flow of ions.

The positive ions move towards the negatively charged cathode, where they are reduced to atoms. The negative ions move towards the positively charged anode, where they are oxidised to atoms. If an electric current is passed through molten sodium chloride, the following reactions take place at the electrodes.

At the cathode (−), reduction occurs as each Na^+ ion gains an electron:

$$Na^+ + e^- \rightarrow Na$$

At the anode (+), oxidation occurs as each Cl^- ion loses an electron:

$$Cl^- \rightarrow \tfrac{1}{2}Cl_2 + e^-$$

The clearest evidence for ionic bonding is when the molten substance, or a solution of it, conducts electricity and new substances are produced at the electrodes. This process is called **electrolysis**.

The movement of ions can be demonstrated by the following experiment.

Some copper(II) chromate, $CuCrO_4$, is dissolved in the minimum of dilute hydrochloric acid. This solution is then saturated with urea, a water-soluble covalent substance. This increases the density of the solution. A U-tube is half-filled with dilute hydrochloric acid and the copper(II) chromate solution is poured carefully down a long-stemmed funnel so that it forms a separate layer beneath the dilute hydrochloric acid (Figure 6.4). Graphite or platinum electrodes are then placed in the dilute hydrochloric acid layer just above the copper(II) chromate layer and an electric current is passed through the solution for at least 30 minutes. The Cu^{2+} ions move towards the negative terminal and the solution around this terminal becomes blue-green because of the copper ions. The solution around the positive terminal becomes yellow because of the migration of the CrO_4^{2-} ions towards it.

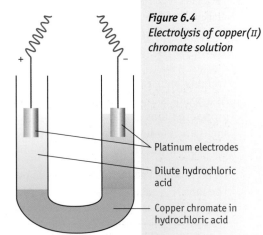

Figure 6.4
Electrolysis of copper(II) chromate solution

Platinum electrodes

Dilute hydrochloric acid

Copper chromate in hydrochloric acid

Melting temperature and solubility

Ionic compounds have very high melting temperatures. Only a few, for example potassium nitrate, can be melted in a test tube in the laboratory. This is because of the strong forces of attraction between the ions.

Many ionic compounds are soluble in water. When the ions become hydrated, the energy released compensates for that required to break the bonds between positive and negative ions in the lattice. Positive ions are surrounded by the δ^- oxygen atoms of water molecules; the negative ions are surrounded by the δ^+ hydrogen atoms.

These properties are not proof of ionic bonding. Some covalent substances also have very high melting temperatures — for example, diamond and silicon dioxide, which have giant covalent structures. In diamond, each carbon atom is covalently bonded to four other carbon atoms, which are covalently bonded to other carbon atoms. In silicon dioxide, each silicon atom is covalently bonded to four oxygen atoms, each of which is bonded to two silicon atoms. These covalent bonds are strong and have to be broken to melt the solid. Therefore, the melting temperature is high. This type of covalent compound is insoluble in water. However, some other covalent substances, such as ethanol, are water-soluble.

This means that a high melting temperature, combined with solubility in water, is strong evidence for ionic bonding.

Electron-density maps

When X-rays are passed through a crystal, the radiation is scattered and a diffraction pattern is obtained. The extent of the scattering of the X-rays depends on the electron density. In this way, the position of atoms and ions in a solid can be determined. This analysis can be extended to measure the electron density in the atoms or ions. The result is a contour map, where the lines join places with the same electron density — just like the contour lines on a geographical map that join places of equal height above sea level.

The electron-density map of sodium chloride is shown in Figure 6.5. It can be seen clearly that the electron density drops to zero between the ions, showing them to be discrete entities that are not joined to specific atoms or ions.

Figure 6.5
Electron-density map of sodium chloride

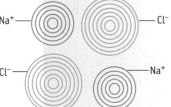

Energetics of ionic bonding

Ion-pair formation

It is often taught at GCSE that a sodium atom 'gives' one electron to a chlorine atom 'so that they both gain the stability of a noble gas'. This is a gross over-simplification. The enthalpy change for the formation of separate sodium and chloride ions from their atoms is the sum of the energies for the endothermic removal of an electron from the sodium atom and for the exothermic addition of an electron to the chlorine atom, thus:

$$Na(g) \rightarrow Na^+(g) + e^- \qquad \Delta H = +494 \text{ kJ mol}^{-1}$$
$$Cl(g) + e^- \rightarrow Cl^-(g) \qquad \Delta H = -364 \text{ kJ mol}^{-1}$$
$$\overline{Na(g) + Cl(g) \rightarrow Na^+(g) + Cl^-(g) \quad \Delta H = +130 \text{ kJ mol}^{-1}}$$

This is very endothermic and, therefore, cannot be the driving force for the reaction. In addition, sodium metal has to be vaporised and chlorine molecules have to be split into atoms, making the whole process even more endothermic.

When sodium and chlorine react, solid sodium chloride is formed and not separate sodium and chloride ions. The energy released when the ions come together and form the solid more than compensates for the endothermic formation of the separate ions. This energy is called the **lattice energy**.

> The lattice energy, ΔH_{latt} or LE, of an ionic compound is the energy change when *1 mol* of the *solid* is formed from its constituent *gaseous ions* that start infinitely far apart.

Born–Haber cycles

The enthalpy of formation of NaCl(s) is the enthalpy change for the reaction of solid sodium with gaseous chlorine molecules. The enthalpy change for this direct one-step reaction will, by Hess's law, be equal to the sum of the enthalpy changes involved if the reaction were to take place in several steps. Born and Haber were the first to think of the formation of an ionic compound in this way. In this case, the steps are:

- turning solid sodium atoms into gaseous atoms
- splitting chlorine molecules into atoms
- removing an electron from each gaseous sodium atom
- adding an electron to each gaseous chlorine atom
- bringing the ions together into an ionic lattice

These processes are shown below:

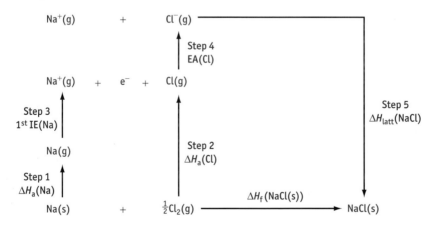

Step 1: the enthalpy of atomisation of sodium, ΔH_a(Na)   (+109 kJ mol^{-1})
Step 2: the enthalpy of atomisation of chlorine, ΔH_a(Cl)   (+121 kJ mol^{-1})
Step 3: the first ionisation energy of sodium, 1st IE(Na)   (+494 kJ mol^{-1})
Step 4: the first electron affinity of chlorine, EA(Cl)   (−364 kJ mol^{-1})
Step 5: the lattice energy of sodium chloride, ΔH_{latt}(NaCl)   (−771 kJ mol^{-1})

ΔH_f(NaCl(s)) = sum of the ΔH values for steps 1–5
$= \Delta H_a$(Na) $+ \Delta H_a$(Cl) $+$ 1st IE(Na) $+$ EA(Cl) $+ \Delta H_{latt}$(NaCl)
$= +109 + 121 + 494 + (−364) + (−771) = −411$ kJ mol^{-1}

The enthalpy of atomisation, ΔH_a, of an element is the enthalpy change when *1 mol* of *gaseous* atoms is made from an element in its standard state.

For example, the enthalpy of atomisation of:
- chlorine is for the change $\frac{1}{2}Cl_2(g) \rightarrow Cl(g)$
- bromine is for the change $\frac{1}{2}Br_2(l) \rightarrow Br(g)$
- sulfur is for the change $\frac{1}{8}S_8(s) \rightarrow S(g)$

If the ionic radii and the arrangement of the ions in the lattice are known, theoretical lattice energies can be calculated. The only way that they can be found experimentally is by using a Born–Haber cycle.

$$\Delta H_f(NaCl(s)) = \Delta H_a(Na) + \Delta H_a(Cl) + 1^{st} IE(Na) + EA(Cl) + \Delta H_{latt}(NaCl)$$
$$\Delta H_{latt}(NaCl) = \Delta H_f(NaCl(s)) - \{\Delta H_a(Na) + \Delta H_a(Cl) + 1^{st} IE(Na) + EA(Cl)\}$$
$$= -411 - \{+109 + (+121) + (+494) + (-364)\} = -771 \text{ kJ mol}^{-1}$$

Calculation of lattice energy from Born–Haber cycle data

Born–Haber cycles can also be drawn as energy-level diagrams. For example, for calcium chloride, $CaCl_2$, the energy-level diagram is:

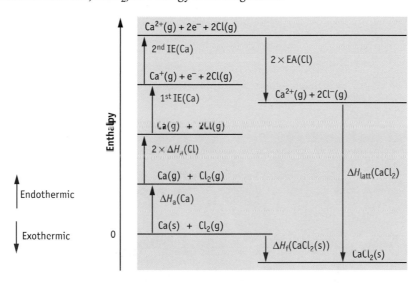

Note that as 2 Cl atoms are formed from Cl_2, the enthalpy change is $2 \times \Delta H_a$. Also, because 2 Cl^- ions are formed, the enthalpy change is $2 \times EA$.

$$\Delta H_f(CaCl_2(s)) = \Delta H_a(Ca) + 2 \times \Delta H_a(Cl) + 1^{st} IE(Ca) + 2^{nd} IE(Ca)$$
$$+ 2 \times EA(Cl) + \Delta H_{latt}(CaCl_2)$$
$$\Delta H_{latt}(CaCl_2) = \Delta H_f(CaCl_2(s)) - \{\Delta H_a(Ca) + 2 \times \Delta H_a(Cl) + 1^{st} IE(Ca)$$
$$+ 2^{nd} IE(Ca) + 2 \times EA(Cl)\}$$
$$= -795 - \{+193 + 2(+121) + (+590) + (+1150) + 2(-364)\}$$
$$= -2242 \text{ kJ mol}^{-1}$$

e The second ionisation energy is the energy change when an electron is removed from a gaseous 1+ ion. It is the energy required to remove the second electron, not the energy for the removal of both electrons.

Factors that affect lattice energy

The magnitude of the lattice energy depends on the strength of the forces acting on the ions. In a lattice, each ion is surrounded by a number of ions of opposite charge, resulting in strong forces of attraction and some forces of repulsion. This is illustrated in Figure 6.6.

The strength of these forces depends on:
- the magnitude of the charges on the ions
- the sum of the radii of the cation and the anion

- the arrangement of the ions in the lattice
- the relative sizes of the ions
- the extent of covalency

In Figure 6.6, the red lines represent forces of attraction; the blue lines represent forces of repulsion.

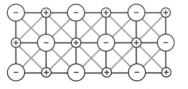

Figure 6.6 The forces acting between the ions in a planar slice through a crystal of sodium chloride

Charge and size of ions

The force between two ions of opposite charge depends upon the value of the ionic charges and how close the centres of the ions are. The stronger the force of attraction between the ions in the solid, the more exothermic is the lattice energy.

The lattice energy is proportional to the product of the charges on the two ions divided by the sum of their radii:

$$\Delta H_{latt} \propto \frac{q(+) \times q(-)}{\{r(+) + r(-)\}}$$

where $q(+)$ = the charge on the cation
$q(-)$ = the charge on the anion
$r(+)$ = the ionic radius of the cation
$r(-)$ = the ionic radius of the anion

ⓔ There are small forces of repulsion between neighbouring ions of the same charge. This is only significant when the anion is much larger than the cation, which causes the anions to be crowded together.

Energy is released in the thermit process because the lattice energy of the aluminium oxide produced is much greater than the lattice energy of the iron(III) oxide reactant.

The exothermic thermit process — an aluminium wrench striking a rusty iron block

CROWN COPYRIGHT/HEALTH & SAFETY LABORATORY/SPL

> **Worked example 1**
>
> Explain why the lattice energy of sodium fluoride is more exothermic than the lattice energy of potassium chloride.
>
> **Answer**
>
> It is because Na^+ has a smaller ionic radius than K^+, and F^- is smaller than Cl^-. Therefore, the forces between sodium ions and fluoride ions are stronger than those between potassium ions and chloride ions.

> **Worked example 2**
>
> Explain why the lattice energy of calcium oxide is approximately four times larger than that of potassium fluoride.
>
> **Answer**
>
> This is because in calcium oxide the product of the ionic charges is 4, whereas in potassium fluoride it is 1. The sums of the ionic radii of the two compounds are not very different. However, the value is smaller in CaO than in KF, so the lattice energy ratio is further increased.

The answer to worked example 2 can be explained more fully by looking at the values involved (Table 6.3).

Substance	ΔH_{latt}/kJ mol^{-1}	Product of ionic charges	Sum of ionic radii/nm
CaO	−3513	4	0.239
KF	−801	1	0.269

Table 6.3 Lattice energy and ionic charge

The lattice energy depends on the product of the charges on the ions divided by the sum of the ionic radii.

$$\text{ratio} = \frac{\text{product of ionic charges divided by sum of ionic radii for CaO}}{\text{product of ionic charges divided by sum of ionic radii for KF}}$$

$$= \frac{4/0.239}{1/0.269} = 4.5$$

$$\text{ratio} = \frac{\Delta H_{latt}(\text{CaO})}{\Delta H_{latt}(\text{KF})}$$

$$= \frac{-3513}{-801} = 4.4$$

The fact that these two values are so similar supports the theory.

Arrangement and relative size

Ions take up an arrangement in the lattice that maximises the lattice energy released. This is the position of minimum potential energy of the ions. If a small cation is surrounded by too many larger anions, a considerable amount of repulsion occurs between the larger anions. This is shown by the structures of ionic compounds of formula AB that are given in Table 6.4.

ⓔ The coordination number of an ion in an ionic lattice is the number of ions of opposite charge that are most closely arranged around it.

In sodium chloride, fitting eight chloride ions around the smaller sodium ion would result in considerable repulsion between the closely packed chloride ions. Each Na^+ ion is surrounded by six Cl^- ions; each Cl^- ion is surrounded by six Na^+ ions. This is called 6:6 coordination. This arrangement minimises repulsion.

Substance	Cationic radius/ nm	Anionic radius/ nm	Ratio of anionic radius: cationic radius	Coordination number
CsCl	0.169	0.181	1.1	8
NaCl	0.095	0.181	1.9	6
ZnS	0.074	0.184	2.5	4

Table 6.4 Ionic radii and coordination numbers

The type of lattice depends upon the ratio of the ionic radii. Compounds with a ratio of approximately 1:1 have a lattice structure similar to that of caesium chloride. Those with a ratio nearer to 2:1 have a structure similar to that of sodium chloride.

(a)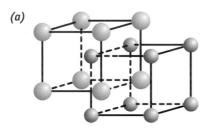
(b)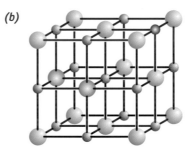

Figure 6.7 Structures of (a) caesium chloride and (b) sodium chloride

Extent of covalency: polarisation of the anion

The positive cation exerts an attraction on the electrons in the negative anion. If the electrons are significantly pulled towards the cation, the anion is **polarised**.
- Cations with a high charge and a small radius have a high polarising power.
- The smaller magnesium ion is more polarising than the larger calcium ion, both of which have the same charge of 2+.
- A magnesium ion is more polarising than a sodium ion, because it is 2+ and it is smaller than the larger, 1+ sodium ion.
- The polarising power is measured by the charge density of the cation.
- Anions with a high charge and a large radius are easily polarised.
- An iodide ion, I^-, is more easily polarised than the smaller chloride ion, Cl^-.
- A sulfide ion, S^{2-}, is more easily polarised than a chloride ion because it has a greater charge and a larger radius (see Table 6.5).

The charge density of a cation is the charge divided by the surface area of the ion. Assuming the ion to be spherical, the charge density is the charge divided by $4\pi r^2$, where r is the ionic radius.

The charge density of a cation can be thought of as its charge divided by the square of its ionic radius.

Cations with a large charge and a small radius are highly polarising. Anions with a large charge and a large radius are highly polarisable. If either type of ion is present, the anion will be significantly polarised, resulting in the bond being partially covalent. This causes the experimental (or Born–Haber) lattice energy to be greater than the theoretical value, which is calculated assuming that the solid is 100% ionic.

Cation	Radius/nm	Anion	Radius/nm
Na^+	0.095	Cl^-	0.181
K^+	0.133	I^-	0.216
Mg^{2+}	0.065	S^{2-}	0.184

Table 6.5 Ionic radii

e If you use the term 'charge density' in an examination, you must explain what it means.

As the polarising power of a cation depends on its charge density, the magnesium ion is the most polarising cation in Table 6.5. The iodide ion is the biggest anion, so magnesium iodide should be the most covalent compound and have the biggest difference between the experimental and the theoretical lattice energies (Table 6.6).

Table 6.6 Lattice energies

Substance	Experimental lattice energy/kJ mol^{-1}	Theoretical lattice energy/kJ mol^{-1}	Extent of covalency
NaCl	−780	−770	Almost none
K_2S	−2052	−1933	Significant
MgI_2	−2327	−1944	Considerable

The effect that covalency has on the experimental value of lattice energy is very much less than the effects of the charges and the radii of the ions.

> ### Worked example
> Explain why calcium chloride is more covalent than potassium chloride.
>
> ### Answer
> The calcium ion is 2+ and has a smaller radius than the 1+ potassium ion. Therefore, its charge density is greater. This means that it polarises the Cl^- ion to a greater extent, causing the bond in calcium chloride to be more covalent than the bond in potassium chloride.

Table 6.7 Lattice energies of some ionic compounds

	Lattice energy/kJ mol^{-1}				
Halides	LiF −1022	NaF −902	KF −801	RbF −767	CsF −716
	LiCl −846	NaCl −771	KCl −701	RbCl −675	CsCl −645
	LiBr −800	NaBr −733	KBr −670	RbBr −647	CsBr −619
	LiI −744	NaI −684	KI −629	RbI −609	CsI −585
	$BeCl_2$ −3006	$MgCl_2$ −2500	$CaCl_2$ −2237	$SrCl_2$ −2112	$BaCl_2$ −2018
Oxides	Li_2O −2814	Na_2O −2478	K_2O −2232	Rb_2O −2161	Cs_2O −2063
	BeO −4444	MgO −3890	CaO −3513	SrO −3310	BaO −3152
Sulfides	Li_2S −2500	Na_2S −2200	K_2S −2052	Rb_2S −1944	Cs_2S −1850
	BeS −3832	MgS −3300	CaS −3013	SrS −2850	BaS −2725
Hydroxides		$Mg(OH)_2$ −2842	$Ca(OH)_2$ −2553	$Sr(OH)_2$ −2354	$Ba(OH)_2$ −2228

Table 6.7 shows that:

- the magnitude of the lattice energy steadily decreases down a group of the periodic table as the size of the cation increases
- the magnitude of the lattice energy steadily decreases down the group as the size of the anion increases
- lattice energy increases as the charge on either or both the cation and the anion increases

Formulae of ionic compounds

The reason why sodium and chlorine form NaCl rather than $NaCl_2$ is not because of the mythical stability of a noble gas electronic configuration. The answer lies in comparing the enthalpy change of the two reactions:

$$Na(s) + \tfrac{1}{2}Cl_2(g) \rightarrow NaCl(s)$$
$$Na(s) + Cl_2(g) \rightarrow NaCl_2(s)$$

ΔH_f for the formation of NaCl(s) can be measured. The value is -411 kJ mol^{-1}.

ΔH_f for the formation of $NaCl_2(s)$ can be calculated from a Born–Haber cycle, assuming that the lattice energy of $NaCl_2$ would be similar to that of $MgCl_2$ (-2500 kJ mol^{-1}).

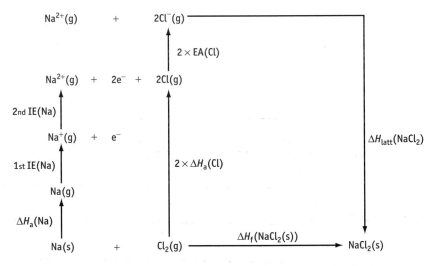

$$\Delta H_f(NaCl_2(s)) = \Delta H_a(Na) + 2 \times \Delta H_a(Cl) + 1^{st} \, IE(Na) + 2^{nd} \, IE(Na)$$
$$+ 2 \times EA(Cl) + \Delta H_{latt}(NaCl_2)$$
$$= +109 + 2(121) + 494 + 4560 + 2(-364) + (-2500)$$
$$= +2177 \text{ kJ mol}^{-1}$$

This process is very endothermic, so $NaCl_2$ is not formed.

The reason why it is so endothermic is that the extra 4560 kJ of energy, required to remove a second electron from sodium, is too great to be compensated for by the increased attraction between a 2+ cation and a 1– anion in the theoretical $NaCl_2$ solid. This huge jump in energy from the first to the second ionisation energy of sodium occurs because the second electron has to be removed from an inner ($2p$) shell. In this shell, the electron is screened from the 11+ nucleus by only the two $1s$-electrons. Therefore, it is held very firmly.

This concept of whether the extra energy required to remove another electron is compensated for by the higher lattice energy (or the hydration energy if the reaction is carried out in aqueous solution) is important in the chemistry of the transition metals.

Simple theories suggested that the driving force behind the formation of ions is gaining the stability of a noble gas electron configuration. However, this ignores the huge number of ionic compounds of the d-block metals that do not gain this so-called stability. A more accurate explanation is that the substances react to give the product with the lowest energy.

> **e** At AS it is insufficient to state that atoms gain or lose electrons in order to reach the stability of the electron configuration of a noble gas.

Magnesium chloride is $MgCl_2$, not $MgCl$, because the energy required to remove the second outer or $3s$-electron is more than compensated for by the extra energy released by the attraction between the 2+ ion and the Cl^- ion. The dot-and-cross diagram for $MgCl_2$ is:

$$\left[\ddot{\underset{\cdot\cdot}{Cl}} \times \right]^{-} \quad \left[\underset{\times\times}{\overset{\times\times}{\times Mg \times}} \right]^{2+} \quad \left[\times \ddot{\underset{\cdot\cdot}{Cl}} \vdots \right]^{-}$$

Covalent bonding

A covalent bond consists of a pair of electrons shared by two atoms. The electrons that are, or could be, involved in bonding are called **valence** electrons. Covalent bonds form between atoms with an electronegativity difference of less than approximately 1.5 units.

In a covalent bond, each atom supplies one electron that is then shared by both atoms. The evidence for this can be found in an electron-density map. Figure 6.8 shows the density of electrons in a hydrogen molecule. Electrons can be seen in the region between the two nuclei.

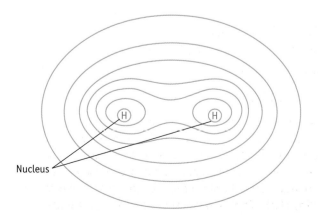

Nucleus

Whether or not a change takes place depends on the value of the change in entropy. This concept is covered in the A2 course.

> **e** The change from atom to cation is endothermic. Ionic bonding only occurs if the ions form a solid or, in solution, are hydrated by water molecules.

Figure 6.8 Electron-density map of a hydrogen molecule

In 1916, G. N. Lewis proposed that atoms share pairs of electrons in order to reach the electron configuration of a noble gas:

■ The halogens are all one electron short of a noble gas configuration. They can gain one electron by sharing and go from $ns^2\, np^5$ to $ns^2\, np^6$ to form one covalent bond, as in H–Cl.

■ Oxygen has the electron configuration [He] $2s^2\, 2p^4$. Therefore, it is two electrons short of the noble gas structure. It forms two covalent bonds, thereby gaining two electrons to have the configuration of neon. An example is in water, H–O–H.

■ Nitrogen is [He] $2s^2\, 2p^3$. It requires three electrons to reach the electron configuration of neon. Thus, it forms three covalent bonds.

■ Carbon is [He] $2s^2\, 2p^2$. It needs four more electrons and so forms four covalent bonds.

Covalent bonding can be shown by dot-and-cross diagrams. The dot-and-cross diagram for hydrogen chloride is:

$$\text{H} \; \overset{\times\times}{\underset{\times\times}{\overset{\bullet}{\times}}} \; \text{Cl} \; \times$$

The electron of hydrogen is shown as a dot and the outer chlorine electrons by crosses.

◀ The electrons are not really different, but it is easier to count them if they are differentiated in this way.

Some people prefer to draw circles around each atom. In this method, the shared electrons are in the overlap of the two circles, as in a Venn diagram where the species in the overlap are common to both sets.

Examples of both types of dot-and-cross diagram are shown in Table 6.8.

Substance	Dot-and-cross diagrams	
Water, H_2O		
Ammonia, NH_3		
Methane, CH_4		

Table 6.8. Dot-and-cross diagrams for water, ammonia and methane

◀ These examples obey the octet rule. This states that atoms proceed, as far as possible, towards having eight electrons in their outer orbit.

Lewis's theory works well for all the period 2 non-metals except boron and for many period 3 and period 4 non-metal compounds. However, it breaks down with such compounds as PCl_5 and SF_6, and particularly with compounds of the noble gases, such as XeF_4.

A more modern theory is that a covalent bond is caused by the overlap of two atomic orbitals, each containing a single unpaired electron. This overlap can happen in several ways:

- The simplest is an overlap of two *s*-orbitals:

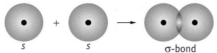

An example is hydrogen.

- Another possibility is the overlap of one *s*-orbital with a *p*-orbital:

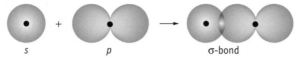

An example is hydrogen chloride.

- Two *p*-orbitals can overlap to form a covalent bond:

An example is chlorine.

These are examples of 'head-on' overlap between two atomic orbitals, in which the overlap lies on the line between the centres of the atoms. This is called a **sigma-bond** (σ-bond).

There is a slight problem with carbon. The electron configuration of carbon is [He] $2s^2\, 2p_x^1\, 2p_y^1\, 2p_z^0$. There are only two unpaired electrons and so carbon might be expected to form only two covalent bonds. However, one of the 2s-electrons is promoted into the empty $2p_z$-orbital, giving it a temporary configuration of [He] $2s^1\, 2p_x^1\, 2p_y^1\, 2p_z^1$, with four unpaired electrons. It can now form four covalent bonds. The energy released in forming four, rather than two, bonds more than compensates for the small amount of energy required to promote an electron from the 2s- into the empty $2p_z$-orbital:

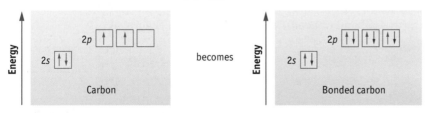

The electrons marked in blue are outer electrons of carbon; those marked in red come from the atoms to which the carbon is bonded.

Expansion of the octet

This concept of promotion explains why phosphorus can form five covalent bonds. Its electron configuration is [Ne] $3s^2\, 3p_x^1\, 3p_y^1\, 3p_z^1$. Therefore, it has three unpaired electrons. When PCl_3 is formed, the three unpaired electrons are used to form three covalent bonds and the configuration of a noble gas is reached. However, phosphorus has five empty 3d-orbitals, with energy close to that of the

3s-orbital. Therefore, one of the 3s-electrons can be promoted into an empty 3d-orbital. The electron configuration becomes [Ne] $3s^1\ 3p_x^1\ 3p_y^1\ 3p_z^1\ 3d^1$ with five unpaired electrons. It can now form five covalent bonds.

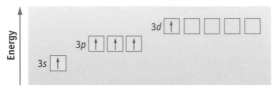

The extra energy released in forming the two extra bonds is much greater than that required to promote one electron from the 3s- to the 3d-orbital. In PCl$_5$, the phosphorus atom has ten electrons in its outer orbit, rather than the octet predicted by the Lewis theory.

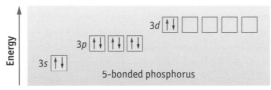

5-bonded phosphorus

Phosphorus does not form PI$_5$ because the P–I bond is not strong enough to compensate for the energy required to promote the 3s-electron of phosphorus into the 3d-orbital.

Single and double bonds

A single bond is a σ-bond between two atoms and is the sharing of one pair of electrons. In carbon dioxide, two double bonds exist between the carbon and the oxygen atoms. The dot-and-cross diagram for carbon dioxide is:

$$O \overset{\times\times}{\underset{\times\times}{}} C \overset{\times\times}{\underset{\times\times}{}} O$$

Here four electrons are shared between the central carbon atom and each oxygen atom. This sharing of two electron pairs is called a **double bond**.

A double bond is the sharing of two pairs of electrons between two atoms.

The orbital overlap theory easily explains a double bond. An orbital in one atom overlaps head-on with an orbital in the other atom, forming a σ-bond. However, a p-orbital, which is at right angles to the σ-bond, overlaps sideways with a p-orbital in the second atom to form a **pi-bond** (π-bond). A π-bond is shown in Figure 6.9.

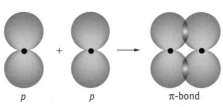

p p π-bond

Figure 6.9 A π-bond

A π-bond is formed by two p-orbitals overlapping, so that the overlap lies above and below the line joining the centres of the two atoms. A double bond consists of a σ-bond and a π-bond between two atoms.

It is not possible to rotate the atoms around a double bond without breaking the π-bond. At room temperature the molecules do not possess enough energy for this to occur. This is why geometric isomers (pages 134–36) are distinct compounds. The rotation is possible if the substance is strongly heated.

In sulfur dioxide, the sulfur atom forms two double bonds. The electron configuration of sulfur is $[Ne]\ 3s^2\ 3p_x^2\ 3p_y^1\ 3p_z^1$. To form two double bonds, one electron must be promoted from the $3p_x$ orbital into an empty $3d$-orbital. There are now four unpaired electrons and it can form two σ-bonds and two π-bonds:

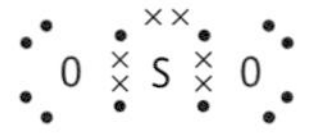

$$\ddot{\underset{\cdot\cdot}{O}} \overset{\times\times}{\underset{\cdot\cdot}{\overset{\times}{S}}} \ddot{\underset{\cdot\cdot}{O}}$$

The sulfur atom has one unused pair of valence electrons. This is called a **lone pair** of electrons.

A lone pair of electrons is a pair of valence electrons that is not used in bonding.

Covalent bond strength

The strength of a covalent bond is measured by the amount of energy needed to break the bond in a gaseous molecule. This energy arises from the attraction between the two nuclei and the shared electrons.

Diamond consists of a regular three-dimensional lattice of carbon atoms each covalently bonded to four other carbon atoms. In order to melt diamond, all these covalent bonds have to be broken. The extremely high melting temperature of nearly 4000 K shows the strength of these covalent bonds. Silicon dioxide, SiO_2, is a giant lattice in which each silicon atom is covalently bonded to four oxygen atoms and each oxygen atom to two silicon atoms. On melting, these covalent bonds have to be broken and, as they are slightly weaker than those in diamond, the melting temperature is around 2000 K. This is still higher than the melting temperatures of most ionic compounds.

The average C–H bond energy in methane is one-quarter of the energy required for the following process:

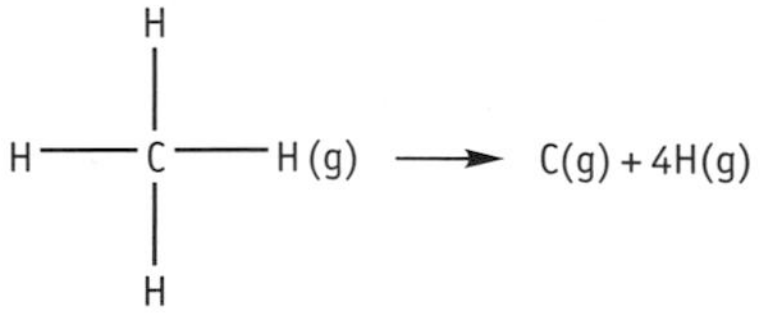

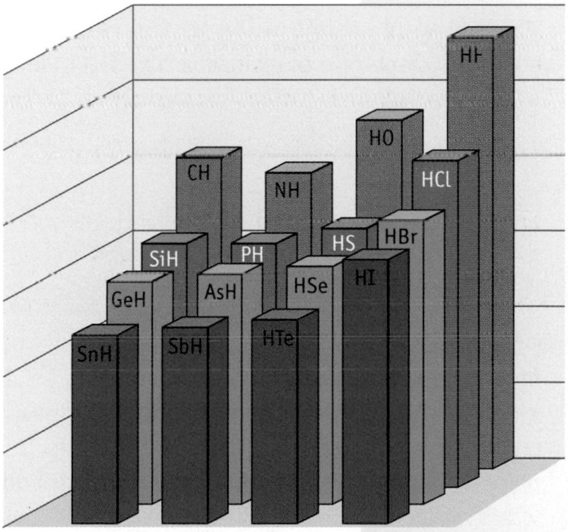

$$H\!-\!\!\underset{\underset{H}{|}}{\overset{\overset{H}{|}}{C}}\!\!-\!H\,(g) \longrightarrow C(g) + 4H(g)$$

Figure 6.10 The relative strength of single bonds

The strength of a covalent bond is determined mainly by:
- the sum of the atomic radii of the two bonded atoms. Small atoms form stronger bonds than larger atoms. Thus the C–Cl bond is shorter and stronger than the C–Br bond because chlorine is a smaller atom than bromine.
- the number of electron pairs being shared. A double bond (two pairs shared) is stronger than a single bond (one pair shared). Nitrogen, N≡N, has the greatest covalent bond strength.

Figure 6.10 shows the relative strength of single bonds between hydrogen and the *p*-block elements.

The strength of a covalent bond is similar to the strength of an ionic bond.

Dative covalent bonds

A dative covalent bond is a covalent bond in which both electrons in the shared pair are provided by one atom.

◀ Dative covalent bonds are sometimes called **coordinate** bonds.

In terms of the overlap of atomic orbitals, a dative covalent bond is the result of an overlap of an empty orbital in one atom and an orbital containing a lone pair of electrons in the other atom. It is represented by an arrow going from the atom with the lone pair to the atom with the empty orbital.

Dative bonds with oxygen

The oxygen atom in a water molecule is able to use one of its two lone pairs to form a dative covalent bond. This occurs with H^+ ions and with many metal cations.

The hydronium ion, H_3O^+

The dot-and-cross diagram and structural formula of a hydronium ion is:

$$\left[H \overset{\times\times}{\underset{\times\times}{\cdot\,O\,\cdot}} H \atop H \right]^+ \quad \text{or} \quad \left[H-O-H \atop \downarrow \atop H \right]^+$$

The empty $1s$-orbital in the H^+ ion overlaps with the orbital in oxygen that contains a lone pair of electrons. The H_3O^+ ion is the ion that makes a solution acidic. For simplicity, it is often written as $H^+(aq)$.

The hydrated magnesium ion, $[Mg(H_2O)_6]^{2+}$

$$\left[\begin{array}{c} \\ H_2O \quad H_2O \qquad OH_2 \\ \searrow \downarrow \swarrow \\ Mg \\ \nearrow \uparrow \nwarrow \\ H_2O \quad\quad OH_2 \\ H_2O \end{array} \right]^{2+}$$

Magnesium has the electron configuration $1s^2\,2s^2\,2p^6\,3s^2$ and that of the Mg^{2+} ion is $1s^2\,2s^2\,2p^6\,3s^0$. Therefore, the Mg^{2+} ion has one empty $3s$-orbital, three empty $3p$-orbitals and five empty $3d$-orbitals. It uses six of these by accepting six pairs of electrons to form six dative covalent bonds with the oxygen atoms in six water molecules.

Hydrated copper(II) ion, $[Cu(H_2O)_4]^{2+}$

- Other metal cations, especially those in the d-block, form hydrated ions in the same way as magnesium.
- The blue colour of hydrated copper sulfate is caused by the $[Cu(H_2O)_4]^{2+}$ ion.
- When hydrated copper sulfate is dissolved in water, two more water molecules bond datively to form the $[Cu(H_2O)_6]^{2+}(aq)$ ion.
- The anhydrous Cu^{2+} ion is colourless. When blue hydrated copper sulfate is heated, it loses the dative-bonded water molecules as steam and the solid turns white as anhydrous copper sulfate is formed.

e If you are asked to label the bonds, make sure that you label the covalent bonds within the water molecules as well as the dative covalent bonds.

e Be careful when drawing the structure of hydrated ions. You must draw the arrow to the metal ion from the oxygen of the water molecule and *not* from the hydrogen.

Hydrated copper sulfate crystals

ANDREW LAMBERT PHOTOGRAPHY/SCIENCE PHOTO LIBRARY

Dative bonds with nitrogen

The nitrogen atom in ammonia, NH_3, has three pairs of covalently bonded electrons and one lone pair.

The ammonium ion, NH_4^+

The lone pair of electrons on the nitrogen atom is used to form a dative covalent bond with an H^+ ion. The result is the ammonium ion, NH_4^+:

$$\left[\begin{array}{c} H \\ \uparrow \\ H - N - H \\ | \\ H \end{array} \right]^+ \quad \text{or} \quad \left[\begin{array}{c} H \\ \times\times \\ H \overset{\bullet}{\times} N \overset{\times}{\bullet} H \\ \bullet\times \\ H \end{array} \right]^+$$

The H^+ ion has no electrons. Its empty $1s$-orbital overlaps with the orbital in the nitrogen atom that contains a lone pair. Once the dative covalent bond has formed, all four N–H bonds are identical.

Ammines

When excess ammonia solution is added to a precipitate of copper hydroxide, a dark blue solution is formed. This colour is due to the $[Cu(NH_3)_4(H_2O)_2]^{2+}$ ion, in which four ammonia molecules and two water molecules are datively bonded to the central copper ion.

Other examples of the formation of soluble datively bonded ions include the 'dissolving' of zinc hydroxide, silver oxide and silver halides in ammonia solution.

Dative bonds with chlorine

Both covalently bonded chlorine and chloride ions are able to use a lone pair to form a dative covalent bond.

Anhydrous aluminium chloride

Anhydrous aluminium chloride is covalent because the difference in the electro-negativities of aluminium and chlorine is only 1.5 (page 161).

■ In the gas phase, just above its boiling temperature, it exists as Al_2Cl_6 molecules.
■ The aluminium in a covalently bonded $AlCl_3$ molecule has only six electrons in its outer orbit and so has an empty orbital.
■ Two $AlCl_3$ molecules bond together. The lone pair on one chlorine atom in one $AlCl_3$ bonds into the empty orbital of the aluminium atom in the other $AlCl_3$ molecule.
■ One chlorine atom from each $AlCl_3$ molecule acts as a bridge, connecting the two molecules with dative covalent bonds.

In three dimensions this has the following structure, with bond angles of 109.5°:

> **e** Do not confuse ammonia, NH_3, with the ammonium ion, NH_4^+.

> Make sure that you have the arrows going the correct way — from the chlorine to the aluminium. There are no Al–Al bonds in Al_2Cl_6.

The PCl_6^- ion

The phosphorus atom in PCl_5 has ten electrons in its outer orbit (pages 117–18), but it also has four empty $3d$-orbitals. Solid phosphorus pentachloride is not PCl_5, but an ionic compound, $PCl_4^+PCl_6^-$. One PCl_5 molecule loses a Cl^- ion, which uses one of its lone pairs of electrons to form a dative covalent bond with an empty orbital of a phosphorus atom in another PCl_5 molecule.

On heating, the dative bond breaks and two molecules of PCl_5 are formed.

Questions

1 Explain why magnesium has a higher melting temperature than sodium.

2 Write equations for the reaction of magnesium with:
   a dilute hydrochloric acid
   b copper sulfate solution

3 a Define ionic bond.
   b Explain, in terms of electrons, how an ionic bond forms between atoms of calcium and atoms of fluorine.
   c Draw electron configuration diagrams for a calcium ion and for a fluoride ion, showing their charges and outer electrons.

4 Arrange the following in order of size of radius, with the largest first and the smallest last:

   Na, Na^+, Mg, Mg^{2+}, Cl, Cl^-

5 Arrange the following in order of polarising power, with the strongest first and the weakest last:

   Na^+, Mg^{2+}, Al^{3+}, K^+

6 Explain the difference between a covalent bond and a dative covalent bond.

7 Carbon monoxide can be thought of as having two covalent bonds and one dative covalent bond between the carbon and oxygen atoms. Draw the dot-and-cross diagram for carbon monoxide.

8 Explain why phosphorus can form five covalent bonds whereas nitrogen can form only three covalent bonds.

9 Explain the difference between a σ-bond and a π-bond.

10 Draw dot-and-cross diagrams for:
   a carbon tetrachloride, CCl_4
   b carbon dioxide, CO_2
   c the sulfate ion, SO_4^{2-}
   d xenon tetrafluoride, XeF_4

11 Suggest which is the strongest and which is the weakest of the following bonds:

   C–Cl, C–C and C–Br

Chapter 7

Introduction to organic chemistry

Nitrogen: northern lights and fertilisers

Atomic number: 7

Electron configuration: $1s^2\ 2s^2\ 2p_x{}^1\ 2p_y{}^1\ 2p_z{}^1$

Symbol: N

Nitrogen is in group 5 of the periodic table. Dry air contains 78.1% nitrogen. The element is also found in all living organisms — in proteins, DNA and other compounds.

Aurora borealis

Northern lights (aurora borealis)

Solar flares are eruptions of the sun that cause millions of protons and electrons to stream into space. These charged particles are concentrated by the Earth's magnetic field to regions above the north and south poles. When they hit the Earth's atmosphere, at a height of 100–250 km, they cause oxygen molecules to split into atoms and they ionise nitrogen molecules, forming excited $N_2{}^+$ ions. Outer electrons in both the oxygen atom and the $N_2{}^+$ ion are in a higher energy level. When they drop down to the ground state, light is emitted.

The oxygen atoms emit crimson and whitish-green light; the nitrogen ions emit violet and blue light. Waving coloured patterns of light can be seen at high latitudes in the northern hemisphere and in the southern hemisphere, where they are called aurora australis.

Fertilisers

Plants require nitrogen for growth, but they cannot use atmospheric nitrogen.

- Some bacteria, particularly those in the roots of leguminous plants, convert nitrogen gas into nitrates. This is called biological fixation.
- The high energy of lightning causes oxygen to react with nitrogen and nitric acid forms. This falls to the ground where it is neutralised and absorbed by plants. About 30 million tonnes of nitrogen are fixed this way annually.
- Nitrogen and hydrogen are converted into ammonia by the Haber process. In the USA, ammonia is drilled directly into the ground as a fertiliser. In Europe, it is converted into urea, ammonium nitrate and ammonium sulfate.

The high temperatures in car engines and in power stations mimic the effect of lightning and produce nitrogen oxides, which are oxidised to nitric acid and fall as acid rain. This can act as a fertiliser. Organic fertilisers, such as animal dung and compost, have to be broken down by bacteria to inorganic nitrates before plants can absorb them.

Introduction to organic chemistry

Introduction

Organic chemistry is the study of compounds containing carbon and hydrogen atoms. Some organic compounds contain other elements, such as oxygen, a halogen or nitrogen. A few contain sulfur. Carbon is unique in that, in compounds, it forms strong and stable bonds with other carbon atoms. Chains of several thousand carbon atoms are possible, giving rise to potentially millions of organic compounds.

Carbon has the electron configuration $1s^2\ 2s^2\ 2p_x^1\ 2p_y^1\ 2p_z^0$. It has only two unpaired electrons, so it might be thought that a carbon atom would only form two bonds. However, it is energetically favourable for one electron from the $2s$-orbital to be promoted into the empty $2p_z$-orbital, which results in four unpaired electrons. The energy required for this promotion is much less than the extra energy released when the atom forms four, rather than two, covalent bonds.

◀ Every carbon atom *must* have four bonds.

Bonding in organic chemistry

The four bonds around every carbon atom can be:

- four single bonds — four σ-bonds

- two single bonds and one double bond — three σ-bonds and one π-bond

◀ σ- and π-bonds are explained on pages 117–18.

- one single bond and one triple bond — two σ-bonds and two π-bonds

- two double bonds — two σ-bonds and two π-bonds

- three single bonds and one ionic bond

Remember that all carbon atoms have a valency of four. This means that each carbon atom has four bonds. In the examples above, note that there are four lines (covalent bonds) coming off each carbon atom, apart from the last type where there are three lines and a charge. In A-level chemistry, you will mostly come across the first two types of bonding.

Four single bonds: four σ-bonds

The four pairs of bonding electrons around the carbon atom repel each other and, as there are no lone pairs, the resulting molecule is tetrahedral in shape with bond angles of 109.5°. The simplest example is methane, CH_4:

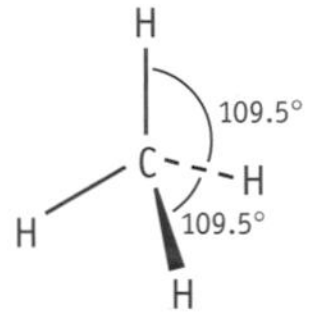

Two single and one double bond: three σ-bonds and one π-bond

A double bond consists of a σ-bond and a π-bond. The σ- and π-electrons in the double bond occur along the same axis. They repel the other two bonded pairs in the two single σ-bonds. Therefore, the three 'lots' of electrons take up a position of maximum separation, which is a planar arrangement with bond angles of 120°. The simplest example is ethene, $H_2C{=}CH_2$:

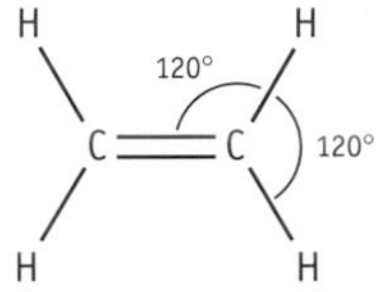

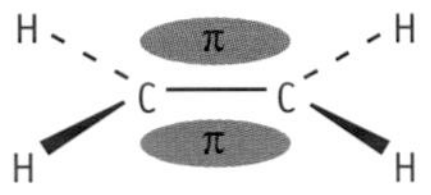

The π-electrons are above and below the σ-bond between the two carbon atoms. If the two CH_2 groups were rotated, the π-bond would break and only reform after a rotation of 180°. This bond breaking would require a large amount of energy (about 250 kJ mol^{-1}) and so does not happen unless the substance is heated very strongly. Therefore, *at room temperature*, there is no rotation about a double bond.

Families of organic compounds

As there are millions of organic compounds, order has to be brought out of chaos. Organic substances are divided into 'families' or homologous series.

A **homologous series** is a series of compounds that contain the same functional group and that have the same general formula. The formula of each member of the series differs from the next by CH_2.

The members of a homologous series show a trend in physical properties, such as melting and boiling temperatures and solubility.

A **functional group** is a small group of atoms or a single halogen atom that gives the compounds in the series particular chemical properties. The functional groups covered at AS are shown in Table 7.1.

Functional group	Name of series	Suffix	General formula	First member
None	Alkanes	-ane	C_nH_{2n+2}	Methane, CH_4
C=C	Alkenes	-ene	C_nH_{2n}	Ethene, $H_2C{=}CH_2$
–X, where X is a halogen	Halogenoalkanes	-ane	$C_nH_{2n+1}X$	Chloromethane, CH_3Cl
–OH	Alcohols	-ol	$C_nH_{2n+1}OH$	Methanol, CH_3OH
(structure: C bonded to H and =O)	Aldehydes	-al	RCHO where R is an alkyl group or H	Methanal, HCHO
(structure: C=O with two alkyl groups)	Ketones	-one	(structure: R–C(=O)–R') where R and R' are alkyl groups	Propanone, CH_3COCH_3
(structure: C with =O and O–H)	Carboxylic acids	-oic acid	(structure: R–C(=O)–O–H) where R is an alkyl group or H	Methanoic acid, HCOOH

Table 7.1 Functional groups

Naming organic compounds

Organic compounds are named according to the IUPAC (International Union of Pure and Applied Chemistry) system.

Hydrocarbons

Hydrocarbons are compounds that contain the elements carbon and hydrogen only.

Alkanes

An alkane is a hydrocarbon in which all the bonds between carbon atoms are single.

- The alkanes are an example of a **homologous series**.
- Alkanes are **saturated** compounds — they have no double or triple bonds.
- The names of all alkanes end in the suffix –*ane*.

The names of individual members of the homologous series of alkanes are found by using three rules.

Rule 1: identify the longest unbranched carbon chain to establish the stem name. The stem names for the alkanes are given in Table 7.2.

Table 7.2 Names of alkanes

Longest unbranched carbon chain	Alkane stem name	Alkane
1	Meth-	Methane
2	Eth-	Ethane
3	Prop-	Propane
4	But-	Butane
5	Pent-	Pentane
6	Hex-	Hexane
7	Hept-	Heptane
8	Oct-	Octane

Formulae of alkanes are traditionally drawn with the carbon atoms in a straight line, but they are really a three-dimensional zigzag.

The compound below has a six-carbon chain. The six carbon atoms are printed in blue.

Rule 2: name any substituent groups that are bonded to a carbon atom (Table 7.3). If there are two or more substituents, they must be named in alphabetical order.

Substituent group	Formula
Methyl	CH_3
Ethyl	CH_3CH_2
Propyl	$CH_3CH_2CH_2$
Butyl	$CH_3CH_2CH_2CH_2$

Table 7.3 Alkyl substituent groups

Rule 3: identify the position of each substituent by a number.
- the numbering is done from the end of the carbon chain that gives the lowest number for the substituent group
- if there are two identical groups on the *same* carbon atom, the number is repeated and the prefix *di-* is added to the name of the substituent group
- if there are two or more identical groups on *different* carbon atoms, the substituent groups are numbered and the prefixes di-, tri- or tetra- (for two, three or four substituents) are used

Worked example
Name the following compound:

2,2,4-trimethyl-pentane is used as a fuel in petrol-driven cars and has an octane number of 100.

> **Answer**
> Rule 1: the longest unbranched carbon chain has five carbon atoms — pentane.
> Rules 2 and 3: there are three methyl groups bonded to this chain — two on carbon atom number 2 and one on carbon atom number 4.
> The name of the compound is 2,2,4-trimethylpentane.

Cycloalkanes

A cycloalkane is a saturated hydrocarbon with the carbon atoms in a ring.

Cycloalkanes have the general formula C_nH_{2n}, which is the same as that of alkenes. The first member of the homologous series is cyclopropane. The most stable is cyclohexane.

Cyclopropane   Cyclohexane

Alkenes

An alkene is a hydrocarbon with *one double* bond between carbon atoms.

Unsaturated compounds have one or more double or triple bonds between carbon atoms, so alkenes are a type of unsaturated compound.

The name of an alkene is derived by replacing the ending *-ane* in the corresponding alkane with the ending *-ene* and, if necessary, specifying the position of the double bond. To do this, number from the end of the longest carbon chain that assigns the lowest number to the first carbon in the double bond.

Alkenes include:
- ethene, $H_2C=CH_2$
- propene, $CH_2=CHCH_3$
- butene

The name of butene depends on the position of the double bond. For instance:
- $H_2C=CHCH_2CH_3$ is but-1-ene
    1 2 3 4
- $H_3CCH=CHCH_3$ is but-2-ene
    1 2 3 4

Naming other organic compounds

Rule 1 has to be modified; rules 2 and 3 are unaltered.

Rule 1: identify the series or class of organic compound to which the formula belongs and ascertain the longest unbranched carbon chain. The suffix of the series is added to the corresponding alkane name (minus the final *e*) of the longest carbon chain. For example, CH_3CH_2OH is ethanol, rather than ethaneol; $CH_3CH_2CH_2CHO$ is butanal, not butaneal.

> **e** Be particularly careful when naming carboxylic acids. The carbon of the –COOH group is part of the carbon chain. Therefore, CH_3CH_2COOH has a three-carbon atom chain and is called *propan*oic acid. The same applies to aldehydes and ketones. CH_3CHO is *ethan*al, and CH_3COCH_3 is *propan*one.

Rule 2: name any substituent groups that are bonded to a carbon atom. If there are two or more substituents, they must be named in alphabetical order.

Rule 3: identify the position of each substituent by a number.

Worked example

Name the following compounds:

a

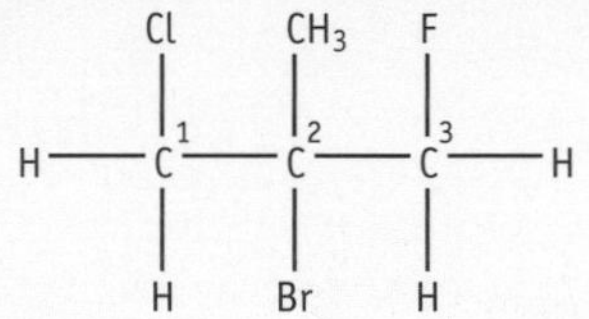

b

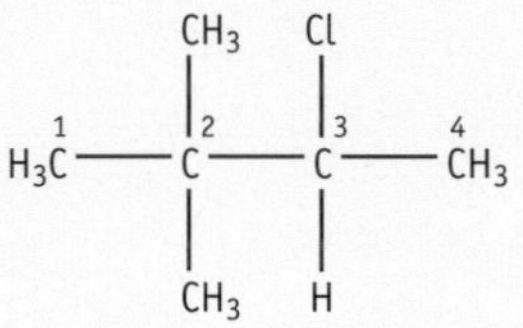

c

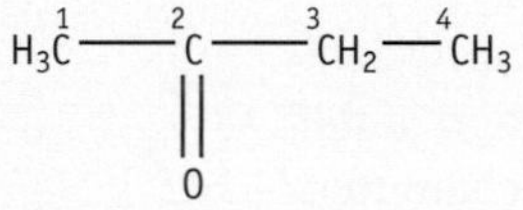

Answers

a Chain length three — propane; substituent groups named in alphabetical order: 2-bromo-1-chloro-3-fluoro-2-methylpropane

b Chain length four — butane; substituents groups named in alphabetical order: 2-chloro-3,3-dimethylbutane

c Compound is a ketone with a chain length of four: butanone

> There is no need to state a number, as the =O must be on one of the inner carbon atoms. Otherwise, the compound would be an aldehyde and not a ketone.

Representing molecules

The **molecular formula** of a compound gives the numbers of atoms of each element present in one molecule of the substance. It does not indicate how the atoms are joined together. The molecular formula of ethanol is C_2H_6O and that of ethene is C_2H_4.

The **empirical formula** is the simplest whole number ratio of the atoms of each element in the compound. The empirical formula of ethanol is the same as its molecular formula; the empirical formula of ethene is CH_2.

Any formula that is used to represent a compound must be unambiguous. Molecules of ethanol and methoxymethane each have two carbon atoms, six hydrogen atoms and one oxygen atom. Therefore, the molecular formula C_2H_6O is ambiguous. The molecular formula of ethene is C_2H_4, which is unambiguous and so can be used to represent ethene in equations.

The use of structural formulae overcomes ambiguous molecular formulae. The **structural formula** of ethanol (molecular formula C_2H_6O) is C_2H_5OH and that of methoxymethane (molecular formula C_2H_6O) is CH_3OCH_3. The structural formula must include any double bonds that are present in the molecule; the structural formula of ethene is $H_2C=CH_2$.

The **displayed** or **full structural formula** must show all the atoms separately and all the bonds between them. The displayed formula for ethene is:

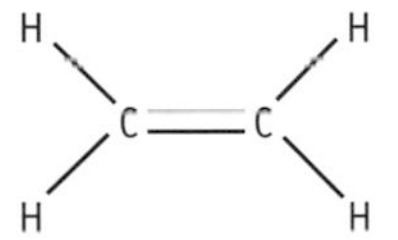

In a **skeletal formula** the carbon atoms are not drawn. A straight line represents a single bond and a double line represents a double bond. A carbon atom is assumed to be at the end of these lines unless another atom, such as oxygen, is shown (see propan-2-ol below). Hydrogen atoms that are bonded to carbon are also omitted. However, remember that each carbon atom has enough hydrogen atoms joined to it to make a total of four bonds.

The following are examples of skeletal formulae:

Propene	Methylpropane	Propan-2-ol	Cyclohexane
$CH_2=CHCH_3$	$CH_3CH(CH_3)CH_3$	$CH_3CH(OH)CH_3$	C_6H_{12}

The skeletal method is particularly useful when drawing the formulae of large molecules, such as cholesterol:

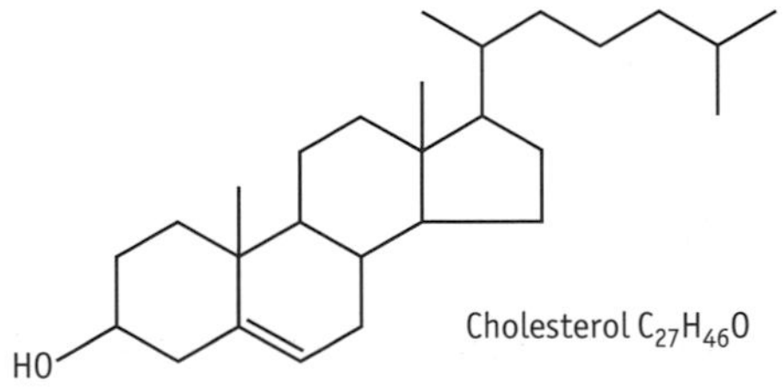

Cholesterol $C_{27}H_{46}O$

The skeletal formula of an unsaturated fat containing an omega-3 acid (A), an omega-6 acid (B) and saturated stearic acid (C) is:

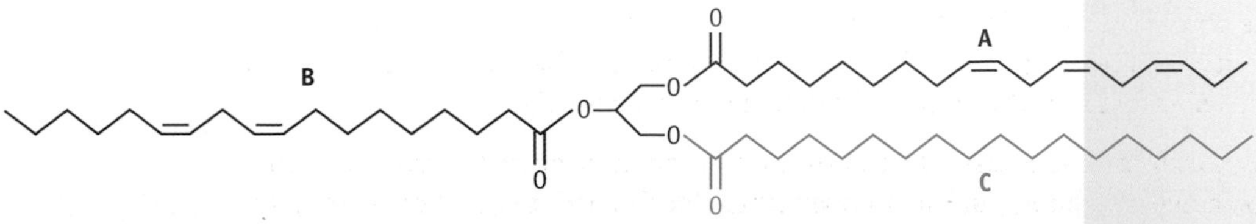

'Omega-3' means that the first double bond is on the *third* carbon atom counting from the non-carboxylic end of the acid. (Omega is the last (end) letter in the Greek alphabet.)

The structural formula of this unsaturated fat is:

$$H_2COOCC_{17}H_{29}$$
$$HCOOCC_{17}H_{31}$$
$$H_2COOCC_{17}H_{35}$$

Isomerism

Isomers are compounds with the same molecular formula but different structural formulae.

There are two main types of isomerism.

Structural isomerism

Structural isomers have the same molecular formula but different molecular structures. This means that the isomers have the same number of atoms of the same elements, but that these atoms are arranged differently.

Structural isomers can be divided into three categories: carbon-chain, positional and functional-group.

Carbon-chain isomers

The only way in which these isomers differ is in the length of the carbon chain. For example, butane and methylpropane both have the molecular formula C_4H_{10}, but butane has a carbon chain of four atoms and methylpropane has a carbon chain of three atoms:

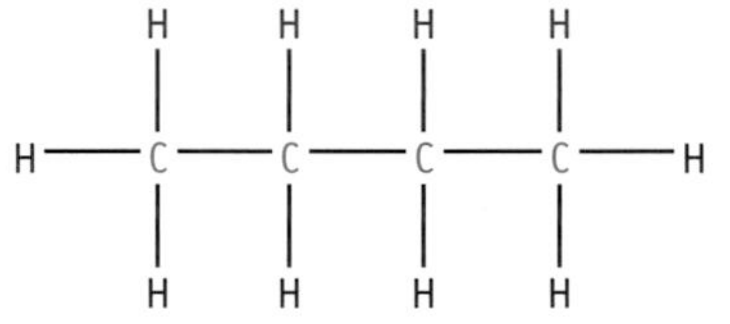

Butane 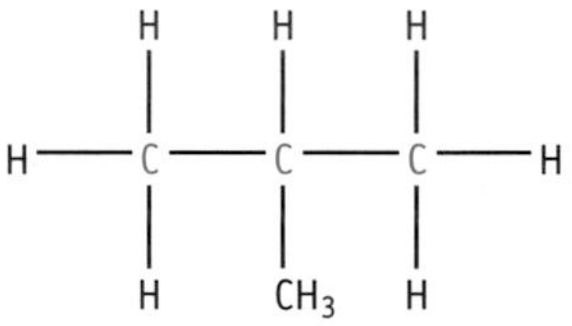   Methylpropane

The skeletal formulae of these two isomers are:

Butane 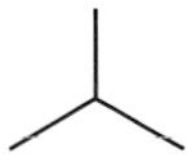   Methylpropane

Positional isomers

Positional isomers have the same functional group in different locations on the carbon framework. A simple example is provided by the alcohols propan-1-ol and propan-2-ol. Both have the molecular formula C_3H_8O, but their structural formulae are $CH_3CH_2CH_2OH$ and $CH_3CH(OH)CH_3$ respectively. Their *full* structural formulae are as follows:

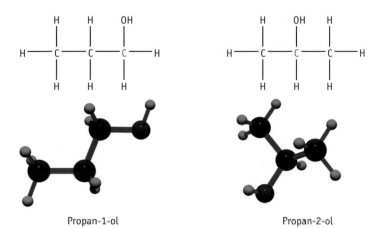

Propan-1-ol   Propan-2-ol

The carbon chains may also be different. Butan-1-ol and butan-2-ol are positional isomers, as are 2-methylpropan-2-ol and 2-methylpropan-1-ol. However, butan-1-ol and 2-methylpropan-1-ol are carbon-chain isomers, as are butan-2-ol and 2-methylpropan-2-ol. All four substances have the molecular formula $C_4H_{10}O$. Their displayed and skeletal formulae are as follows:

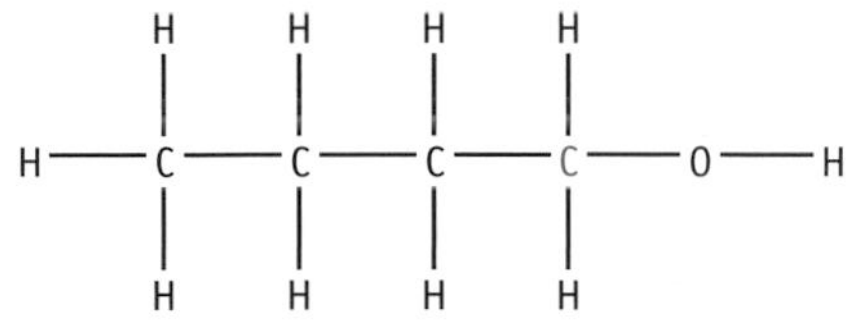

or

$CH_3CH_2CH_2CH_2OH$

Butan-1-ol

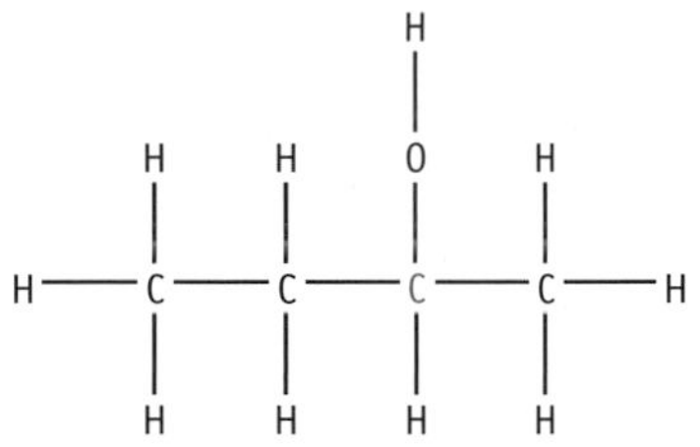

or

$CH_3CH_2CH(OH)CH_3$

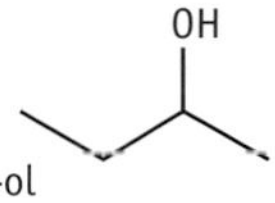

Butan-2-ol

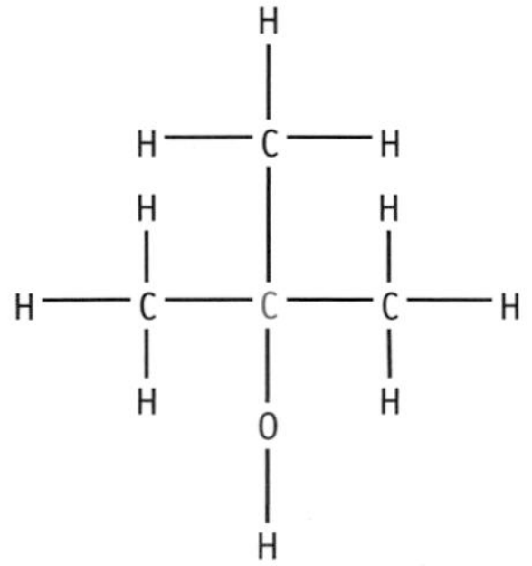

or

$(CH_3)_3COH$

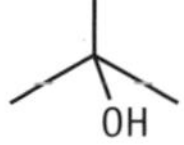

2-methylpropan-2-ol

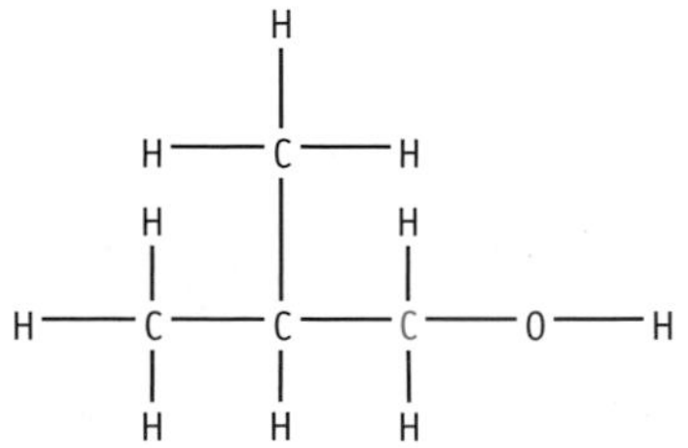

or

$CH_3CH(CH_3)CH_2OH$

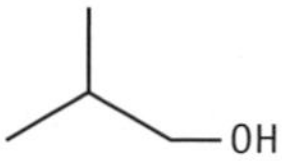

2-methylpropan-1-ol

Functional-group isomers

Functional-group isomerism occurs when two compounds with the same molecular formula are members of different homologous series and have different functional groups. For example, the ester methyl methanoate, $HCOOCH_3$, is an isomer of ethanoic acid, CH_3COOH:

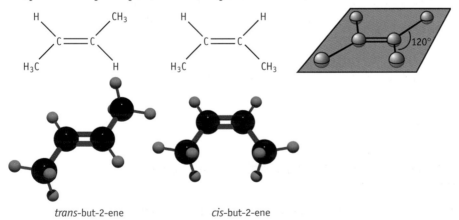

Methyl methanoate   Ethanoic acid

Another example is hex-1-ene and cyclohexane, both of which have the molecular formula C_6H_{12}. However, a molecule of hex-1-ene has a double bond whereas a cyclohexane molecule is cyclic with no double bond.

Stereoisomerism

There are two quite separate types of stereoisomers — geometric and optical. In both types, the carbon chain, the position of the groups on the chain and the nature of the functional groups are the same. However, the arrangement of the atoms in space is different.

Geometric isomers

> Geometric isomers are isomers that differ only in the spatial arrangement of atoms in the planar part of the molecule or above and below a ring.

The most common way that this occurs is when a molecule has a C=C group and each of these carbon atoms has two different groups or atoms joined to it. For example, but-2-ene exists as two geometric isomers. The two double-bonded carbon atoms and the four atoms joined to them all lie in a plane and can take up different spatial positions in that plane.

trans-but-2-ene   *cis*-but-2-ene

The geometry of the two isomers is different. In the left-hand isomer, the two $-CH_3$ groups are staggered at 180° to each other and are on opposite sides of the double bond. This compound is called the *trans* isomer. In the right-hand isomer, the two $-CH_3$ groups are on the same side of the double bond. This compound is called the *cis* isomer.

e This type of isomerism is the least important in A-level chemistry.

But-1-ene does not have geometric isomers because one of the carbon atoms in the double bond has two identical atoms attached to it:

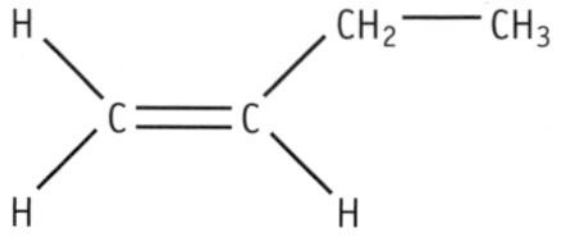

But-2-enoic acid does exhibit geometric isomerism. One carbon atom has a $-CH_3$ group and a hydrogen atom attached to it and the other has a $-COOH$ group and a hydrogen atom. Geometric isomerism arises because each carbon atom of the double bond has two different atoms or groups attached to it.

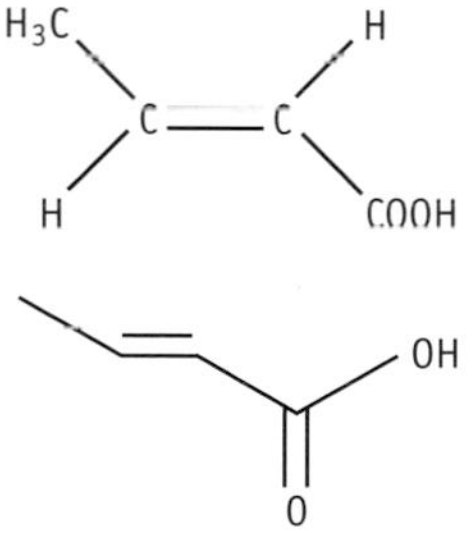

trans-but-2-enoic acid 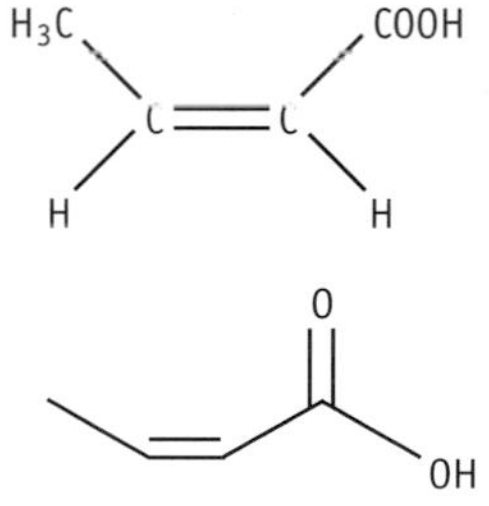   cis-but-2-enoic acid

Geometric isomers exist because there is restricted rotation around the double bond. Therefore, one isomer cannot spontaneously convert into the other unless sufficient energy is supplied (page 118).

The exact shape of a molecule is often important in biochemistry. For example, vision in human beings depends upon the geometric isomerism of retinal, which is found in the retina in the eye. When light hits a cell containing *cis*-retinal, the energy of the light breaks the π-bond. The molecule rotates and reforms as the *trans*-isomer. This shape change causes a signal to be sent, via the optic nerve, to the brain.

This method of naming geometric isomers breaks down in compounds such as 2-chlorobut-2-ene. One of its two geometric isomers is shown below:

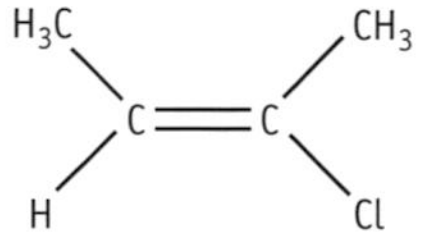

The $-CH_3$ group on the left is *cis* to the $-CH_3$ group on the right but *trans* to the chlorine atom on the right.

Priorities have to be assigned to the atoms attached to each of the C=C carbon atoms. The higher the atomic number, the higher is the priority assigned to that atom.

The left-hand carbon atom of the double bond has a carbon atom (atomic number 6) and a hydrogen atom (atomic number 1) attached. The carbon has the higher priority.

e Do not state that there is no rotation about the double bond; the rotation is restricted and does not occur at room temperature.

The right-hand carbon atom has a carbon atom (atomic number 6) and a chlorine atom (atomic number 17) attached. The chlorine has the higher priority.

If the two higher priority atoms are on the opposite sides of the double bond, the isomer is labelled E (for *entgegen*, the German for opposite). If they are on the same side, the isomer is labelled Z (for *zusammen*, the German for together).

A way to remember this is that E is a vowel and 'opposite' starts with a vowel. Similarly, Z is a consonant and 'together' starts with a consonant. (*Z* meanz on ze zame zide!)

The names and structural formulae of the two isomers of 2-chlorobut-2-ene are:

(E)-2-chlorobut-2-ene   (Z)-2-chlorobut-2-ene

The higher priority atoms are marked in red.

Cis- and trans-*fatty acids*
Oils and fats are the esters of propane-1,2,3-triol and long-chain fatty acids, some of which are unsaturated. In natural oils and fats, the fatty acids are all *cis*-isomers, in which a hydrogen atom is attached to each carbon atom in each double bond. When an unsaturated fat is partially hydrogenated in the presence of a nickel catalyst, some of the double bonds are broken. Rotation occurs and the *trans*-isomer is formed. *Trans*-fatty acids are more harmful than the *cis*-acids and may cause heart disease.

e In *E*- and *Z*-notation, the *cis*-acids are called *Z*-acids.

The skeletal formula of a polyunsaturated fat on page 131 shows that the omega–3 and the omega-6 fatty acids are *cis*- (or *Z*-) isomers.

Optical isomers
Optical isomers are molecules that have non-superimposable mirror images.

> **e** Knowledge of optical isomers is only required at A2 and so is not covered in this book.

Alkanes

Alkanes are saturated hydrocarbons that have the general formula C_nH_{2n+2}.

'Saturated' means that the molecules do not have any double bonds, so they cannot react with hydrogen.

Physical properties

Boiling temperature
The first four members of the homologous series of alkanes are gases at room temperature. The remainder are liquids up to a value of about $n = 30$; subsequently they are waxy solids. A graph of boiling temperatures at 1 atm pressure is shown in Figure 7.1.

The boiling temperature rises because of the increasing number of electrons in the molecule. This causes stronger instantaneous induced dipole–induced dipole forces of attraction (dispersion forces — page 164). Stronger intermolecular forces require more energy to separate the molecules, resulting in a higher boiling temperature.

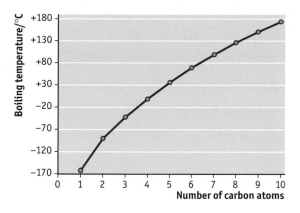

Figure 7.1 Boiling temperatures of the first ten unbranched alkanes

Propane and butane can easily be liquefied at room temperature by increasing the pressure. The liquid in bottles of domestic gas and camping gas and in cigarette lighters is compressed butane and propane. LPG (liquefied petroleum gas or Autogas) is sold as an alternative fuel to petrol (gasoline) and is also a mixture of propane and butane. However, it produces about 25% less energy per litre than petrol.

Branched alkanes have lower boiling temperatures than their straight-chain isomers. This is because there are fewer points of contact between adjacent molecules — they do not pack so well together (Table 7.4).

Name	Number of carbon atoms	Boiling temperature/°C
Butane	4	−0.5
Methylpropane	4	−12
Pentane	5	36
2-methylbutane	5	28
2,2-dimethylpropane	5	10

Table 7.4 Boiling temperatures of butane and pentane and their branched isomers

Solubility
Alkanes are insoluble in water, but dissolve readily in each other. Crude oil is a mixture of over 100 different alkanes dissolved in each other.

Density
Liquid alkanes are less dense than water. Their density increases as the molar mass of the alkane increases.

Occurrence
The major source of methane is natural gas. Crude oil is the source of many other alkanes. It is first fractionally distilled. The heavier fractions are either used as fuels or are cracked to give a mixture of alkenes and smaller alkanes (page 141). These alkanes, together with the gasoline and naphtha fractions, are reformed to give a fuel with a higher octane rating. On heating with a suitable metal catalyst, the straight-chain alkanes turn into branched-chain isomers. These do not explode at the high pressures inside the cylinder of a car engine, until they are ignited with a spark. Therefore, they have higher octane ratings than unbranched alkanes. The gasoline and naphtha fractions are reformed to give a fuel with a higher octane rating.

Chemical reactions

Reaction with halogens

In the presence of bright white or ultraviolet light, alkanes react with chlorine and bromine. The reaction involves the replacement of the hydrogen atoms in the alkane by halogen atoms. There is normally a mixture of products as the hydrogen atoms are successively replaced:

$$CH_4 + Cl_2 \rightarrow HCl + CH_3Cl$$
$$CH_3Cl + Cl_2 \rightarrow HCl + CH_2Cl_2 \text{ and so on}$$

The products are hydrogen chloride and a mixture of chloromethane, dichloromethane, trichloromethane and tetrachloromethane (carbon tetra-chloride).

This is an example of a **free-radical substitution** reaction. A **free radical** is an uncharged species with an unpaired electron that is used to form a covalent bond. An example is a chlorine atom, which has the electronic configuration $1s^2\, 2s^2\, 2p^6\, 3s^2\, 3p_x^2\, 3p_y^2\, 3p_z^1$.

> A substitution reaction is a reaction in which one atom or group is replaced by another atom or group. There are always two reactants and two products.

Reaction mechanism

The mechanism of an organic reaction shows the pathway from the reactants, via the transition state or any intermediates, to the products. It also shows the movement of electrons during the reaction by the use of arrows.

A **half-headed** or **fishhook** arrow represents the movement of a *single* electron:

A half-headed arrow can start:
- from a covalent bond. In this case, it goes either to an atom (thus forming a radical) or to another radical (thus forming a covalent bond).
- from the single electron in a free radical. In this case, it goes to a radical to form a covalent bond.

A free radical is represented by a *single* dot beside the symbol of the atom. For instance, Cl• represents a chlorine radical.

Initiation

Light energy splits a chlorine molecule into chlorine radicals.

The light must be of a high enough frequency to cause the σ-bond to break. It does so homolytically with one of the bonding electrons going to each chlorine atom:

$$Cl\text{—}Cl \xrightarrow{\ UV\ } Cl\bullet \ + \ Cl\bullet$$

> This process is called photolysis, because the bond is broken (-*lysis*) by light (*photo*-).

Propagation

Chlorine radicals are very reactive. When a chlorine radical collides with a methane molecule with sufficient energy, a hydrogen atom is removed to form a methyl radical, •CH_3, and a molecule of hydrogen chloride:

$$CH_4 + Cl\bullet \rightarrow \bullet CH_3 + HCl$$

> **e** Remember that the fishhook arrows show that one electron goes to each atom as the bond breaks.

Methyl radicals are also very reactive and will remove a chlorine atom from a chlorine molecule, forming a chloromethane molecule and another chlorine radical:

$$CH_3 + Cl_2 \rightarrow CH_3Cl + Cl\bullet$$

The chlorine radical then removes a hydrogen atom from another methane molecule and so on. This type of reaction is called a **chain reaction** as a small amount of UV light results in many propagation reactions. It has been estimated that, in the reaction of methane with chlorine, a single photon of light causes 10^6 propagation steps.

> **e** The H–Br bond is weaker than the H–Cl bond, so less energy is given out in its formation. This makes the first propagation step endothermic ($\Delta H = +46$ kJ mol^{-1}) for bromine, whereas it is exothermic ($\Delta H = -19$ kJ mol^{-1}) for chlorine. The result is that bromination is much slower than chlorination.

As the concentration of CH_3Cl builds up, an alternative propagation step becomes increasingly likely — a chlorine radical collides with a CH_3Cl molecule and removes a hydrogen radical from it:

$$CH_3Cl + Cl\bullet \rightarrow \bullet CH_2Cl + HCl$$

The $\bullet CH_2Cl$ radical then removes a chlorine atom from a chlorine molecule, thus continuing the chain reaction:

$$\bullet CH_2Cl + Cl_2 \rightarrow CH_2Cl_2 + Cl\bullet$$

Termination (chain breaking)

The chain reaction can be broken when two radicals collide. As their concentration is low, the chance of this happening is slight. The second propagation step has the highest activation energy, so the $\bullet CH_3$ radicals have the longest life and are the most likely to be involved in chain breaking. Examples of chain-breaking reactions are:

$$\bullet CH_3 + \bullet CH_3 \rightarrow CH_3CH_3 \text{ (ethane)}$$
$$\bullet CH_3 + Cl\bullet \rightarrow CH_3Cl$$
$$Cl\bullet + Cl\bullet \rightarrow Cl_2$$

Any of these reactions breaks the chain and slows down the reaction.

When there are no radicals left, the reaction stops.

Products of the reaction

The reaction between methane and chlorine in the presence of UV light produces a mixture of mono- and poly-substituted chloromethanes, hydrogen chloride and some ethane.

It is the presence of the ethane that provides conclusive evidence for the mechanism given above. The absence of any H_2 molecules in the product is evidence that no hydrogen radicals are formed in the propagation steps.

Evidence for the free-radical mechanism includes:

- the production of ethane as a minor by-product
- the need for high-frequency light to produce chlorine radicals from chlorine molecules; intense red light does not initiate the reaction

The chlorine radical cannot bond to the carbon atom until a hydrogen atom has been removed, so the propagation step is *not* $CH_4 + Cl\bullet \rightarrow CH_3Cl + H\bullet$. Also, the energetics of this reaction are much less favourable than for the reaction in which a chlorine radical removes a hydrogen atom.

e In a chain reaction, a radical reacts with a molecule and produces another radical, which then continues the process.

In focused sunlight or strong ultraviolet light, chlorine and methane react explosively.

The reaction of alkanes with bromine is much slower because the second propagation step has a very high activation energy (page 236). The reaction can be demonstrated in the laboratory by placing a few drops of hexane into each of two small conical flasks, very carefully adding two drops of liquid bromine and stoppering both flasks. One flask is placed on a windowsill and the other in a lightproof cupboard.

The next day, the contents of the flask in the cupboard will still show the brown colour of bromine but the contents of the flask on the windowsill will be colourless. If this flask is carefully opened in a fume cupboard, steamy fumes of hydrogen bromide will be observed.

Combustion

Alkanes, like almost all organic compounds, burn when ignited in air. If the air is in excess, the products are water and carbon dioxide. Combustion is highly exothermic and is the major source of energy in gas-fired power stations:

$$CH_4(g) + 2O_2(g) \rightarrow CO_2(g) + 2H_2O(g) \qquad \Delta H^\ominus = -802 \text{ kJ mol}^{-1}$$

Combustion of gasoline provides the power in petrol-driven cars:

$$C_8H_{18}(g) + 12\tfrac{1}{2}O_2(g) \rightarrow 8CO_2(g) + 9H_2O(g) \qquad \Delta H^\ominus = -5074 \text{ kJ mol}^{-1}$$

In excess air, some of the nitrogen in the air reacts at the very high temperatures in the flame to produce nitric oxide pollutant. This pollutant can cause photo-chemical smog as well as trigger asthma attacks.

$$N_2 + O_2 \rightarrow 2NO$$

When alkanes burn in a limited amount of air, incomplete combustion takes place. Carbon monoxide and even some carbon may be produced as well as some cracked alkanes. The hydrogen in the alkanes is always oxidised to water.

Sustainability and climate change

Alkanes are examples of fossil fuels. Over many millions of years, dead marine organisms have been converted to a mixture of hydrocarbons. Some of these became trapped underground as natural gas (mostly methane) and oil. However, these resources are limited.

Although new gas and oil fields are being discovered, eventually there will not be enough oil and gas in the world to satisfy demand. In addition, the combustion of fossil fuels releases carbon dioxide into the air. The increase in atmospheric carbon dioxide is a well-established fact. It is also known that carbon dioxide is a greenhouse gas (see page 291) and contributes to climate change.

Carbon dioxide emissions could be reduced by a number of methods, for example:
- by using alternative sources of energy such as nuclear, wind or wave power
- by conserving energy in the home and workplace. For example, recycling old aluminium cans requires much less energy than making new aluminium from bauxite ore.

A possible new source of methane is methane hydrate. This is found on the continental shelves of oceans at a depth of more than 300 feet and in permafrost (Figure 7.2). The methane molecules are trapped in a cage of ice crystals. It has been estimated that the amount of methane hydrate on Earth is twice that of all previously known gas and oil reserves.

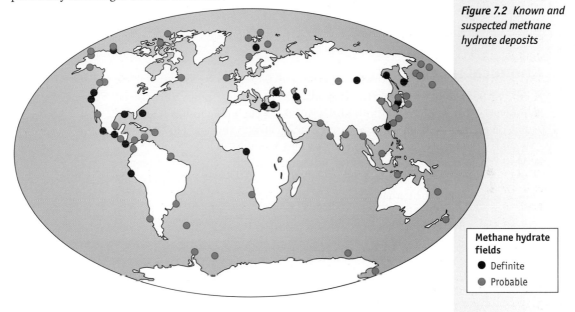

Figure 7.2 *Known and suspected methane hydrate deposits*

Methane hydrate fields
- ● Definite
- ● Probable

Alkenes

Alkenes are **unsaturated** compounds that have the general formula C_nH_{2n}.

Alkenes contain one C=C group. Compounds containing two such groups are called alkadienes. They react in the same way as alkenes.

Manufacture

Alkenes do not occur naturally. They are made from alkanes by the process of **cracking**. The naphtha fraction from the primary distillation of crude oil is mixed with steam and rapidly heated to 900°C. The yield of ethene is about 30%, with a smaller amount of propene and about 25% high-grade petrol. A temperature of 700°C gives less ethene but more high-grade petrol. Alkenes can also be made by cracking propane and butane mixtures (LPG) and by the catalytic cracking of heavier fractions from the primary distillation of oil.

Physical properties

- Alkenes have lower melting and boiling temperatures than alkanes. This is because the rigidity of the double bond does not allow the molecules to pack together as efficiently as alkanes. This effect of molecules packing together is shown by the difference in the melting points of fats and oils. Fats are saturated esters and pack well together, so they are soft solids at room temperature. Oils, such as corn or sunflower oil, are esters of unsaturated acids that are always geometric *cis*-isomers. These molecules do not pack well together and so are liquid. Fish oils are highly unsaturated and remain liquid even in the very cold temperatures of arctic waters. Unsaturated oils are metabolised more easily than saturated fats. Fish oils are particularly good at lowering cholesterol levels.

- Ethene, propene and all three butenes are gases at room temperature. The other alkenes are liquids or waxy solids.
- Alkenes are insoluble in water because they cannot form hydrogen bonds with water molecules.

Chemical reactions

- Alkenes are much more reactive than alkanes.
- Alkenes burn in air when ignited. However, as each alkene has a smaller percentage of hydrogen than the corresponding alkane, they tend to burn with a smoky flame, producing carbon and water vapour.
- Most of the reactions of alkenes are **addition** reactions. The π-bond breaks, leaving a σ-bond between the two carbon atoms. Two atoms or groups add on — one to each of the carbon atoms:

An addition reaction occurs when two substances react together to form a single substance.

A π-bond breaks and two new σ-bonds form. A π-bond is weaker than a σ-bond. Alkanes have no π-bond, so they are much less reactive than alkenes. Some bond strengths are given in Table 7.5.

Bond type	Bond enthalpy/kJ mol^{-1}
σ-bond in C–C	348
σ-bond + π-bond in C=C	612
π-bond only (difference between C=C and C–C)	264

Table 7.5 Strengths of carbon–carbon bonds

The reactions of the alkenes are illustrated below using ethene as the example. The other alkenes react similarly.

Combustion
Like almost all organic compounds, alkenes burn when ignited:

$$H_2C=CH_2 + 3O_2 \rightarrow 2CO_2 + 2H_2O$$

However, they are more useful as chemicals, so they are not used as fuels.

Reaction with hydrogen
When ethene is mixed with hydrogen and passed over a suitable catalyst, an addition reaction occurs. The product is ethane:

$$H_2C=CH_2 + H_2 \rightarrow H_3C–CH_3$$

The π-bond breaks, leaving the σ-bond between the two carbon atoms.

Reagent: hydrogen
Conditions: catalyst of nickel at 150°C or platinum at room temperature
Product: ethane
Reaction type: addition

The catalyst loosens the bonds between the hydrogen atoms, which then add on across the double bond.

This reaction is used in the hardening of polyunsaturated vegetable oils to form margarine.

> **Worked example**
>
> 0.747 g of an unsaturated hydrocarbon, C_6H_8, reacts with 448 cm^3 of hydrogen gas. Calculate the molar ratio of hydrogen to hydrocarbon and hence the number of carbon-to-carbon double bonds in the hydrocarbon. Under the conditions of the experiment, the molar volume of gas is 24.0 dm^3 mol^{-1}.
>
> **Answer**
>
> molar mass of $C_6H_8 = (6 \times 12.0) + 8.0 = 80.0$ g mol^{-1}
>
> amount of $C_6H_8 = \dfrac{0.747\,g}{80.0\,g\,mol^{-1}}$
>
> $= 0.00934$ mol
>
> volume of hydrogen $= 448$ cm^3
>
> amount of hydrogen $= \dfrac{448\,cm^3}{24\,000\,cm^3\,mol^{-1}}$
>
> $= 0.01867$ mol
>
> ratio (by moles) of hydrogen to hydrocarbon $= 0.01867:0.00934 \approx 2:1$
>
> Therefore, as 2 moles of hydrogen react with each mole of C_6H_8, there must be two C=C double bonds in the molecule.

Reaction with halogens

When ethene is passed into liquid bromine or bromine dissolved in an inert solvent such as tetrachloromethane, an addition reaction takes place. The π-bond breaks and a bromine atom adds on to each carbon atom:

$$H_2C=CH_2 + Br_2 \rightarrow H_2CBr–CH_2Br$$

Reagent: bromine
Conditions: mix at room temperature
Product: 1,2-dibromoethane
Reaction type: electrophilic addition

Bromine attacks the electron-rich π-bond in the ethene molecule and forms a σ-bond with one of the carbon atoms. The bromine is described as an **electrophile** and the reaction as **electrophilic addition**.

> An electrophile is a species that bonds to an electron-rich site in a molecule. It accepts a pair of electrons from that site and forms a new covalent bond.

The mechanism for this reaction is shown on page 144.

Alkenes react in a similar way with chlorine, but not with iodine.

Chapter 7: Introduction to organic chemistry 143

This is a test for compounds containing one or more C=C bonds.

Test for alkenes

If an alkene is shaken with bromine dissolved in water (bromine water), an addition reaction takes place and the brown colour of bromine disappears.

Reaction with hydrogen halides

When a gaseous hydrogen halide is mixed with ethene, an electrophilic addition reaction takes place:

$$H_2C{=}CH_2 \ + \ H{-}Br \rightarrow H_3C{-}CH_2Br$$

The reaction is fastest with hydrogen iodide and slowest with hydrogen chloride. This is because the H–I bond is weaker than the H–Br bond, which is weaker than the H–Cl bond.

Reagent: hydrogen bromide
Conditions: mix gases at room temperature
Product: bromoethane
Reaction type: electrophilic addition

The δ^+ hydrogen in hydrogen bromide attacks the electron-rich π-bond in ethene and forms a σ-bond using the two electrons from the π-bond. The hydrogen atom is the electrophile.

Mechanism of electrophilic addition

Alkenes react with reagents such as halogens and hydrogen halides by **electrophilic addition.**

The π-bond between the carbon atoms is an area of high electron density. The first step in electrophilic addition is the movement of these π-electrons towards the electrophile to make a new covalent σ-bond. This movement of a pair of electrons is shown by a full-headed curly arrow ($\frown\!\!\searrow$). A curly arrow must:

■ start from a π bond and go towards an atom forming a new bond
■ start from a σ or π bond and go to an atom forming a negative ion
■ start from an atom or ion and go towards another atom forming a new bond

In the examples below, the alkene is represented by the formula $RCH{=}CH_2$, where R can be a hydrogen atom (ethene) or a group, for example $-CH_3$ (propene). R can also represent much more complex groups in, for example complex natural products such as limonene.

Addition of hydrogen bromide

Addition of hydrogen bromide across a double bond is a two-step reaction.

The HBr molecule is polar with the hydrogen atom δ^+ and the more electro-negative bromine, δ^-. The δ^+ hydrogen is the electrophile. It adds on to one of the carbon atoms by attracting the π-electrons from the double bond and forming a covalent bond. As this happens, the Br–Br σ-bond breaks and the two electrons both go to the bromine atom, forming a Br^- ion. This uneven breaking of the H–Br bond is called **heterolytic fission**.

Step 1:

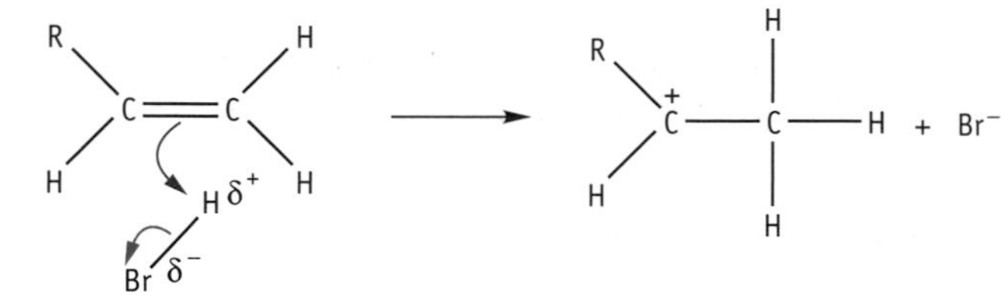

The red curly arrow shows the movement of the π-electrons and the blue curly arrow shows the movement of the H–Br σ-electrons.

Step 2: The Br⁻ ion then adds on to the positive carbon atom.

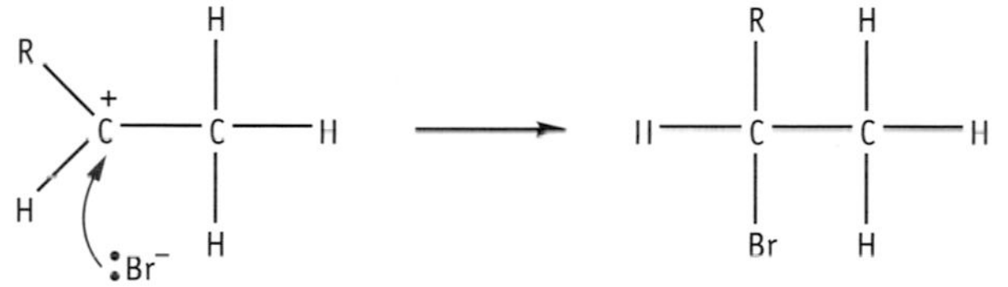

The red curly arrow shows the movement of the lone pair of electrons on the Br⁻ ion as it forms a covalent bond with the carbon atom.

Hydrogen chloride, HCl, and hydrogen iodide, HI, also undergo addition reactions with this type of mechanism.

Addition to asymmetrical alkenes: Markovnikoff's rule

If the alkene is asymmetrical, such as propene, two addition products are possible. For example, the reaction between propene, $CH_3CH=CH_2$, and hydrogen bromide can give:

- $CH_3CHBrCH_3$, 2-bromopropane
- $CH_3CH_2CH_2Br$, 1-bromopropane

The *major* product is 2-bromopropane. This is predicted, but *not* explained, by Markovnikoff's rule:

> When HX adds to an asymmetrical alkene, the hydrogen atom goes to the carbon which already has most hydrogen atoms directly attached.

The explanation for this is that a secondary carbocation, $CH_3C^+HCH_3$ is formed preferentially to a primary carbocation, $CH_3CH_2C^+H_2$ or that a tertiary carbocation is more stable than a secondary carbocation.

The secondary carbocation is stabilised by the electron-pushing effect of the $-CH_3$ groups, whereas the primary carbocation is less stabilised.

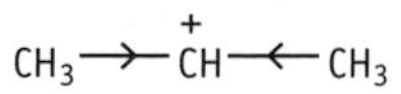

$$CH_3 \longrightarrow \overset{+}{C}H \longleftarrow CH_3$$

The arrowhead on a σ-covalent bond represents a movement of the bonding electrons in the direction the arrowhead is pointing. In the above example, it is pointing towards the positively charged carbon atom, reducing its charge and hence stabilising the cation. A $-CH_2$ group does not do this to the same extent as a $-CH_3$ group.

e Do not forget to include the positive charge on the carbon atom in the intermediate.

A carbocation is an organic ion that has a positive charge on a carbon atom.

The pushing effect of a $-CH_3$ group is explained by the difference in electronegativities of carbon (2.5) and hydrogen (2.1). The carbon becomes δ⁻ and so pushes electrons towards the neighbouring carbon atom.

Similarly with 2-methylbut-2-ene, the hydrogen goes to the CH carbon and not to the $C(CH_3)_2$ carbon. The major product is 2-chloro-2-methylbutane ($CH_3CH_2CCl(CH_3)_2$), and *not* 3-chloro-2-methylbutane ($CH_3ClCH(Cl)CH(CH_3)_2$):

$$CH_3CH=C(CH_3)_2 + HCl \rightarrow CH_3CH_2CCl(CH_3)_2$$
2-methylbut-2-ene   2-chloro-2-methylbutane

> **e** Do not state that the secondary halogenoalkane is more stable than the primary halogenoalkane. They are equally stable. It is the *intermediate* secondary carbocation that is more stable.

Addition of bromine (bromination)

Addition of bromine across a double bond is a two-step reaction.

The Br_2 molecule is non-polar. However, on its approach to the high electron density of the π-bond, a $δ^+$ charge is induced in the nearer bromine. This atom is the electrophile. It attracts the π-electrons and forms a covalent bond with one of the carbon atoms. At the same time, the Br–Br σ-bond breaks heterolytically and a Br^- ion is formed.

Step 1:

The red curly arrow shows the movement of the π-electrons and the blue curly arrow shows the movement of the Br–Br σ-electrons.

Step 2:

The red curly arrow shows the movement of the lone pair of electrons on the Br^- ion as it forms a covalent bond with the positively charged carbon atom.

The same mechanism occurs when an alkene reacts with chlorine.

Iodine is not a strong enough electrophile to undergo this reaction, unless the C=C group is activated by an oxygen atom on one of the carbon atoms.

Evidence for this mechanism is the formation of CH_2ClCH_2Br when ethene reacts with bromine in the presence of sodium chloride. Step 1 is the same, with the formation of the intermediate carbocation $H_2C^+CH_2Br$. This can then be attacked by either a Br^- ion (thus forming CH_2BrCH_2Br) or by a Cl^- ion (thus forming CH_2ClCH_2Br).

There is a partial bond between the bromine atom that has added on and the other carbon atom. This is represented in the intermediate by a dotted line.

The Br^- ion attacks from the side away from the bromine atom. This has stereochemical implications for alkenes such as RCH=CHR because stereoisomers are possible in the product.

A carbocation is an unstable ion with a positive charge on a carbon atom. The carbon atom has three covalent bonds and the positive charge. It is surrounded by three bond pairs of electrons and has an empty orbital, which it uses to form a bond with a nucleophile, such as Br^- or Cl^-.

The compound CH_2ClCH_2Cl does not form because there is no δ^+ chlorine atom in sodium chloride.

Addition with bromine water

The reaction between $RCH=CH_2$ and bromine water produces $RCH(OH)CH_2Br$ as the major product, rather than $RCHBrCH_2Br$. This is because the intermediate, RC^+HCH_2Br is more likely to collide and react with a water molecule than with a Br^- ion.

$$RCH=CH_2 + Br_2 \rightarrow RC^+H-CH_2Br + Br^-$$
$$RC^+H-CH_2Br + H_2O \rightarrow RCH(OH)CH_2Br + H^+$$

This reaction is used as a test for the presence of a carbon–carbon double bond. The decolourisation of brown bromine water to a colourless solution, without any precipitate being formed, is proof of the presence of a $C-C$ group.

Reaction with potassium manganate(VII) solution

When ethene is shaken with a neutral solution of potassium manganate(VII), oxidation with addition takes place. The purple manganate(VII) ions are reduced to a precipitate of brown manganese(IV) oxide. As the exact stoichiometry is not fully known, [O] is used in the equation as the symbol for the oxidising agent:

$$H_2C=CH_2 + [O] + H_2O \rightarrow HOCH_2-CH_2OH$$

Reagent: neutral potassium manganate(VII) solution
Conditions: shake together at room temperature
Product: ethane-1,2-diol
Reaction type: oxidation

Polymerisation reactions

Alkene molecules can react together in an **addition polymerisation** reaction. For example, one ethene molecule can break the π-bond in another ethene molecule and become attached. This is repeated with other ethene molecules so that, eventually, a long hydrocarbon chain containing thousands of carbon atoms is formed. This chain is called a **polymer**.

Other alkenes, such as propene, and substituted alkenes, such as chloroethene, $ClCH=CH_2$, tetrafluoroethene, $CF_2=CF_2$, and phenylethene, $C_6H_5CH=CH_2$, also form polymers.

Polymer production

The usual procedure is to cause radicals to be formed that initiate the polymerisation reaction.

One method of polymerising ethene to poly(ethene) is to heat it under a pressure of 1000 atm in the presence of a trace of oxygen. The oxygen produces radicals that initiate the polymerisation:

The electrophile, Br^+, attaches to the carbon atom that has more hydrogen atoms bonded to it, thus forming the secondary carbocation. This is more stable than a primary carbocation.

A polymer is a chain of covalently bonded molecules.

e A common error when writing the structural formula of the chain in poly(propene) is to put the three carbon atoms in a line thus:
$-(CH_2CH_2CH_2)_n-$
This is quite wrong, as the structure has CH_3 groups on alternate carbon atoms.

Monomer	Polymer — showing *two* repeat units	Name	Uses
Ethene, $H_2C=CH_2$	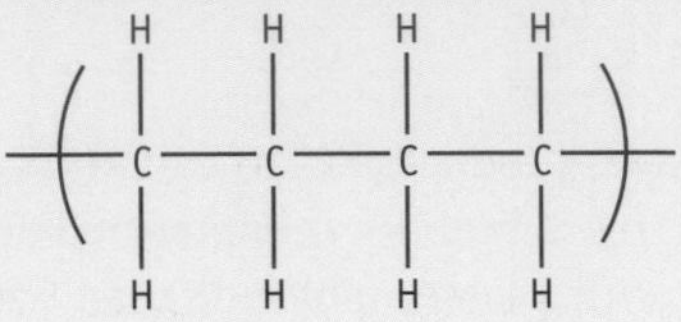	Poly(ethene)	Low density — plastic bags; high density — water pipes, bottles, washing-up bowls, buckets
Propene, $CH_3CH=CH_2$	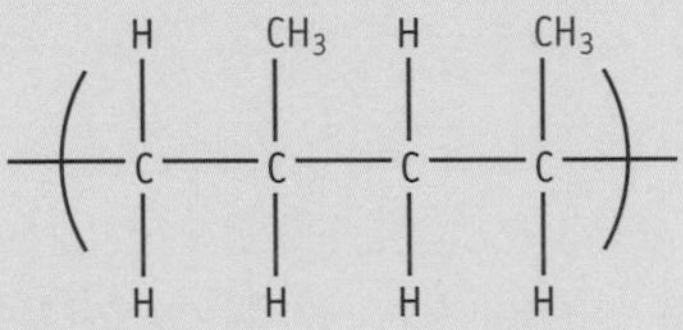	Poly(propene)	Ropes, containers that have to withstand boiling water
Chloroethene, $CHCl=CH_2$	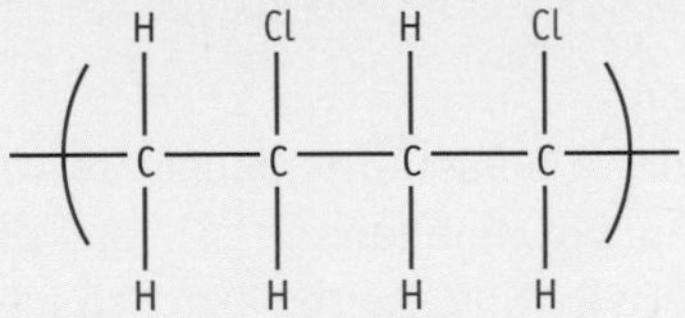	Poly(chloroethene) or PVC	Window frames, guttering, drain pipes, electrical insulation
Tetrafluoroethene, $CF_2=CF_2$	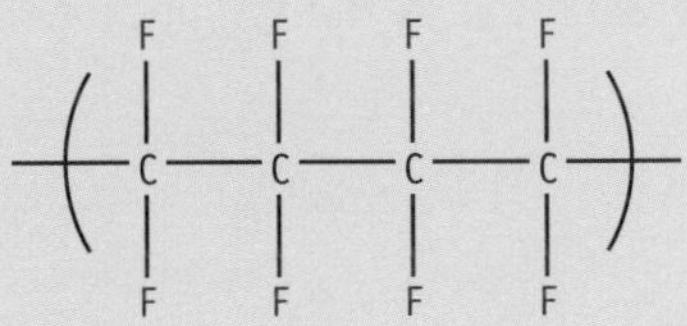	Poly(tetrafluoroethene) PTFE or Teflon	Non-stick coatings, low-friction bearings (burette taps, plumber's tape), electrical insulation
Propenenitrile, $H_2C=CHCN$	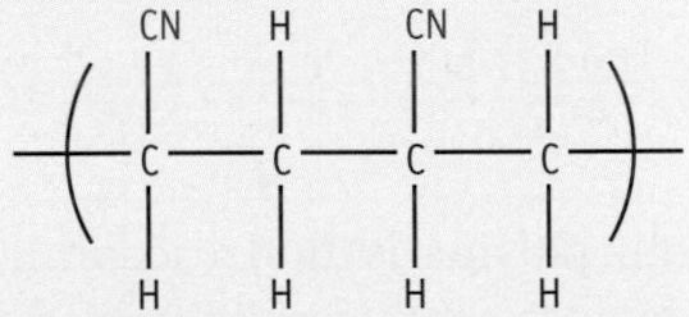	Acrilan	Fibres for clothing
Phenylethene, $C_6H_5HC=CH_2$	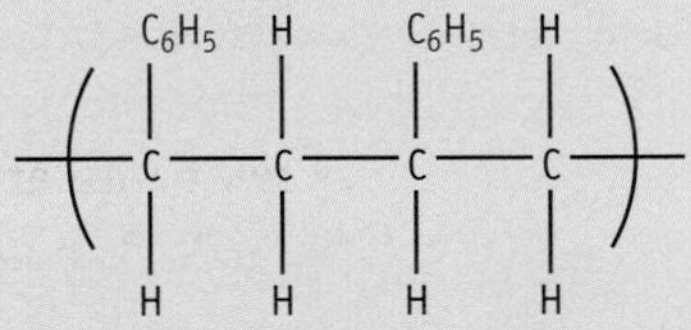	Polystyrene	Thermal insulation, packaging for fragile articles
Ethenol, $H_2C=CHOH$	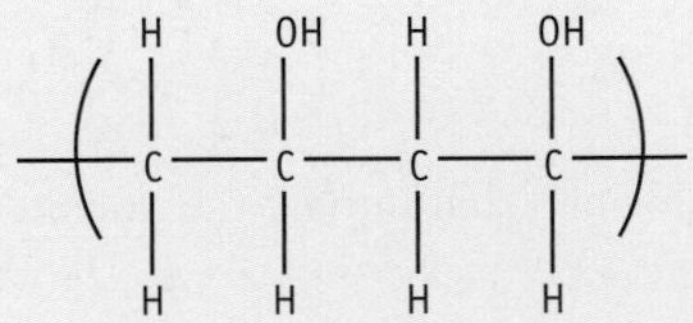	Poly(ethanol) or polyvinyl alcohol	Soluble laundry bags and liquitab capsules

Table 7.6 Some common addition polymers

The other method of producing poly(ethene) is to mix ethene with a solution containing alkylaluminium and titanium chloride catalysts. This method produces a harder, higher-density polymer with a higher melting temperature.

The reaction on page 147 for polymerising ethene is not as simple as implied by the equation given. A significant amount of branching occurs and the lengths of the polymer chains vary considerably. This results in the polymer softening over a range of temperatures before it becomes molten, rather than it having a definite melting temperature.

Disposal of polymers

Because of the large size of the molecules and the strength of the C–C and C–halogen bonds, addition polymers are resistant to chemical and biological attack. They are not naturally occurring substances, so no enzymes have evolved to break them down. This is both a disadvantage and an advantage. Resistance to corrosion and chemical attack are useful properties, in that they confer a long life to plastic objects. However, discarded plastic bags can litter the environment and bulky plastic objects can fill waste disposal sites. Disposal by combustion, especially of halogenated polymers, causes toxic fumes.

Summary of alkene reactions

Propene is used as the example.

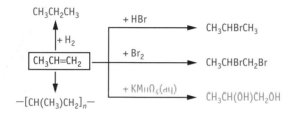

- The reactions in blue are free radical addition reactions.
- The reactions in red are electrophilic addition reactions.
- The reaction in green is oxidation.

Hazard and risk

Many organic chemicals and some of the substances that they react with are hazardous. This means that they are potentially dangerous for anyone using them. Possible hazards include:

- toxicity (e.g. lead compounds)
- absorption through the skin (e.g. 1-bromobutane)
- irritation if inhaled (e.g. hydrogen chloride)
- corrosive compounds (e.g. concentrated sulfuric acid)
- high flammability (e.g. ethoxyethane, $C_2H_5OC_2H_5$)
- carcinogenic compounds (e.g. benzene, C_6H_6)

A reaction might be hazardous if it is rapid and highly exothermic or if it produces a volatile hazardous product.

The risk relates to how a hazardous material is used. For instance, using small quantities or taking specific precautions (e.g. wearing gloves or using a water bath for heating) may reduce the risk. Hazardous materials used in small quantities with proper containment pose little risk.

e When answering a question about hazard and risk, *never* give standard safety precautions, such as wearing safety glasses and a laboratory coat, as methods of minimising risk.

It is important to understand the possible hazards when carrying out an experiment and to take precautions to reduce the risk. For example, alkenes react with bromine (a hazardous chemical) and the reaction between pure liquid bromine and a liquid alkene is fast and extremely exothermic. The risk is reduced by using a dilute solution of bromine in a suitable solvent. The reaction is then much slower. Therefore, the temperature does not rise dangerously quickly and the risk to the user is reduced considerably. However, bromine fumes are still present. Performing the reaction in a fume cupboard further reduces the risk.

Questions

1 Name the following compounds:
  a $CH_3CH(OH)CH_2Br$
  b CH_2ClCH_2COOH
  c $CH_2=CHC(CH_3)_3$

2 Give the formulae of the following compounds:
  a 1,2-dichloro-1,2-difluoroethene
  b 1-hydroxybutanone
  c 2-amino-3-chloropropanoic acid

3 Identify the functional groups in the following:
  a $CH_2OHCOCH(NH_2)COOH$
  b $CH_2=CHCH(OH)CHO$

4 Draw and name all the structural isomers of $C_3H_6Br_2$.

5 Draw and name all the structural isomers of C_4H_9Cl.

6 Draw and name five isomers of the alkene C_5H_{10}.

7 Identify the major organic product when ethane and *excess* chlorine are exposed to white light for a long period of time.

8 Write the equation for the combustion of butane in excess air.

9 Write the equation and give the conditions for the reaction between hydrogen iodide and 2-methylbut-2-ene.

10 What would you observe when but-2-ene and a neutral solution of potassium manganate(VII) are shaken together? Name the organic product and give its formula.

11 Explain why bromine reacts rapidly with alkenes, but only reacts slowly with alkanes, even in the presence of light.

12 But-1-ene could be polymerised. Draw a section of the polymer, showing two repeat units.

Unit 2 Application of core principles of chemistry

Chapter 8

Shapes of molecules and ions, polarity and intermolecular forces

Oxygen: life-saving gas, ozone and magnetic liquid

Atomic number: 8

Electron configuration: $1s^2\ 2s^2\ 2p_x^2\ 2p_y^1\ 2p_z^1$

Symbol: O

Oxygen is in group 6 of the periodic table. Dry air contains 21.0% oxygen. It is found in many rocks as aluminosilicates and carbonates and is the most abundant element in the Earth's crust.

Life-saving gas

All animals need oxygen for aerobic respiration, in which sugars are oxidised to carbon dioxide and water. Hospital patients who have difficulty breathing are given oxygen, as are premature babies in incubators. There are cells in our lungs that detect the partial pressure of oxygen in the blood and other, more sensitive, cells that detect the level of carbon dioxide. If the carbon dioxide level in blood rises or the oxygen level falls, these cells trigger an increase in breathing rate. This is why the oxygen used by paramedics contains a little carbon dioxide and why expired breath (mouth to mouth) resuscitation is effective in stimulating breathing.

Ozone

Ozone is a form of oxygen — it has the formula O_3. It is formed in the stratosphere when oxygen absorbs UV radiation of wavelengths below 240 nm. Ozone is decomposed by UV radiation of wavelength 200–300 nm. The oxygen–ozone equilibrium protects us from UV radiation, cutting out most of the harmful rays. The natural balance between oxygen and ozone has been affected by the release of CFCs into the atmosphere. These compounds of carbon, fluorine and chlorine are so unreactive that they slowly diffuse up to the stratosphere, where they are broken down by UV radiation and produce chlorine radicals. These chlorine radicals catalyse the decomposition of ozone in a chain reaction. It is estimated that one chlorine radical can cause the decomposition of 100 000 ozone molecules. The release of CFCs has caused a thinning of the ozone layer all over the world, allowing more harmful UV radiation to reach the Earth's surface.

Magnetic liquid oxygen

Oxygen boils at −183°C. The liquid is blue and is magnetic. Liquid oxygen can be held between the poles of a very strong magnet. The bonding in the O_2 molecule is not a simple double bond. Modern molecular orbital theory shows that there are two unpaired electrons in antibonding π^*-orbitals. These give rise to the colour of liquid oxygen and to its magnetic properties.

Liquid oxygen suspended between the poles of a magnet

Shapes of molecules and ions, polarity and intermolecular forces

Shapes of molecules and ions

In biochemistry, knowledge of the shapes of molecules is important in understanding many reactions. For example, when light strikes a rod cell in the retina of the eye, the light energy causes a molecule of 11-*cis*-retinal to change its shape, which triggers an impulse along the optic nerve to the brain. Animal fats and vegetable oils are esters of fatty acids, some of which are unsaturated. The *cis*-isomers of these fatty acids are less harmful than the *trans*-isomers.

The shapes of molecules such as DNA, proteins and enzymes are extremely complicated. However, the rules that govern the shapes of such molecules are the same as those used to predict the shapes of simple molecules and ions, such as water, H_2O, and ammonium, NH_4^+. The shape is determined according to the valence shell electron pair repulsion (VSEPR) theory. This states that the shape of a molecule or ion is caused by repulsion between the pairs of electrons, both bond pairs and lone pairs, that surround the central atom.

The electron pairs repel each other to the position of minimum repulsion, which is also the position of maximum separation.

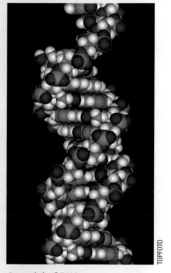

A model of DNA

Molecules with single bonds only

The method for predicting the shape is:

Step 1: count the number of bond pairs around the central atom. This is equal to the number of bonds formed by the central atom.

Step 2: count the number of lone pairs on the central atom. The number of non-bonding electrons is found by subtracting the number of electrons used in bonding from the group number of the element. The number of lone pairs is half this number.

$$\text{number of lone pairs} = \frac{\text{group number} - \text{number of electrons used in bonding}}{2}$$

Step 3: add these numbers together to get the total number of electron pairs; this determines the arrangement of the electron pairs.

Step 4: the name of the shape depends on the position of the *atoms* around the central atom.

> **Worked example**
> How many bond pairs and lone pairs are there around a sulfur atom in:
> **a** hydrogen sulfide, H_2S?
> **b** sulfur hexafluoride, SF_6?

> ### Answer
> Sulfur is in group 6, so has six outer electrons.
> a Sulfur forms two bonds in H_2S, so has two bond pairs.
>   There are $\frac{1}{2} \times (6-2) = 2$ lone pairs.
> b Sulfur forms six bonds, so has six bond pairs.
>   There are $\frac{1}{2} \times (6-6) = 0$ lone pairs.

Molecules and ions with double bonds

A double bond consists of a σ-bond and a π-bond and, therefore, can be regarded as one 'lot' of electrons.

> *e* Stereochemically, a double bond is equivalent to a single bond because they are between the same two atoms. Either count a double bond as one bond or count the number of σ-bonds around the central atom. Remember that each σ-bond is *one* pair of electrons.

The number of lone pairs is calculated as in steps 1 and 2 above. However, the arrangement of electrons around the central atom is determined by the number of σ-bond pairs and the number of lone pairs.

> ### Worked example 1
> How many σ-bonds are there around the sulfur atom in the sulfate ion, SO_4^{2-}, and how many lone pairs?
>
> ### Answer
> The structural formula is:
>
> $$^-O - \overset{\overset{O}{\|}}{\underset{\underset{O}{\|}}{S}} - O^-$$
>
> It uses six electrons to form two single and two double bonds, which is four σ- (and two π-) bonds. Sulfur is in group 6, so the number of lone pairs $= \frac{1}{2} \times (6-6) = 0$.

> ### Worked example 2
> Calculate the number of lone pairs around the sulfur in sulfur dioxide, SO_2 (O=S=O).
>
> ### Answer
> Sulfur is in group 6. The sulfur atom forms two double bonds and, therefore, uses four electrons in bonding. The number of non-bonding electrons is $6-4=2$, which is one lone pair.

e The number of electrons not used in bonding is the group number of the element minus the number of electrons that have been used in bonding.

Species with two electron pairs

- The two electron pairs repel each other to a position of minimum repulsion (maximum separation) and so take up a **linear** arrangement.
- Beryllium chloride, $BeCl_2$, is covalent. Its structural formula is Cl–Be–Cl. There are two bond pairs of electrons around the central beryllium atom. It is in

group 2, so the number of lone pairs = $\frac{1}{2} \times (2 - 2) = 0$. The two electron pairs repel each other to a position of maximum separation, so the molecule is linear.

$$\underset{\times\times}{\overset{\times\times}{\times}}\, Cl\, \underset{\times\times}{\overset{\times\times}{\times}}\, \underset{}{\overset{}{\bullet}}\, Be\, \underset{}{\overset{}{\bullet}}\, \underset{\times\times}{\overset{\times\times}{\times}}\, Cl\, \underset{\times\times}{\overset{\times\times}{\times}}$$

The Cl–Be–Cl bond angle is 180°.

- Carbon dioxide is O=C=O and has two σ-bonds and two π-bonds. All four electrons of the carbon atom are used in bonding. There are two σ-bond pairs and no lone pairs, so the molecule is also linear, with a O–C–O bond angle of 180°.

Species with three electron pairs

The position of minimum repulsion caused by three pairs of electrons is a planar triangle around the central atom, with bond angles of 120°. This arrangement is called **trigonal planar**.

Three bond pairs

Boron trichloride, BCl_3, is covalent. Boron is in group 3 and has three valence electrons which are used to form three bonds. There are no lone pairs around the boron atom, so the position of minimum repulsion of the electron pairs is trigonal planar (triangular) with a bond angle of 120°.

In sulfur trioxide, SO_3, there are three σ- and three π-bonds. These use all six valence electrons, so there are no lone pairs. The three σ-pairs of electrons move as far apart as possible and take up the trigonal planar shape.

The carbonate ion is more complicated. As with all anions containing oxygen, the minus charges are on oxygen atoms. The structural formula is:

The carbon atom has used all *four* of its electrons in bonding and has no lone pairs (group $4 - 4 = 0$). It has formed three σ-bonds (and one π-bond). Therefore, the ion is trigonal planar because the three σ-bond electron pairs repel each other to a position of maximum separation.

Alkenes are planar around the C=C group. Each carbon atom in the C=C group forms three σ-bonds and one π-bond. The three σ-bond pairs repel each other to a position of maximum separation and, therefore, the two carbon atoms and the four atoms directly attached to them lie in the same plane. Ethene is planar.

Note that the beryllium atom does not have an octet of electrons.

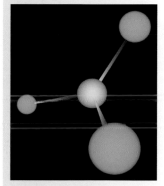

A model of BCl_3

Cis-but-2-ene is planar around the C=C group, so the –CH₃ groups and the hydrogen atoms should be drawn with 120° bond angles to the C=C group.

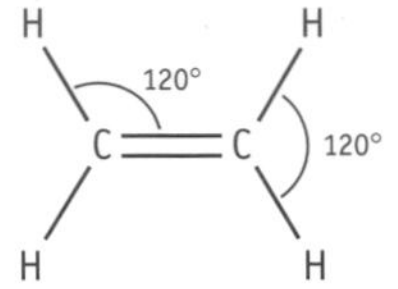

Ethene

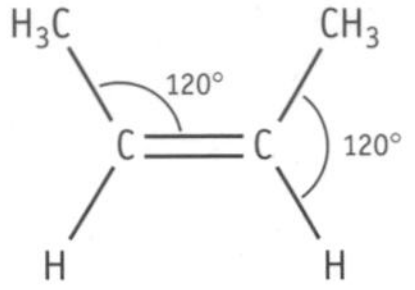

cis-but-2-ene

Two bond pairs and one lone pair

Sulfur dioxide, SO_2, has the structural formula O=S=O. Four of the six valence electrons are used in bonding, so there is one lone pair (group $6 - 4 = 2$ electrons $= 1$ lone pair). The three electron pairs (two σ-bonds and the lone pair) are in a trigonal planar arrangement. However, the shape of the molecule is described by the position of the *atoms* around the central sulfur atom. The sulfur dioxide molecule is V-shaped or bent:

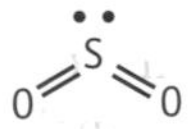

Species with four electron pairs

The electron pairs are in a three-dimensional **tetrahedral** arrangement, the internal angle of which is 109.5°.

Four bond pairs

A model of methane, CH_4

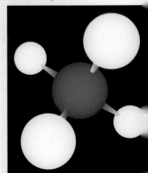

- A simple example is methane, CH_4. The carbon forms four σ-bonds and, as all the valence electrons are used in bonding, it has no lone pairs. The four bond pairs repel each other and the molecule has a **tetrahedral** shape, with H–C–H bond angles of 109.5°.

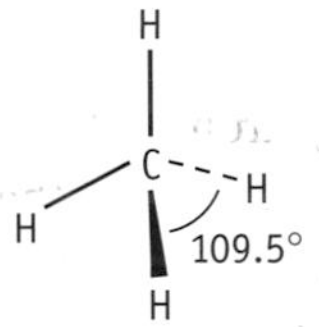

All covalent singly bonded carbon compounds have this shape around the carbon atom.

Carbon atoms in alkanes are tetrahedrally bonded. A skeletal formula shows the bond angles more correctly than the displayed formula, as is illustrated by the skeletal and displayed formulae of pentane:

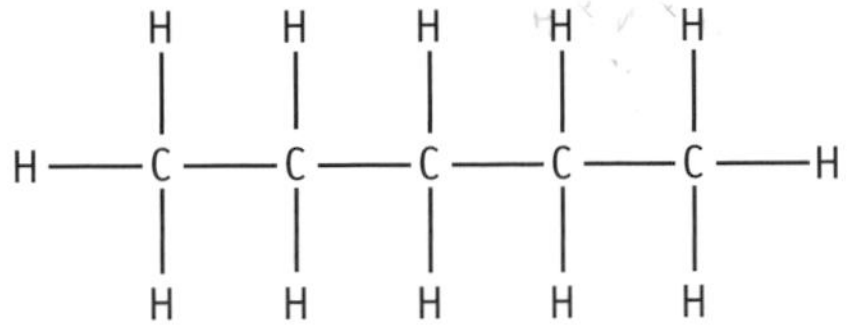

Skeletal

Displayed

C_5H_{12}

Molecular

- Silicon is also in group 4, so compounds such as SiH_4 and $SiCl_4$ are tetrahedral.
- The ammonium ion, NH_4^+, is tetrahedral. Three of the five valence electrons of nitrogen are used in forming single covalent bonds with three hydrogen

atoms and two are used in forming a dative covalent bond with an H^+ ion. The four bond pairs of electrons repel each other and the ion is tetrahedral.

- The dimer, Al_2Cl_6, is a complicated example. Four bond pairs of electrons surround each aluminium, so the four chlorine atoms are arranged in a tetrahedron around the aluminium atoms, with a bond angle of 109.5°.

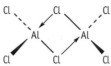

- The sulfate ion has the following structural formula and shape:

There are four σ-bonds and two π-bonds. Therefore, all six valence electrons of sulfur are used in bonding. The four σ-bonds repel equally and the ion is tetrahedral.

> **e** The best way to draw a tetrahedral molecule is by using two ordinary lines, one wedge and one dashed line coming from the central atom. A double bond is represented by a pair of lines.
>
> A wedge indicates a bond coming out of the paper and a dashed line indicates a bond going behind the paper, thus creating a three-dimensional effect.

Three bond pairs and one lone pair

The four pairs of electrons are arranged tetrahedrally around the central atom. However, the name of the shape is determined by the position of the atoms. The shape is **pyramidal** — a triangular-based pyramid around the central atom, with the lone pair on the opposite side of the central atom to the three bonded atoms.

Ammonia, NH_3, molecules are this shape. Nitrogen is in group 5, so the nitrogen atom has five valence electrons. Three of these are used in forming covalent bonds with the three hydrogen atoms. This leaves a lone pair. The electrons are arranged tetrahedrally, but the shape of the molecule is pyramidal.

A model of ammonia, NH_3

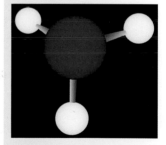

> **e** The best way to draw a pyramidal molecule is with one ordinary line, one wedge and one dashed line.

The repulsion between the lone pair and the bond pairs is *greater* than the repulsion between bond pairs. This reduces the H–N–H bond angle from the tetrahedral angle of 109.5° to 107°.

◀ Do not confuse the shape of the *ammonia* molecule, NH_3 (pyramidal), with that of the *ammonium* ion, NH_4^+ (tetrahedral). Ammonia has three bond pairs and one lone pair; ammonium has four bond pairs and no lone pairs.

The sulfite ion, SO_3^{2-}, is also pyramidal. The sulfur atom forms two single bonds with two oxygen atoms and one double bond with the third oxygen atom. Therefore, four of the six valence electrons are used and there is one lone pair (group $6 - 4 = 2$ electrons = 1 lone pair).

Two bond pairs and two lone pairs

The four pairs of electrons are arranged tetrahedrally. However, as there are only two atoms bonded to the central atom, the molecule is **V-shaped** or bent.

An example of this is water. Two of the six valence electrons of oxygen are used in bonding, so there are two lone pairs ($6 - 2 = 4$ unused electrons = 2 lone pairs). The four electron pairs repel to a position of minimum repulsion, so the molecule is V-shaped.

The repulsion between the two lone pairs is greater than that between the bond pairs, so the tetrahedral angle is further compressed to 104.5°.

In ethanol, CH_3CH_2OH, the oxygen atom has two σ-bonds and two lone pairs. This means that the C–O–H bond angle is the same as that in water at 104.5°. The other bond angles in the molecule are the normal tetrahedral angles of 109.5°.

Species with five electron pairs

The examples covered by the A-level specification all have five bond pairs and no lone pairs. The position of maximum separation is a planar triangle with one atom above and one below the plane. This shape is called **trigonal bipyramidal**. For example, phosphorus pentachloride, PCl_5, has five bonding pairs and no lone pairs of electrons around the phosphorus atom:

>
> ℮ The best way to draw this complex shape is to have three ordinary lines at 90° to each other with one wedge and one dashed line on the opposite side.

The bond angles between the chlorine atoms and the phosphorus atom in the triangular plane are all 120°. The angle between the top or bottom chlorine atom and those in the plane is 90°.

Models of (a) water, H_2O and (b) ethanol, C_2H_5OH

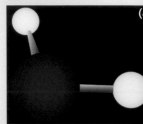

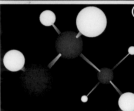

◄ Lone pair/lone pair repulsion is greater than lone pair/bond pair, which is greater than bond pair/bond pair repulsion.

℮ Make sure that you can draw a good three-dimensional diagram of this shape, with the bond angles clearly marked.

Species with six electron pairs

Six electron pairs repel and take up an octahedral arrangement with 90° bond angles between the bonding atoms and the central atom.

In SF_6, all six valence electrons of sulfur are used in bonding, so the shape of the molecule is octahedral.

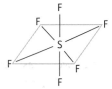

e Another way of drawing this shape is to have one line going vertically upwards from the central atom, one line going vertically downwards, two wedges bottom left and right, and two dashed lines top left and right.

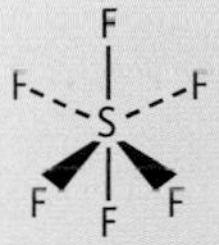

Many metal ions are hydrated, with the oxygen of six water molecules dative-covalently bonded to the central metal ion. These ions are octahedral, for example $[Cr(H_2O)_6]^{3+}$ and $[Mg(H_2O)_6]^{2+}$:

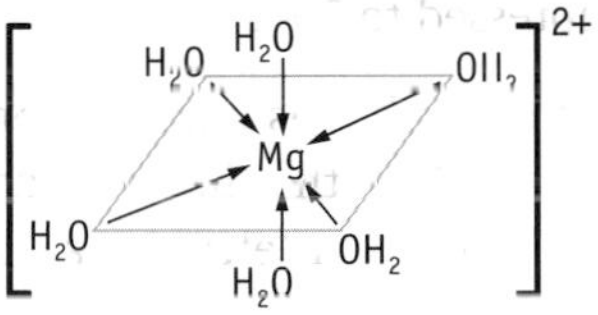

Summary

- Work out the number of σ-bonds and lone pairs of electrons around the central atom. This gives the *arrangement* of electrons.
- The *shape* of the molecule is this arrangement, modified by any lone pairs.
- Repulsion of the bond pairs by any lone pairs reduces the bond angle.

Examples are given in Table 8.1.

Polarity of molecules

In a covalent bond between two atoms of the same element (e.g. H_2 or Cl_2), the average position of the bonding electrons is exactly halfway between the centres of the two atoms. However, in the bond between atoms of different elements (e.g. HF or H_2O) this is not the case. In the H–F bond, the electrons are nearer to the fluorine atom. The ability of an atom in a covalent bond to draw the shared electrons nearer to itself is measured by the **electronegativity** of the element.

Electronegativity is the ability of an element to attract the pair of electrons in a covalent bond.

◄ Fluorine is the most electronegative element.

Total number of electron pairs	Number of σ-bond pairs	Number of lone pairs	Arrangement of electron pairs (lone pairs in red)	Shape of molecule	Bond angle	Examples
2	2	0	Linear 	Linear 	180°	$BeCl_2$ CO_2
3	3	0	Trigonal planar 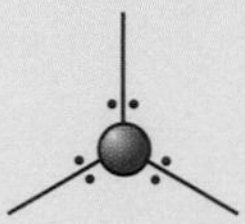	Trigonal planar 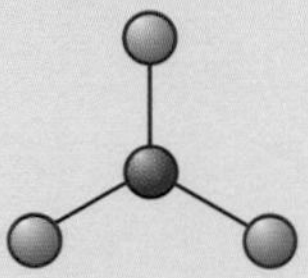	120°	BCl_3 CO_3^{2-} SO_3 NO_3^-
	2	1	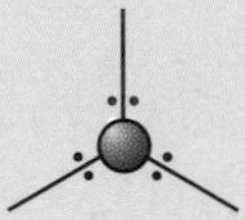	V-shaped 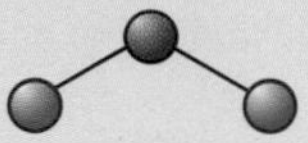	<120°	SO_2 NO_2^-
4	4	0	Tetrahedral 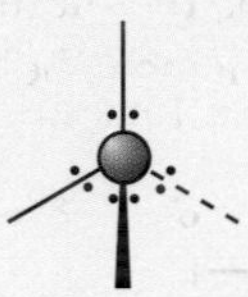	Tetrahedral 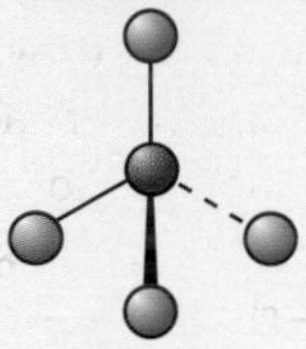	109.5°	NH_4^+ CH_4 CCl_4 $SiCl_4$ PCl_4^+ SO_4^{2-}
	3	1	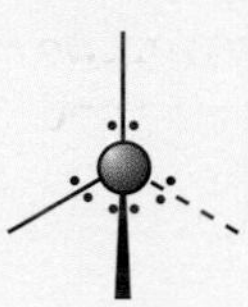	Pyramidal 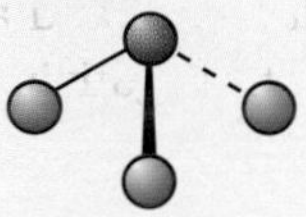	<107°	NH_3 SO_3^{2-} H_3O^+
	2	2	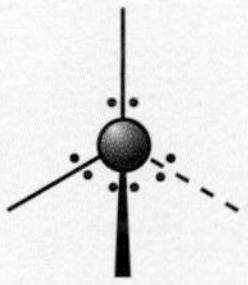	V-shaped 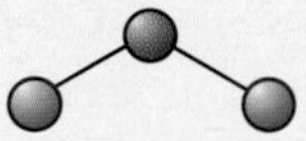	<104.5°	H_2O
5	5	0	Trigonal bipyramidal 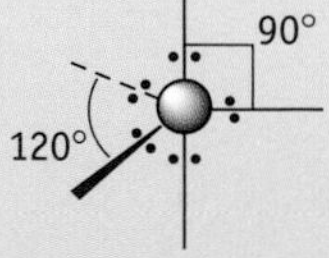	Trigonal bipyramidal 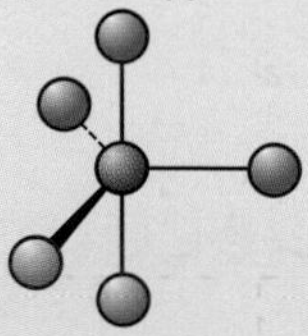	120° and 90°	$PCl_5(g)$
6	6	0	Octahedral 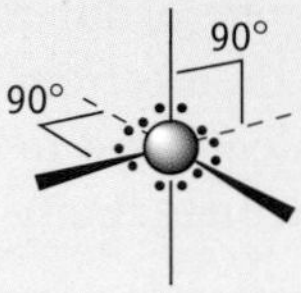	Octahedral 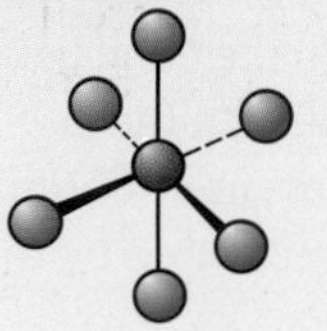	90°	SF_6 $[Mg(H_2O)_6]^{2+}$ and other hydrated cations

Table 8.1 Shapes of molecules and ions

In the periodic table, electronegativity increases across a period and decreases down a group. The electronegativities (estimated using the Pauling scale) of some elements are shown in Table 8.2.

H 2.1						
Li 1.0	Be 1.5	B 2.0	C 2.5	N 3.0	O 3.5	F 4.0
Na 0.9	Mg 1.2	Al 1.5	Si 1.8	P 2.1	S 2.5	Cl 3.0
K 0.8						Br 2.8
Rb 0.8						I 2.5

Table 8.2
Electronegativities of some elements

When two covalently bonded elements have different electronegativities, small positive (δ^+) and small negative (δ^-) charges are present and the bond is said to be polar. The bonding electrons are drawn towards the more electronegative element, making it δ^-. The less electronegative element becomes δ^+. Some examples of polar bonds are shown below.

$$
\begin{array}{cc cc cc cc cc}
\delta^+ & \delta^- & \delta^+ & \delta^- & \delta^+ & \delta^- & \delta^- & \delta^+ & \delta^- & \delta^+ \\
H\!-\!O & & H\!-\!Cl & & C\!-\!Cl & & N\!-\!H & & C\!-\!H &
\end{array}
$$

Polar covalent bonds can be regarded as being between two ideals — a 100% covalent and hence non-polar bond and a 100% ionic bond. The relationship between the difference in electronegativity and the percentage ionic character is shown in Figure 8.1.

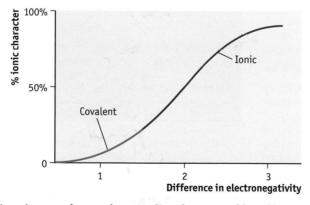

Figure 8.1 Relationship between difference in electronegativity and percentage ionic character

An approximation can be used to predict the type of bonding — an electronegativity difference of more than 1.5 results in an ionic bond.

Shape and polarity

The bonds in both CH_3Cl and CCl_4 are polar. The CH_3Cl molecule is polar, because the polarity of the C–Cl bond is not cancelled out. However, the CCl_4 molecule is not polar because the polarities of the four C–Cl bonds cancel since the molecule is symmetrical.

e It is important to distinguish between a polar *molecule* and a polar *bond*.

- Tetrahedral molecules of formula AB$_4$, in which all four B atoms are the same, are not polar.
- If one of the B atoms is replaced by a different atom — if the bonds are polar — or by a lone pair, the molecule is polar. Ammonia, NH$_3$, has three polar bonds and one lone pair. The polarities do not cancel, so the molecule is polar.
- Trigonal planar molecules, such as SO$_3$ and BCl$_3$, are not polar because the bond polarities cancel.
- Linear molecules, such as CO$_2$, are also non-polar since the dipoles cancel. However, H$_2$O is V-shaped, so the polarities do not cancel.

A simple test for polarity

A simple test to find whether or not a liquid consists of polar molecules is to run a stream of the liquid from a burette close to a charged rod or to a balloon that has been rubbed on wool and become charged with static electricity. A polar substance will be attracted towards the high charge of the rod or balloon, whereas a non-polar substance will not be deflected.

The polarity of a molecule is measured by its **dipole moment**. For a diatomic molecule such as hydrogen chloride, HCl, this is defined as the charge difference (how δ^+ and δ^- the atoms are) multiplied by the distance between the hydrogen and chlorine atoms. For a polyatomic molecule, it is more complex because the polarities of each bond have to be taken into account, as well as any lone pairs.

The polarity of a bond is sometimes shown as an arrow with a plus at one end. The arrow points to the δ^- atom in the bond, for example:

The polarity of the *molecule* is a combination of all the bond polarities and those due to any lone pairs.

Molecule	Dipole moment	Molecule	Dipole moment
HF	1.91	O_3	0.53
HCl	1.08	CH_4	0
HBr	0.80	BF_3	0
HI	0.42	CO_2	0
NH_3	1.47	trans-CHCl=CHCl	0
H_2O	1.85	cis-CHCl=CHCl	1.90

Table 8.3 Dipole moments of some covalent molecules

Forces between ions, atoms and molecules

The seven types of interaction that we need to consider are summarised in Table 8.4.

Type of interaction	Typical strength/ kJ mol^{-1}	Example
Ion–ion	250	Between two ions
Atom–atom	400	Between two atoms in a covalent bond
Ion–electron	Very varied	Metals
Ion–water molecule	100	Hydrated ions
Permanent dipole–dipole	1.5	Between polar molecules
Instantaneous induced dipole–induced dipole	10	Between all types of molecules
Hydrogen bonding	20	Between δ^+ H and δ^- N, O or F in other molecules

Table 8.4 Forces between ions, atoms and molecules

The interactions in green are chemical bonds; those in red are intermolecular forces.

Ion–ion bonds

- An ion–ion bond is a simple electrostatic force of attraction between a positive and a negative ion.
- The strength is proportional to the product of the charges and inversely proportional to the sum of the ionic radii.

Atom–atom bonds

- Atom–atom bonds are covalent bonds.
- A covalent bond is the attraction of the two nuclei for the shared electrons in the bond.
- Short bonds are stronger than long bonds.

Ion–electron bonds

- Ion–electron bonds are the forces that act in a metal.
- Small metal ions bond more strongly than large ones.
- Metals that have more delocalised electrons have stronger bonds than those that release fewer electrons.

Ion–water molecule bonds

- Water is a polar molecule that consists of δ^+ hydrogen atoms and δ^- oxygen atoms.
- The δ^- oxygen atoms are attracted to positive metal ions and the δ^+ hydrogen atoms to negative anions.
- One of the lone pairs of electrons in the oxygen forms a dative covalent bond with an empty orbital in a metal cation, producing a hydrated ion.

Forces between covalently bonded molecules

Intermolecular forces also exist between two covalently bonded molecules. These are of three types:

- **Permanent dipole forces** exist between polar molecules.
- **Instantaneous induced dipole forces** exist between all molecules.
- **Hydrogen bonds** exist between a δ^+ hydrogen atom in one molecule and a δ^- oxygen, nitrogen or fluorine atom in another molecule.

Permanent dipole–dipole forces

- Molecules with a permanent dipole can attract neighbouring molecules.
- The dipoles line up so that the δ^+ end of one molecule is next to the δ^- end of another molecule.
- This is an example of an intermolecular force.

◄ An intermolecular force exists *between* separate covalent molecules.

Instantaneous induced dipole–induced dipole forces

Instantaneous induced dipole–induced dipole forces are also called **dispersion** or **London forces.**

The electrons in a covalent molecule oscillate within their atomic orbitals or covalent bonds. This random motion causes a temporary dipole in the molecule. The electrons in neighbouring molecules oscillate in phase, each inducing a dipole in the molecule next to it so that, at any instant, the δ^+ end of one molecule is next to the δ^- end of a neighbouring molecule. A fraction of a second later the polarity will change, but again the new δ^+ end will induce a δ^- charge on the end of the molecule next to it.

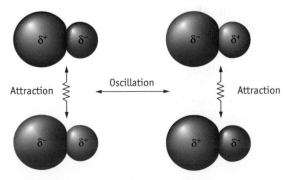

e The term 'van der Waals force' is sometimes used to describe induced dipole (dispersion) forces. However, it can also be used to mean permanent dipole forces. Sometimes it is used for both. For this reason, the term should not be used without explanation.

Instantaneous induced dipole–induced dipole forces are quite strong — stronger than permanent dipole–dipole attractions. The strength depends mainly upon the number of electrons in the molecule. For example, at room temperature, iodine is a solid and chlorine is a gas because the dispersion forces between iodine molecules (106 electrons) are much stronger than the dispersion forces between chlorine molecules (34 electrons).

It has been estimated that the energy required to overcome the permanent dipole forces between hydrogen chloride molecules at room temperature is about 1 kJ mol^{-1} and that required to overcome the London forces between the non-polar methane molecules is nearly 5 kJ mol^{-1}. The London forces between hydrogen chloride molecules are stronger than those between methane molecules as they have more electrons, so the strongest intermolecular force between hydrogen chloride molecules are the London forces.

Permanent dipole forces are not necessarily stronger than induced dipole forces.

The strength of dispersion forces between multi-atom molecules also depends upon their shape and hence the number of points of contact between neighbouring molecules. For example, the almost spherical molecule $C(CH_3)_4$ has fewer points of contact with other molecules than the zigzag molecule $CH_3CH_2CH_2CH_2CH_3$. The first, 2,2-dimethylpropane, boils at 10°C and the other, pentane, at 36°C. Both are non-polar molecules with the same number of electrons.

London forces exist between *all* covalent molecules. Even atoms of noble gases have dispersion forces between them. However, as non-bonding electrons are less able than bonding electrons to form instantaneous dipoles, the boiling temperatures of noble gases are very low.

Dispersion and permanent dipole–dipole forces are sometimes collectively known as van der Waals forces.

Hydrogen bonds

Although the word 'bond' implies a chemical bond, a hydrogen bond is a type of intermolecular force. A hydrogen bond forms between a δ^+ hydrogen atom in one molecule and a δ^- fluorine, oxygen or nitrogen atom in another molecule. This only happens when the hydrogen atom is covalently bonded to a fluorine, oxygen or nitrogen atom.

A hydrogen bond is a relatively strong intermolecular force. Hydrogen is the smallest atom and has no shielding electrons. Therefore, a δ^+ hydrogen atom is like a partially exposed nucleus and can form a strong interaction with a small δ^- atom, such as fluorine, oxygen or nitrogen. A chlorine atom is too large to get close enough to a hydrogen atom to form a hydrogen bond. Therefore, although chlorine in hydrogen chloride is as δ^- as nitrogen is in ammonia, hydrogen chloride molecules do not form hydrogen bonds with each other.

Hydrogen bonding through fluorine

The only fluorine compound with intermolecular hydrogen bonds is hydrogen fluoride. The δ^+ hydrogen atoms in one molecule form hydrogen bonds with δ^- fluorine atoms in an adjacent molecule. The F–H–F bond angle is 180°; the H–F–H bond angle is 109.5°. The result is a zigzag chain of HF molecules.

Hydrogen bonding can only occur between a δ^+ hydrogen atom, which is extremely small, and the next smallest atoms (fluorine, oxygen or nitrogen), which must be δ^-. Chlorine atoms are too large to form hydrogen bonds.

Hydrogen bonding through oxygen

- All compounds containing an –OH group form hydrogen bonds.
- The most important example is water. The hydrogen atoms are δ^+ and form hydrogen bonds with the δ^- oxygen atoms in other water molecules.

Hydrogen bonding between water molecules is stronger than that between hydrogen fluoride molecules. Although a given hydrogen bond is stronger between a δ^+ hydrogen atom and a δ^- fluorine atom than between a δ^+ hydrogen and a δ^- oxygen atom, there are more hydrogen bonds per molecule in water than in hydrogen fluoride. Water boils at 100°C, whereas hydrogen fluoride boils at 20°C. This shows that the sum of the intermolecular forces in water is higher than the sum of those in hydrogen fluoride.

Water has the unusual property of the solid being less dense than the liquid. This is because ice crystals contain interlocking rings of six water molecules held together by hydrogen bonds. The distances between molecules opposite each other in the ring is quite large relative to the close approach of neighbouring molecules:

- oxygen
- hydrogen

When ice melts, the ring structure is destroyed and the *average* distance apart of the molecules decreases, causing the density to increase.

Alcohol molecules contain –OH groups. Therefore, they can form hydrogen bonds with other alcohol molecules or, when dissolved in water, with water molecules:

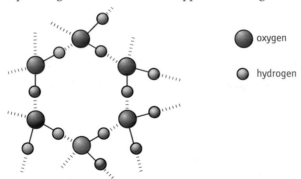

Carboxylic acid molecules also contain –OH groups and behave similarly.

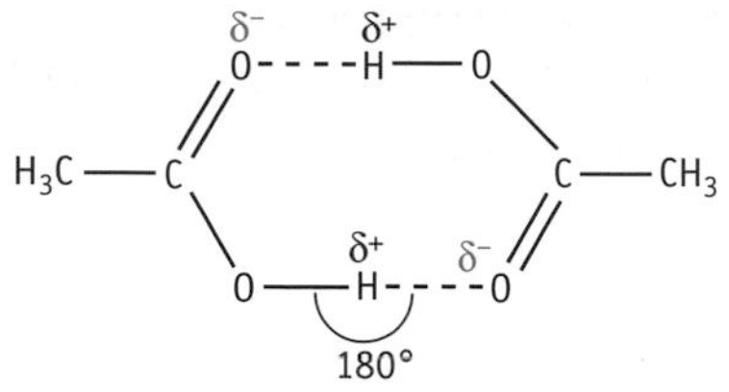

Hydrogen bonding through nitrogen

When nitrogen forms three bonds, it has a lone pair. If nitrogen is bonded to hydrogen, it becomes δ^-; the hydrogen becomes δ^+. This allows hydrogen bonding to occur between molecules. This is true for all nitrogen compounds with an –NH group. An example is ammonia, NH_3, in which the δ^+ hydrogen atom in one molecule hydrogen bonds to the δ^- nitrogen atom in another molecule:

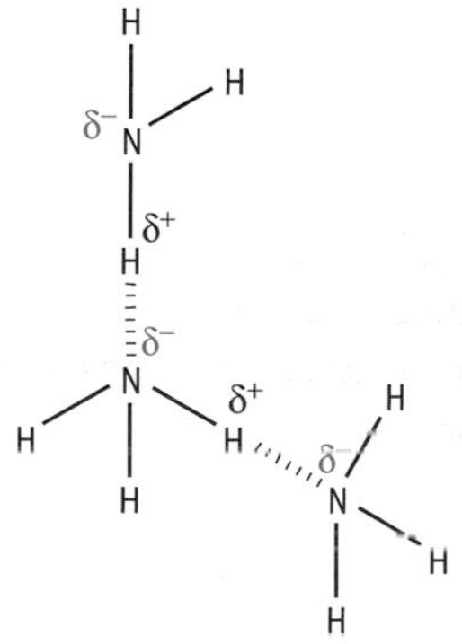

Organic compounds containing the primary amine group, $-NH_2$, and those with the $>$NH group form intermolecular hydrogen bonds. This includes amino acids, such as glycine, NH_2CH_2COOH.

Proteins and DNA

Proteins are long chains of amino acids arranged either in a helix or in a pleated sheet. Both structures are held in shape by hydrogen bonds between the δ^+ hydrogen atom in an –NH group and the δ^- oxygen in a C=O group, which is four amino acid residues away on the chain in a helix, or on an adjacent chain in a pleated sheet.

A DNA molecule is a double helix. Each strand of the helix is made up of a pattern of base–sugar–phosphate units. Each strand is held to the other strand by hydrogen bonds. The base adenine in one strand is linked to thymine in the other strand by two hydrogen bonds and the base cytosine to guanine by three hydrogen bonds.

Physical properties

The physical properties of an element or compound depend upon the nature of the forces between the particles.

Melting a solid

- A solid consists of a regular arrangement of particles (ions, atoms or molecules), which vibrate about a fixed position and cannot move through the solid.

- When a solid is heated, the vibration increases and the temperature rises. When the melting temperature is reached, the structure of the solid begins to break down and it starts to melt.
- As the heating continues, more solid melts. However, the temperature stays constant until all the solid has melted.
- When molten, the particles are free to move around through the liquid; they are not fixed in position.
- If the forces between the particles are strong, much energy is required to break down the structure of the solid. Therefore, it will have a high melting temperature.

Boiling a liquid
- In a liquid, the particles (ions, atoms or molecules) are close together, but their positions are random and they tumble over each other. Particles move at different speeds and continually collide with each other.
- When a liquid is heated, the particles gain kinetic energy and their average speed increases.
- Some particles move fast enough to escape from the surface as a gas. This is called **evaporation**.
- When the boiling temperature is reached, some particles have the necessary energy to escape from the middle of the liquid. Bubbles of gas are produced and the liquid boils.
- If the forces between the particles are strong, much energy is required to separate the particles. Therefore, the liquid will have a high boiling temperature.

Electrical conductivity
- An electric current is a flow of charge (involving electrons or ions).

Types of solid structure

The types of solid structure are:
- ionic
- giant atomic
- hydrogen bonded molecular
- metallic
- simple molecular
- polymeric

Ionic solids

Ionic solids consist of a regular arrangement of charged ions. This arrangement is called a **lattice**.

Melting
- The melting temperature is dependent on the strength of the ion–ion forces.
- Ions with a 2+ or 2– charge exhibit stronger forces than those that are singly charged.
- Ions with a small radius give rise to stronger forces of attraction than those with a large radius.
- The melting temperature also depends upon the lattice structure. Lattices in which the ion ratio is 2:1 melt at a lower temperature than that expected from their charge and radius values.

e Do *not* say that the particles *start* to vibrate; they vibrate at all temperatures.

e You must be able to predict the type of solid structure when given the identity of a substance.

The melting temperatures of some ionic substances are shown in Table 8.5.

	Ions	Ionic radius/nm	Melting temperature/°C
Sodium chloride	Na^+ Cl^-	0.095 0.181	801
Potassium chloride	K^+ Cl^-	0.133 0.181	776
Calcium chloride	Ca^{2+} Cl^-	0.010 0.181	782
Calcium oxide	Ca^{2+} O^{2-}	0.099 0.140	2614
Magnesium oxide	Mg^{2+} O^{2-}	0.065 0.140	2800

Table 8.5 Melting temperatures of some ionic substances

Electrical conductivity

Ionic solids do not conduct electricity because the ions are fixed in the lattice and cannot move through the solid.

When molten or dissolved, the lattice is broken down and the ions are able to move freely, allowing electricity to be conducted.

Ionic compounds conduct electricity by the movement of charged ions; metals conduct by the movement of electrons.

> **Worked example 1**
> Explain why calcium oxide has a lower melting temperature than magnesium oxide.
>
> **Answer**
> Calcium and magnesium ions both have a charge of 2+. However, the radius of an Mg^{2+} ion is less than that of a Ca^{2+} ion. Thus, the force of attraction between the cation and the oxide ion is less with the bigger Ca^{2+} ion. Therefore, calcium oxide has a lower melting temperature.

In a comparison question such as this, make sure that you say something about both compounds.

> **Worked example 2**
> Why does molten calcium oxide conduct electricity whereas solid calcium oxide does not?
>
> **Answer**
> In the solid state the ions are fixed in position, so there are no charge carriers that can move. When molten, the solid lattice breaks down. The ions become free to move and are able to carry the current.

Do not answer this question in terms of the movement of electrons. That applies only to metals and graphite.

Metals

The structure and properties of metals are described earlier in Chapter 6 (pages 101–104).

Giant atomic substances

Giant atomic substances are sometimes called network covalent. They are solids that consist of a giant network of atoms linked to each other by covalent bonds.

Diamond and graphite

Carbon exists in two main forms:

- In diamond, each carbon atom has four σ-bonds to four other carbon atoms, in a giant three-dimensional tetrahedral arrangement, with all bond angles 109.5°.

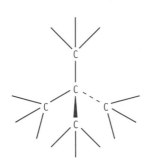
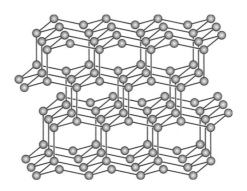

Diamond has an extremely high melting temperature because each carbon atom is covalently bonded to four others. Millions of strong covalent bonds have to be broken in order to melt it.

- Graphite has a layered structure.

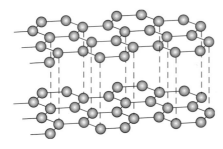

Each carbon atom is bonded to three others by σ-bonds, forming inter-locking hexagonal rings. The fourth electron is in a *p*-orbital, which forms a delocalised cloud above and below the plane of the rings. The planar rings are weakly bonded to the ones above and below by dispersion forces, so they can slide over each other easily. This is why a pencil leaves a mark of graphite on paper. The delocalised electrons are the reason why graphite conducts electricity when solid, as they can move across the surface of the planes.

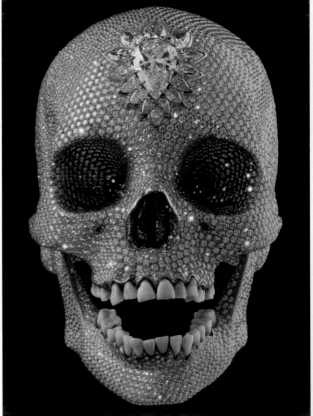

Damien Hirst's diamond-encrusted skull

JAY JOPLING/WHITE CUBE (LONDON)

There are other types of solid carbon. Sir Harry Kroto of the University of Sussex and Richard Smalley of Rice University, Texas, discovered the C_{60} molecule by firing a laser at a disc of carbon and condensing the vapour formed. The molecule consists of 20 interlocking hexagons and twelve pentagons. Each carbon atom has three σ-bonds and the fourth electron is delocalised in the structure in a similar way to in graphite. The molecule has the same basic geometry as the geodesic dome designed by Richard Buckminster Fuller and the C_{60} molecule is colloquially called a buckyball (Figure 8.2).

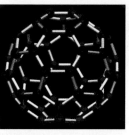

Figure 8.2 A buckyball

Further studies have shown the existence of a C_{70} molecule. This has 25 hexagons and the molecule is shaped like a rugby ball. Evaporation of a solution of C_{60} or C_{70} in benzene produces red crystals.

Two C_{60} buckyballs can join together via a four-membered carbon ring to make a C_{120} molecule.

Molecules such as C_{60} and C_{70} are called **fullerenes**. Their chemistry is similar to that of alkenes — for example, they undergo addition reactions with hydrogen and halogens; C_{60} can be polymerised. An atom or an ion can even be trapped inside the carbon ball. This has an important medical application. Radioactive radon-224 atoms are trapped inside C_{60} molecules, which are coated with tumour-targeting antibodies. The molecules are injected into a patient suffering from cancer. The treated C_{60} is adsorbed by the cancer cells. The radon atoms emit α-rays, which destroy the cancer cells in the tumour but do not affect other organs in the body because the range of α-rays is short.

A further development was the discovery of carbon **nanotubes.** A nanotube consists of a cylinder of interlocking hexagons of carbon atoms (Figure 8.3). One end of the cylinder is sealed with a fullerene molecule. The internal radius of carbon nanotubes is a few nanometres. Nanotubes can be used to deliver drugs to specific parts of the body.

*Figure 8.3
A nanotube*

Clusters of carbon nanotubes are one hundred times as strong as steel wire. They contain delocalised electrons and can conduct electricity.

Silicon
- Silicon has a structure similar to that of diamond.
- Silicon carbide, SiC, has a diamond-like structure in which each silicon atom is surrounded tetrahedrally by four carbon atoms, which, in turn, are surrounded by four silicon atoms. The whole is a giant covalently bonded network of alternating silicon and carbon atoms.
- Quartz, SiO_2, is also a giant atomic lattice. Each silicon atom is bonded to four oxygen atoms and each oxygen atom to two silicon atoms.

Melting

Since covalent bonds have to be broken when the solid is melted or sublimed, all giant atomic substances have very high melting or sublimation temperatures. Diamond and graphite sublime at over $3500°C$ and quartz melts at $1610°C$.

Sublimation is when a solid is heated and turns straight into a gas without melting.

Hardness

Giant atomic substances are very hard, because of the strong covalent bonds in the lattice. This makes them useful as abrasives. For example, sandpaper is coated with grains of quartz, while the abrasive carborundum is made of silicon carbide.

Simple molecular substances

Most covalent substances belong to this category.

Many non-metallic elements form simple molecules. They include:
- nitrogen, N_2, and phosphorus, P_4, in group 5
- oxygen, O_2, and sulfur, S_8, in group 6
- the halogens, in group 7, which are all diatomic

The forces between elemental molecules are dispersion forces only.

Compounds of non-metallic elements also form simple molecular structures. Examples include:
- organic compounds, such as alkanes, alkenes and halogenoalkanes
- hydrogen sulfide, H_2S
- phosphorus trichloride, PCl_3, and phosphorus pentachloride, PCl_5
- many halogen compounds, such as SF_6

The forces between molecules of a compound are dispersion forces and, if the molecule is polar, permanent dipole–dipole forces.

Melting and boiling

The intermolecular forces are weak, so simple molecular substances have low melting and boiling temperatures.

The boiling temperatures of the noble gases increase down the group. Helium has the lowest value and radon the highest. This is because there are dispersion forces between the atoms, and the strength of these forces increases as the number of electrons in the atom increases.

The boiling temperature of members of the homologous series of alkanes increases as the number of carbon atoms increases. The larger molecules have more points of contact and therefore more dispersion forces, so the total dispersion force is greater. The variation in boiling temperature is shown in Figure 8.4.

The boiling temperatures of the hydrogen halides, from hydrogen chloride, HCl, to hydrogen iodide, HI, increase because the strength of the dispersion forces increases as the number of electrons increases. This outweighs the decrease in the strength of the permanent dipole–dipole forces. However, hydrogen fluoride is anomalous because it forms intermolecular hydrogen bonds. The variation is shown in Figure 8.5.

e The only time *covalent* bonds are broken on melting is when a giant atomic solid is heated.

e Covalent bonds are *not* broken when a molecular substance is melted or boiled. The strength of the covalent bonds within the molecules has nothing to do with the boiling temperature of the substance. This is a common misunderstanding in answers at A-level.

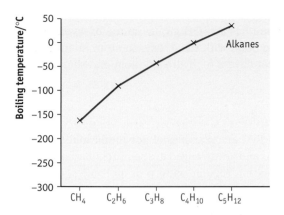

Figure 8.4 *Variation in boiling temperatures*

Hydrogen-bonded molecular substances

Hydrogen fluoride, water, alcohols, carboxylic acids, ammonia and organic amines are examples of hydrogen-bonded molecular substances.

Boiling

For a hydrogen-bonded substance to boil, the molecules have to be separated. This involves breaking the hydrogen bonds *and* the dispersion forces between the molecules. Hydrogen bonds are quite strong. Therefore, the boiling temperatures of hydrogen-bonded substances are much higher than those of similar compounds that are not hydrogen-bonded.

Hydrogen fluoride has intermolecular hydrogen bonds. These are much stronger than the dispersion and dipole–dipole forces that exist between molecules of the other group 7 hydrogen halides. This means that the boiling temperature of hydrogen fluoride is higher than the boiling temperatures of the other hydrogen halides.

The boiling temperature of water is also anomalous, being higher than that of other group 6 compounds that contain hydrogen. Similarly, the boiling temperature of ammonia, NH_3, is higher than that of phosphine, PH_3.

However, the group 4 hydrides do not show this anomalous behaviour of the first member. The carbon atom in methane is only slightly δ^- and the hydrogen atoms are only slightly δ^+, so hydrogen bonding is not possible. This is shown by the graphs in Figure 8.5.

Referring to Figure 8.5, the anomalous high values for NH_3, H_2O and HF are due to hydrogen bonding. The steady increase in the boiling temperatures of the other compounds is caused by an increase in the strength of the dispersion forces because of the increase in the number of electrons in the molecule.

Figure 8.5 *Variation in boiling temperature down groups and across periods*

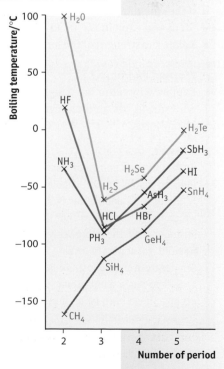

Ethanol, CH_3CH_2OH, has a much higher boiling temperature than its isomer methoxymethane, CH_3OCH_3. Ethanol has an –OH group, so the δ^+ hydrogen atom in one molecule can hydrogen bond with the δ^- oxygen atom in another molecule. Methoxymethane does not have a δ^+ hydrogen atom and so cannot form any hydrogen bonds.

Polymeric substances

Polymers such as poly(ethene) and PVC are examples of polymeric solids. The molecules have varying chain lengths and are arranged in a semi-regular fashion, which is in-between a true solid, with a regular pattern of molecules, and a liquid, with a random arrangement.

Melting
- The chain length of a polymer is variable and the solid is not fully crystalline, so polymers soften rather than melt at a fixed temperature.
- Thermosetting polymers, which are highly cross-linked, decompose rather than melt. Examples include melamine and polyurethane, which are used as surfaces that must not deform when a hot object is placed on them.

Answering questions on melting and boiling temperatures

To answer a question comparing the melting or boiling temperatures of different substances, adopt the following procedure:
- Work out the type of structure.
- List all the forces between the particles in the substances being compared.
- If they are different, explain why one has one type of force and the others do not.
- State and explain the relative strengths of the forces.
- Relate the strength of the forces to the energy required to separate the particles.

Worked example 1

Explain why sulfur has a higher melting temperature than phosphorus, and why silicon has a much higher melting temperature than both sulfur and phosphorus.

Answer

Both sulfur and phosphorus have simple molecular structures, with instantaneous induced dipole–induced dipole (dispersion) forces between the molecules.

Sulfur is S_8 and so has more electrons in a molecule than does phosphorus, which is P_4. This means that the dispersion forces between sulfur molecules are stronger than those between phosphorus molecules. Therefore, more energy is needed to separate them. This results in sulfur having a higher melting temperature than phosphorus.

Silicon has a giant atomic structure with strong covalent bonds between the atoms. These have to be broken for silicon to melt and so a very much higher temperature is required.

Worked example 2

Explain why, in comparison with the other group 6 hydrides, water has an anomalous boiling temperature.

Answer

All the group 6 hydrides are covalent molecules. However, only oxygen atoms are both small and δ^-. In water, hydrogen bonds form between the δ^- oxygen and δ^+ hydrogen atoms. Therefore, the intermolecular forces between water molecules are hydrogen bonds and dispersion forces.

The atoms of other group 6 elements are too large and not sufficiently electronegative for hydrogen bonding to be possible. The intermolecular forces between molecules of these hydrides are dispersion forces, which increase as the number of electrons in the molecule increase.

Hydrogen bonds are stronger than dispersion forces, so more energy has to be supplied to separate water molecules than is needed to separate molecules of the other group 6 hydrides. Therefore, water has a much higher boiling temperature than the other hydrides, the boiling temperatures of which steadily increase as the strengths of the dispersion forces increase.

Solubility

- For a substance to dissolve, the solute particles must be separated from each other and the individual solute particles become surrounded by solvent particles.
- The forces between solute and solvent must be strong enough to overcome the solvent–solvent and solute–solute forces

Solubility of ionic solids

Many ionic solids dissolve in water. The energy required to separate the ions in the solid is compensated for by the energy released by the hydration of the ions.

The positive cations are surrounded by the δ^- oxygen atoms of water molecules and the negative anions are surrounded by the δ^+ hydrogen atoms (Figure 8.6). This process is called **hydration** and the energy released is called the **hydration energy**.

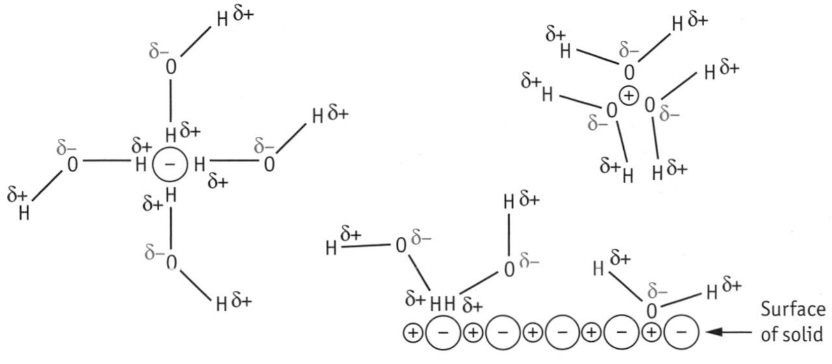

Figure 8.6 An ionic solid dissolving

Ionic solids dissolve in other ionic compounds when they are molten. The extraction of aluminium metal from aluminium oxide uses a solution of aluminium oxide in molten cryolite, which is an ionic compound of formula Na_3AlF_6.

Solubility of compounds that can form hydrogen bonds with the solvent

Compounds that can form hydrogen bonds are soluble in water. The δ^+ hydrogen atoms in the compound form hydrogen bonds with the δ^- oxygen atoms in the water molecules. The δ^- oxygen or nitrogen atoms in the compound form hydrogen bonds with the δ^+ hydrogen atoms in the water molecules.

Both ethanol, CH_3CH_2OH, and ethanoic acid, CH_3COOH, mix in all proportions (are **miscible**) with water.

The δ^- oxygen atom in ethanol forms a hydrogen bond with the δ^+ hydrogen atom in water. The δ^+ hydrogen atom in ethanol forms a hydrogen bond with the δ^- oxygen atom in water.

Molecules of propanone, CH_3COCH_3, cannot hydrogen-bond with each other because they do not have a δ^+ hydrogen atom. However, propanone is very water-soluble because its δ^- oxygen atom can form hydrogen bonds with the δ^+ hydrogen atoms in the water:

Note that the O–H–O angle around the hydrogen atom that is hydrogen-bonded is 180°. This is because it is surrounded by one pair of σ-bonded electrons and one lone pair from the δ^- oxygen atom forming the hydrogen bond.

Ethanoic acid, CH_3COOH, is miscible with water in all proportions. However, pentanoic acid, $CH_3CH_2CH_2CH_2COOH$, is almost insoluble because of its hydrocarbon chain of four carbon atoms. There are only two sites (C=O and OH) for hydrogen bonding and there are four other non-polar groups.

Solubility of non-hydrogen-bonding substances

If a substance can neither form hydrogen bonds with water nor react with it, it is likely to be insoluble, whether or not the molecule is polar. The highly polar fluoromethane, CH_3F, is insoluble, as are the non-polar alkanes and alkenes.

Solubility in non-aqueous solvents

This depends on the similarities between the intermolecular forces of the solvent and the solute. Thus, non-polar solvents tend to dissolve non-polar solutes and polar solvents dissolve polar solutes.

Alkanes are soluble in each other. Crude oil is a complex mixture of alkanes dissolved in each other. Non-polar hexane dissolves non-polar bromine.

Ethoxyethane, $C_2H_5OC_2H_5$, is polar and is a good solvent for many organic substances.

The solubility of drugs in water and in the lipids that make up the membranes of animal cells is important. Balms rubbed on the skin must be lipid-soluble; those injected into the bloodstream or ingested orally must be water-soluble. The presence of –OH, >NH and >C=O groups increases water solubility. The presence of hydrocarbon chains and rings increases lipid solubility.

Other solubilities

Solubility can also result from a chemical reaction. All the hydrogen halides dissolve in water because they form ions, which then become hydrated:

$$HX(g) + H_2O(l) \rightarrow H_3O^+(aq) + X^-(aq)$$

Halogenoalkanes 'dissolve' in hot aqueous sodium hydroxide because they react to form an alcohol, which then hydrogen-bonds with the water:

$$CH_3CHBrCH_3(l) + OH^-(aq) \rightarrow CH_3CH(OH)(aq) + Br^-(aq)$$

Questions

1 State the number of lone pairs of electrons on:
   a the iodine atom in ICl_3
   b the oxygen atom in F_2O
   c the carbon atom in CO_2

2 Draw and explain the shapes of the following:
   a PH_3
   b SCl_2
   c PO_4^{3-}
   d AsF_5
   e HCN (H–C≡N)

3 Give the bond angles of species a–e in Question 2.

4 Decide whether the following bonds are polar or non-polar. If the bond is polar, state which is the δ^+ atom, and explain whether or not the molecule is polar.
   a I–Cl as in ICl_3
   b F–O as in F_2O
   c C=O as in CO_2
   d C–I as in CH_3I

5 State all the forces operating between molecules of:
   a ammonia, NH_3
   b methane, CH_4
   c oxygen fluoride, OF_2

6 Explain why molecules of hydrogen fluoride form intermolecular hydrogen bonds whereas molecules of hydrogen chloride do not.

7 Identify the type of solid structure in:
   a chlorine
   b vanadium
   c lithium fluoride
   d glucose, $CHO(CHOH)_4CH_2OH$
   e silicon tetrachloride, $SiCl_4$

8 a Explain why graphite conducts electricity whereas diamond does not.
   b Would you expect the fullerene C_{60} to conduct electricity? Justify your answer.

9 Other than the delivery of radon to a tumour, describe a use of either a fullerene or carbon nanotubes.

10 Explain the following:
   a Silicon and phosphorus are both covalent substances, but silicon has a much higher melting temperature than phosphorus.
   b Solid sodium and solid sodium chloride both have lattice structures involving ions. Solid sodium conducts electricity, but solid sodium chloride does not.

11 Explain the variation of boiling temperatures among the hydrides in group 5 (NH_3 to BiH_3).

12 Explain why propanone, CH_3COCH_3, has a higher boiling temperature than butane, C_4H_{10}.

13 Explain why hydrogen bromide has a lower boiling temperature than hydrogen iodide.

14 Explain why ammonia is soluble in water whereas chloromethane is not. Both are gases. Chloromethane is more polar (dipole moment 1.9 D) than ammonia (dipole moment 1.5 D).

15 Explain why magnesium chloride is soluble in water, even though its lattice energy is $-526\,kJ\,mol^{-1}$.

16 Discuss the solubility of aspirin and ibuprofen in water and in lipids.

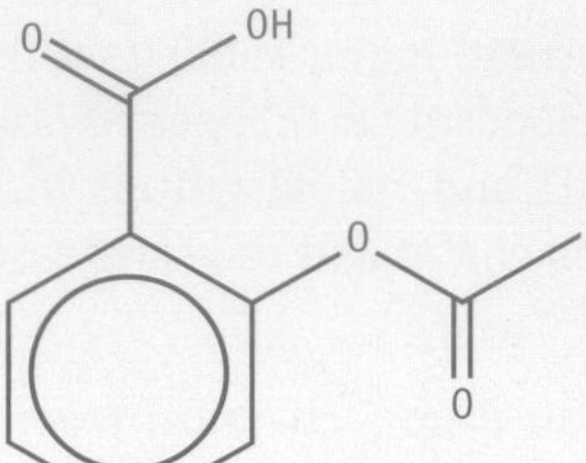

Aspirin

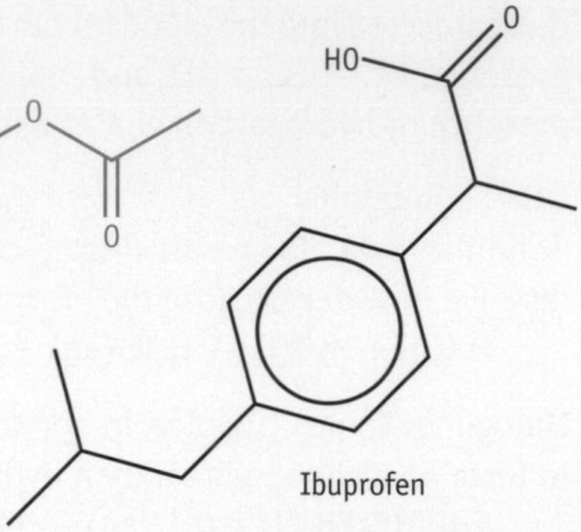

Ibuprofen

Chapter 9

Oxidation and reduction: redox

Fluorine: etchings, toothpaste and nuclear power

Atomic number: 9

Electron configuration: $1s^2\ 2s^2\ 2p_x^2\ 2p_y^2\ 2p_z^1$

Symbol: F

Fluorine is a halogen and is in group 7 of the periodic table. It is the most electronegative element.

An etched vase

Hydrogen fluoride

Hydrogen fluoride is strongly hydrogen-bonded and is a better solvent for ionic compounds than water. It has a boiling point of 20°C but cannot be kept in glass bottles because it attacks glass. This property is exploited in etching. To etch glass, it is first covered in wax. A design is scratched through the wax, exposing the glass beneath. Then the glass is either lowered into a concentrated solution of hydrogen fluoride or placed in damp hydrogen fluoride vapour. A reaction takes place between hydrogen fluoride and the silicates in the glass, which dissolve, marking the glass where the hydrogen fluoride was able to attack it.

Toothpaste

Tooth enamel contains a small proportion of fluoride. The ingestion of small quantities of fluoride during childhood, when teeth develop, reduces the occurrence of tooth decay. Some water companies add 1 ppm of sodium fluoride to the water supply. Most toothpaste contains sodium fluoride, typically about 0.22%. The addition of fluoride to water and toothpaste has greatly reduced tooth decay in the population. Too much fluoride, however, causes teeth to turn brown.

Uranium enrichment

Nuclear fission reactors normally use uranium-235 as the fuel. Natural uranium contains only 0.72% of this fissile isotope. It has to be enriched to 3% before it can sustain the chain reaction in the reactor. This involves separating the ^{235}U and ^{238}U isotopes. Isotopes have identical chemical properties, so a physical method has to be used. One way is to convert the uranium into the volatile compound uranium hexafluoride, UF_6, and then carry out fractional diffusion on the gas. The lighter $^{235}UF_6$ diffuses 1.004 times faster than $^{238}UF_6$. After many separate diffusions, the uranium is enriched sufficiently for it to be used as a nuclear fuel. Another method that partially separates the lighter $^{235}UF_6$ from the heavier $^{238}UF_6$ is to use a gas ultracentrifuge.

Oxidation and reduction: redox

Introduction

At first sight, chemistry appears to be a huge collection of reactions. To bring order out of this chaos, chemical reactions are categorised. One such category includes **oxidation**, together with its opposite, **reduction**.

Oxidation in action

TOM TARRANT

Everyday examples of oxidation include the burning of fuels, the rusting of iron and the bleaching of dyes.

The extraction of a metal, such as iron, from its ore is an example of reduction.

Neither oxidation nor reduction can take place on its own. They always happen together. When a fuel, such as methane, is burnt in oxygen, methane is oxidised and oxygen is reduced. Such a reaction is called a **redox reaction**.

In a redox reaction, one substance is oxidised and another is reduced.

Oxidation was originally defined as the addition of oxygen to a species and reduction as the removal of oxygen from a species. The reaction between carbon monoxide and iron(III) oxide that takes place in a blast furnace in the extraction of iron is an example:

$$Fe_2O_3 + 3CO \rightarrow 2Fe + 3CO_2$$

The iron(III) oxide loses oxygen and is therefore reduced. The carbon monoxide gains oxygen and is therefore oxidised.

The definition was expanded to include the removal of hydrogen as an example of oxidation. When chlorine gas is bubbled into aqueous hydrogen sulfide, the hydrogen sulfide is oxidised to sulfur by the removal of hydrogen. The chlorine is reduced to HCl by the addition of hydrogen:

$$H_2S + Cl_2 \rightarrow S + 2HCl$$

When it became accepted that atoms consist of protons, neutrons and electrons, chemists understood what was happening in oxidation reactions and the electron transfer theory was put forward. This is the approach that is used by chemists today.

Electron transfer in redox reactions

The combustion of magnesium in oxygen is an oxidation reaction:

$$Mg + \tfrac{1}{2}O_2 \rightarrow MgO$$

This involves the addition of oxygen.

Magnesium has the electron configuration [Ne] $3s^2$. In this reaction, each magnesium atom loses its two outer electrons, becoming [Ne] $3s^0$. The electrons are gained by an oxygen atom. The result is the formation of an Mg^{2+} ion and an O^{2-} ion.

$$\overset{2e^-}{\underset{}{Mg(s) + \tfrac{1}{2}O_2(g) \rightarrow Mg^{2+}O^{2-}(s)}}$$

What happens to the magnesium in this reaction is identical to what happens to it in the reaction with chlorine:

$$\overset{2e^-}{\underset{}{Mg(s) + Cl_2(g) \rightarrow Mg^{2+}2Cl^-(s)}}$$

The magnesium is oxidised because it has lost electrons. Oxygen in the first equation, and chlorine in the second, are reduced because they have gained electrons.

The definitions are as follows:

Oxidation occurs when a species loses one or more electrons.

Reduction occurs when a species gains one or more electrons.

The reduction of iron oxide in the blast furnace can be seen as the gain of three electrons by the Fe^{3+} ions in the oxide:

$$Fe^{3+} + 3e^- \rightarrow Fe$$

When one substance is reduced by gaining electrons, the electrons have been supplied by another species, which lost them and so became oxidised. When chlorine gas is passed through a solution of potassium bromide, the bromide ions lose electrons (are oxidised) and become bromine. The chlorine molecule gains the electrons (is reduced) and becomes chloride ions:

OIL RIG — oxidation is loss, reduction is gain

$$\overset{\displaystyle 2e^-}{\underset{}{2Br^- + Cl_2 \rightarrow Br_2 + 2Cl^-}}$$

Bromide ions are oxidised to bromine and chlorine is reduced to chloride ions.

Oxidising and reducing agents

In the example above, chlorine has oxidised bromide ions. Therefore, chlorine is an **oxidising agent**. The bromide ions have reduced the chlorine. Therefore, the bromide ion is a **reducing agent**.

- An oxidising agent is a species (atom, molecule or ion) that oxidises another species by removing one or more electrons. When an oxidising agent reacts, it is itself reduced.
- A reducing agent reduces another species by giving it one or more electrons. When a reducing reagent reacts, it is itself oxidised.

Consider a redox reaction in which substance X oxidises substance Y:

$$\overset{\displaystyle e^-}{\underset{}{X + Y \rightarrow products}}$$

Substance X has removed an electron from substance Y and so has oxidised it. Substance X is the oxidising agent. It has gained an electron and has itself been reduced. Substance Y has given an electron to substance X and so has reduced it. Substance Y is the reducing agent. It has lost an electron and has itself been oxidised.

Some common oxidising and reducing agents and the products formed when they react are given in Table 9.1.

Oxidising agents	Product when reduced	Reducing reagents	Product when oxidised
Chlorine, Cl_2	Chloride ions, Cl^-	Iodide ions, I^-, or hydrogen iodide, HI	Iodine, I_2
Bromine, Br_2	Bromide ions, Br^-	Hydrogen sulfide, H_2S	Sulfur, S
*Manganate(VII) ions, MnO_4^-	Manganese(II) ions, Mn^{2+}	Sulfur dioxide, SO_2	Sulfate ions, SO_4^{2-}
*Dichromate(VI) ions, $Cr_2O_7^{2-}$	Chromium(III) ions, Cr^{3+}	Iron(II) ions, Fe^{2+}	Iron(III) ions, Fe^{3+}
*Hydrogen peroxide, H_2O_2	Water, H_2O	Hydrogen peroxide, H_2O_2	Oxygen, O_2
Iron(III) ions, Fe^{3+}	Iron(II) ions, Fe^{2+}	Tin(II) ions, Sn^{2+}	Tin(IV) ions, Sn^{4+}
Concentrated sulfuric acid, H_2SO_4	Sulfur dioxide, SO_2	Carbon	Carbon monoxide or carbon dioxide
Concentrated nitric acid, HNO_3	Nitrogen dioxide, NO_2	Carbon monoxide, CO	Carbon dioxide, CO_2
Hydrogen ions, H^+, in dilute acid	Hydrogen, H_2	Metals (e.g. Mg)	Metal ions (e.g. Mg^{2+})

*The solution must be made acidic with dilute sulfuric acid.

Table 9.1 Common oxidising and reducing agents

Ionic half-equations

Oxidation reactions can be written as half-equations, which show the loss of electrons from a single species and the oxidation product. For example, the oxidation half-equation for the oxidation of zinc atoms to zinc ions is:

$$Zn(s) \rightarrow Zn^{2+}(aq) + 2e^-$$

There are four points to note about a half-equation such as this:

- This is an oxidation reaction, so the electrons are on the right-hand side of the equation.
- The equation must balance for charge as well as for numbers of atoms. In this example, both sides add up to a charge of zero ($0 = +2 - 2$).
- State symbols are usually included in half-equations.
- The reactant must always be on the left-hand side of the half-equation.

Reduction reactions can also be written as half-equations. When zinc is added to dilute hydrochloric acid, hydrogen ions are reduced to hydrogen. The half-equation is:

$$2H^+(aq) + 2e^- \rightarrow H_2(g)$$

This is a reduction reaction, so the electrons are on the left-hand side of the half-equation.

Half-equations can be combined to give the overall equation for the reaction (page 188).

(page 188).

Worked example 1

Write an ionic half-equation for the oxidation of Fe^{2+} ions.

Answer

Fe^{2+} ions each lose an electron and become Fe^{3+} ions. The half-equation is:

$$Fe^{2+}(aq) \rightarrow Fe^{3+}(aq) + e^-$$

Worked example 2

Write an ionic half-equation for the reduction of chlorine.

Answer

Chlorine atoms each gain two electrons and are reduced to chloride ions. The half-equation is:

$$Cl_2(g) + 2e^- \rightarrow 2Cl^-(aq)$$

This equation can be halved:

$$\tfrac{1}{2}Cl_2(g) + e^- \rightarrow Cl^-(aq)$$

---

e Remember that :
- *reducing* agents are *oxidised* when they react, so their half-equations must have electrons on the *right*
- *oxidising* agents are *reduced* when they react, so their half-equations must have the electrons on the *left*

Ionic half-equations always have electrons on either the left-hand side or the right-hand side.

e Remember that electrons are negative. It is advisable to put the charge as a superscript in the same way as the charge on an ion is shown.

As the Fe^{2+} ions have been oxidised, the electron appears on the right. The charges on both sides of the half-equation add up to +2.

The chlorine has been reduced, so the electrons appear on the left. The charges on both sides add up to −2.

More complex half-equations

Some oxidising agents require the presence of acid. For example, in the presence of dilute sulfuric acid, potassium manganate(VII) solution is a strong oxidising agent. The manganese in the MnO_4^- ion is reduced to Mn^{2+} ions. A first, but totally wrong, attempt at a half-equation would be:

$$MnO_4^- \rightarrow Mn^{2+} + 3e^-$$

This equation balances for charge but cannot be correct because the electrons are on the wrong side. They have to be on the left-hand side because MnO_4^- ions are being reduced.

The equation also does not balance for atoms, as there are four oxygen atoms on the left and none on the right. In this reaction, oxygen gas is not produced, so the atom balance cannot be made by adding $2O_2$ to the right-hand side. The answer is to have hydrogen ions on the left-hand side. These pick up the oxygen atoms from the MnO_4^- ions and form water. There are four oxygen atoms, so eight hydrogen ions are needed and four water molecules are formed.

A second attempt at the half-equation is:

$$MnO_4^-(aq) + 8H^+(aq) + xe^- \rightarrow Mn^{2+}(aq) + 4H_2O(l)$$

where x is the number of electrons.

The value of x can be worked out, because the equation must balance for charge. The charge on the right-hand side is $+2$. The charge on the left-hand side is:

$$-1 + 8 - x = 7 - x$$

The charge on the left-hand side must equal the charge on the right-hand side, i.e. $+2$. Therefore:

$$7 - x = +2$$
$$x = 7 - 2 = 5$$

There are five electrons on the left-hand side, so the correct ionic half-equation is:

$$MnO_4^-(aq) + 8H^+(aq) + 5e^- \rightarrow Mn^{2+}(aq) + 4H_2O(l)$$

In acid solution, hydrogen peroxide, H_2O_2, acts as an oxidising agent. It is reduced to water (Table 6.1). The half-equation has H^+ and electrons on the left-hand side:

$$H_2O_2(aq) + 2H^+(aq) + 2e^- \rightarrow 2H_2O(l)$$

Worked example

Write the ionic half-equation for the reduction of dichromate(VI) ions to Cr^{3+} ions in acid solution.

Answer

Electrons must be on the left-hand side because the $Cr_2O_7^{2-}$ ions are being reduced. The solution is acidic and so H^+ ions are also on the left. The half-equation is:

$$Cr_2O_7^{2-}(aq) + 14H^+(aq) + 6e^- \rightarrow 2Cr^{3+}(aq) + 7H_2O(l)$$

---

e If there is more oxygen in a molecule or ion on the left-hand side of a half-equation than on the right-hand side, you must add H^+ ions to the left-hand side and H_2O molecules to the right-hand side.

e Remember that an oxidising agent becomes reduced, so electrons are on the left-hand side in the half-equation.

ⓔ There are seven oxygen atoms in the $Cr_2O_7^{2-}$ ion. Therefore, seven water molecules are produced and 14 hydrogen ions are needed on the left-hand side of the equation. The charge on the right-hand side is +6, so the left-hand side must also be +6:

$$-2 + 14 - x = 6$$
$$x = -2 + 14 - 6$$
$$= 14 - 8$$
$$= 6$$

There are six electrons on the left.

Always work out the number of electrons last. The number is such that the equation balances for charge.

The rules for writing half-equations are:

Rule 1: the electrons are on the left-hand side for a reduction equation and on the right-hand side for an oxidation equation.

Rule 2: if the reaction takes place in acid solution, add H^+ ions to the left-hand side and water molecules to the right-hand side.

Rule 3: make sure that the equation balances for atoms.

Rule 4: change the number of electrons so that the equation balances for charge.

Overall redox equations

Half-equations can be combined to give the overall equation for a redox reaction. This is done in two steps:

Step 1: multiply one or both half-equations by integers so that the number of electrons becomes the same in both.

Step 2: add the two half-equations together to obtain the overall equation.

Note that the electrons on the left-hand side cancel the electrons on the right-hand side.

The reaction between aqueous silver nitrate and copper metal is a redox reaction. The silver ions are reduced to silver atoms and the copper atoms are oxidised to copper ions. The two half equations are:

$$Ag^+(aq) + e^- \rightarrow Ag(s)$$
$$Cu(s) \rightarrow Cu^{2+}(aq) + 2e^-$$

The overall equation is found by multiplying the first equation by 2 and then adding it to the second equation:

$$2Ag^+(aq) + 2e^- + Cu(s) \rightarrow 2Ag(s) + Cu^{2+}(aq) + 2e^-$$

The electrons cancel, so the result is:

$$2Ag^+(aq) + Cu(s) \rightarrow 2Ag(s) + Cu^{2+}(aq)$$

The rules for deriving the overall equation from half-equations are:

Rule 1: check that one half-equation has the electrons on the left-hand side and the other has them on the right-hand side.

Rule 2: multiply the half-equations by integers so that the number of electrons is the same in both.

Rule 3: add the two equations and cancel the electrons (and any spectator ions).

Rule 4: check that both reactants are on the left-hand side of the overall equation.

(a)
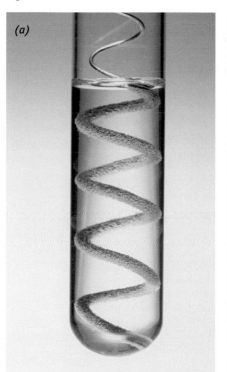

(b)

PETICOLAS/MEGNA/FUNDAMENTAL PHOTOS/SCIENCE PHOTO LIBRARY

Copper wire suspended in silver nitrate solution: (a) after 3 minutes; (b) after 24 hours. The silver ions preferentially replace the copper, leaving crystalline silver deposits and forming a solution of copper nitrate.

Worked example 1

Acidified dichromate(VI) ions oxidise Fe^{2+} ions to Fe^{3+} ions. In this reaction, the acidified dichromate(VI) is itself reduced to Cr^{3+} ions. Write the two ionic half-equations and hence derive the overall equation.

Answer

The half-equations are:

$Fe^{2+}(aq) \rightarrow Fe^{3+}(aq) + e^-$

$Cr_2O_7^{2-}(aq) + 14H^+(aq) + 6e^- \rightarrow 2Cr^{3+}(aq) + 7H_2O(l)$

Multiply the first equation by 6 to obtain the same number of electrons in both equations. Then, add the two equations together and cancel the electrons. The overall equation is:

$Cr_2O_7^{2-}(aq) + 14H^+(aq) + 6Fe^{2+}(aq) \rightarrow 2Cr^{3+}(aq) + 7H_2O(l) + 6Fe^{3+}(aq)$

◄ The Fe^{2+} ions are oxidised, so they lose electrons. As the dichromate(VI) ions are the oxidising agent, they are reduced and so the electrons are on the left.

Worked example 2

In the presence of dilute sulfuric acid, potassium manganate(VII) oxidises Sn^{2+} ions to Sn^{4+} ions and is itself reduced to Mn^{2+} ions. Write the two ionic half-equations and use them to derive the overall equation.

Answer

The half equations are:

$$Sn^{2+}(aq) \rightarrow Sn^{4+}(aq) + 2e^-$$
$$MnO_4^-(aq) + 8H^+(aq) + 5e^- \rightarrow Mn^{2+}(aq) + 4H_2O(l)$$

Multiply the first equation by 5 and the second equation by 2 to obtain the same number of electrons in each equation. Then, add the two equations and cancel the electrons. The overall equation is:

$$2MnO_4^-(aq) + 16H^+(aq) + 5Sn^{2+}(aq) \rightarrow 2Mn^{2+}(aq) + 8H_2O(l) + 5Sn^{4+}(aq)$$

Deduction of half-equations from the overall equation

The half-equations can be deduced from the overall equation because the latter shows the formulae of all the reactants and products. The redox reaction between nitric acid, HNO_3, and hydrogen sulfide, H_2S, is a suitable example:

$$2HNO_3 + H_2S \rightarrow 2NO_2 + S + 2H_2O$$

First, work out which substance has been oxidised and write electrons on the right-hand side of the half-equation.

The sulfur in H_2S loses hydrogen and so is oxidised. The half-equation is:

$$H_2S \rightarrow 2H^+ + S + 2e^-$$

The hydrogen sulfide is oxidised and therefore the nitric acid must be reduced. The number of electrons on the left-hand side in the half-equation for nitric acid reduction must be the same as the number on the right-hand side in the half-equation for hydrogen sulfide oxidation. Therefore, the half-equation for the reduction of nitric acid to nitrogen dioxide is:

$$2HNO_3 + 2H^+ + 2e^- \rightarrow 2NO_2 + 2H_2O$$

On addition of the two half-equations, the $2H^+$ on the left-hand side of the nitric acid reduction equation cancels with the $2H^+$ on the right-hand side of the hydrogen sulfide oxidation equation.

As the oxidation number of sulfur changes by two, there must be two electrons in the equation. They will be on the *right* as H_2S has been *oxidised*. The hydrogen in H_2S has been neither oxidised nor reduced. Therefore, it must still be in the +1 state and so H^+ and not H_2 is the product in the half-equation. Note that the equation balances for charge.

Note that nitric acid is the oxidising agent and hydrogen sulfide is the reducing agent.

Oxidation numbers

Redox reactions can be described in terms of the **oxidation numbers** of the elements concerned. The oxidation number of the species in a simple ionic compound is the charge on the ion. For example, in magnesium chloride, the magnesium ions are 2+, so the oxidation number of magnesium is +2. The charge on each chloride ion is 1−, so the oxidation number of each chlorine in $MgCl_2$ is −1. This concept can be extended to covalent substances and to polyatomic ions such as sulfate, SO_4^{2-}.

The oxidation number of an element in a compound or ion is the charge that the element would have if the compound were fully ionic.

The oxidation number is calculated on the basis that the bonding electrons are assigned to the more electronegative atom in a covalent bond. Consider hydrogen sulfide, H_2S. The bonding is covalent and the sulfur atom shares a pair of electrons with each hydrogen atom. However, sulfur is more electronegative than hydrogen, so the bond pair is assigned to the sulfur in both H–S bonds. In this way, the sulfur atom 'gains' one electron from each of the two H–S bonds, giving it an assigned charge of 2–. The compound is not ionic, but the sulfur is δ^- and is said to have an oxidation number of –2, or to be in the –2 oxidation state.

Deduction of oxidation numbers

Oxidation numbers can be deduced using a series of rules:

- The oxidation number of an uncombined element is zero. For example, the bonding electrons in chlorine, Cl_2, 'go' one to each chlorine atom. Neither gains an electron, so both chlorine atoms have an oxidation number of 0.
- The oxidation number of the element in a monatomic ion is the charge on the ion. For example, the oxidation number of copper in the Cu^{2+} ion is +2 — it is in the +2 state. The name of a compound, such as copper(II) oxide, indicates that the copper is in the +2 oxidation state.
- All group 1 metals have an oxidation number of +1 in their compounds; all group 2 metals have an oxidation number of +2 in their compounds.
- Fluorine always has the oxidation number –1 in its compounds.
- Hydrogen has the oxidation number +1 in its compounds, apart from when it is combined with a metal, when the oxidation number is –1.
- Oxygen has the oxidation number –2 in its compounds, apart from in peroxides and superoxides or when it is combined with fluorine.
- The sum of the oxidation numbers of the atoms in a neutral compound is zero. For example, in ammonia, NH_3, the oxidation number of nitrogen (–3) plus three times the oxidation number of hydrogen (+1) equals zero.
- The sum of the oxidation numbers in a polyatomic ion is the charge on that ion. For example, in the MnO_4^- ion, the oxidation number of manganese (+7) plus four times the oxidation number of oxygen (–2) equals –1.

The possible oxidation numbers of some elements in their compounds are shown in Table 9.2.

Element	Oxidation number
Lithium, sodium and potassium	+1
Magnesium, calcium, strontium and barium	+2
Aluminium	+3
Nitrogen (except in some oxides)	–3, +3 or +5
Oxygen (except in peroxides and superoxides or when combined with fluorine)	–2
Sulfur	–2, +1, +2, $+2\frac{1}{2}$, +4 or +6
Fluorine	–1
Chlorine, bromine and iodine	–1, +1, +3, +5 or +7

e Remember that the most electronegative element is fluorine, followed by oxygen and then chlorine. Electronegativity *increases* across a period and *decreases* down a group in the periodic table.

Table 9.2 Oxidation numbers

e Remember that if an element is joined to a *more* electronegative element, it will have a positive oxidation number. So, bromine in BrO_3^- is in a positive oxidation state.

Worked example 1

Deduce the oxidation number of chlorine in:

a NaCl

b ClO^-

c ClO_3^-

Answer

a The oxidation number of Na is +1. The two oxidation numbers must add up to zero. Therefore, Cl must be −1.

b Let the oxidation number of chlorine = y.
The oxidation number of oxygen is −2 and the charge on the ion is −1, so:

$$-2 + y = -1$$
$$y = +1$$

The oxidation number of chlorine in the ClO^- ion is +1.

c Let the oxidation number of chlorine = z.
The oxidation number of oxygen is −2 and the charge on the ion is −1, so:

$$(3 \times -2) + z = -1$$
$$z = -1 + 6 = +5$$

The oxidation number of chlorine in the ClO_3^- ion is +5.

---

e Remember that you must always put the sign of the oxidation number *before* the number. The sign must *never* be omitted. If it is omitted, the answer may be marked as wrong.

Worked example 2

Deduce the oxidation number of sulfur in:

a SCl_2

b SO_2

c SO_4^{2-}

Answer

a Chlorine is more electronegative than sulfur. Therefore, (2×-1) and the oxidation number of sulfur add up to zero.
The oxidation number of sulfur in SCl_2 is +2.

b The oxidation number of sulfur and 2 × the oxidation number of oxygen add up to zero. The oxidation number of oxygen is −2.
Therefore, the oxidation number of sulfur in SO_2 is +4.

c Let the oxidation number of sulfur = z.
The oxidation number of sulfur and 4 × the oxidation number of oxygen add up to −2.

$$z + (4 \times -2) = -2$$
$$z = -2 + 8 = +6$$

The oxidation number of sulfur in the SO_4^{2-} ion is +6.

Worked example 3

Deduce the oxidation number of manganese in:

a MnO_4^-

b MnO_4^{2-}

Answer

Let the oxidation number of manganese $= z$.

a $z + (4 \times -2) = -1$

       $z = -1 + 8 = +7$

The oxidation number of manganese in the MnO_4^- ion is +7.

b $z + (4 \times -2) = -2$

       $z = -2 + 8 = +6$

The oxidation number of manganese in the MnO_4^{2-} ion is +6.

Worked example 4

Deduce the oxidation number of vanadium in VO_2^+.

Answer

    $z + (2 \times -2) = +1$

    $z = 1 + 4 = +5$

The oxidation state of vanadium in the VO_2^+ ion is +5.

Oxidation and reduction in terms of oxidation number

Consider the following oxidation half-equations:

$Fe^{2+}(aq) \rightarrow Fe^{3+}(aq) + e^-$

$H_2S(g) \rightarrow S(s) + 2H^+(aq) + 2e^-$

$2I^-(aq) \rightarrow I_2(s) + 2e^-$

In all three examples, notice that the oxidation number of the oxidised species has increased. The iron has gone from the +2 state to +3, sulfur from −2 to 0 and iodine from −1 to 0. This leads to another definition of oxidation.

> An increase in the oxidation number of an element means that it has been oxidised.

Consider the following reduction half-equations:

$MnO_4^-(aq) + 8H^+(aq) + 5e^- \rightarrow Mn^{2+}(aq) + 4H_2O(l)$

$PbO_2(s) + 4H^+(aq) + 2e^- \rightarrow Pb^{2+}(aq) + 2H_2O(l)$

In each case, the oxidation number of the reduced species has decreased. The manganese has gone from the +7 state to +2 and the lead from +4 to +2. Reduction can also be defined in terms of oxidation number.

> A decrease in the oxidation number of an element means that it has been reduced.

Names of compounds in terms of oxidation number

The names of compounds and ions that contain an element that can have more than one oxidation number include the oxidation number. This is written as a Roman numeral in brackets:

- Manganese in the MnO_4^- ion is in the $+7$ state. Therefore, the ion is manganate(VII).
- Chromium in the $Cr_2O_7^{2-}$ ion is in the $+6$ state. Therefore, the ion is dichromate(VI).
- $FeCl_3$ is iron(III) chloride; $FeSO_4$ is iron(II) sulfate.

> **Worked example**
>
> Name the following ions:
>
> **a** ClO^-
>
> **b** ClO_3^-
>
> **Answer**
>
> **a** The chlorine is in the +1 oxidation state, so the ion is chlorate(I).
>
> **b** The chlorine is in the +5 oxidation state, so the ion is chlorate(V).

Balancing equations using oxidation numbers

Half-equations

The number of electrons in a half-equation is equal to the total change in oxidation number of the element.

In acid solution, ferrate(VI) ions, FeO_4^{2-}, can be reduced to Fe^{2+} ions. The oxidation number of iron changes from $+6$ to $+2$. This is a decrease of four, so there are four electrons on the left-hand side of the reduction half-equation. The half-equation is:

$$FeO_4^{2-}(aq) + 8H^+(aq) + 4e^- \rightarrow Fe^{2+}(aq) + 4H_2O(l)$$

When potassium dichromate(VI), $Cr_2O_7^{2-}$, in dilute sulfuric acid, is reduced to Cr^{3+} ions, the oxidation number of each chromium decreases by three. However, there are two chromium atoms in the equation, so the total change is six. Therefore, the half-equation has six electrons on the left-hand side:

$$Cr_2O_7^{2-}(aq) + 14H^+(aq) + 6e^- \rightarrow 2Cr^{3+}(aq) + 7H_2O(l)$$

Overall redox equations

The total change in oxidation number of the element being reduced is equal to the total change in oxidation number of the element being oxidised.

When potassium manganate(VII) oxidises Sn^{2+} ions to Sn^{4+} ions, the oxidation number of the manganese changes from $+7$ to $+2$ — a decrease of five. The oxidation number of tin increases by two. To balance the change in oxidation numbers, $2MnO_4^-$ and $5Sn^{2+}$ are needed. In this way, the total change in oxidation number of both manganese and tin is 10. The overall equation is:

$$2MnO_4^-(aq) + 16H^+(aq) + 5Sn^{2+}(aq) \rightarrow 2Mn^{2+}(aq) + 8H_2O(l) + 5Sn^{4+}(aq)$$

◄ Polyatomic anions containing oxygen have names ending in -ate. The exception is the OH^- ion, which is called the hydroxide ion.

ⓔ Remember that oxidising agents have electrons on the left-hand side in half-equations (they become reduced) and reducing agents have electrons on the right-hand side (they become oxidised).

Disproportionation reactions

Disproportionation is a redox reaction in which an element in a *single species* is simultaneously oxidised and reduced.

When chlorine gas is bubbled into aqueous sodium hydroxide, a disproportionation reaction takes place. One atom in the Cl_2 molecule is oxidised to NaOCl and one is reduced to NaCl. The equation, with the oxidation numbers of chlorine, is:

$$Cl_2 + 2NaOH \rightarrow NaOCl + NaCl + H_2O$$
$$0 \qquad\qquad\qquad +1 \quad -1$$

> To be involved in a disproportionation reaction, an element must have at least three oxidation states — the initial one, one higher and one lower.

Worked example 1

When potassium bromate(I) is heated, it decomposes according to the equation:

$$3KOBr \rightarrow 2KBr + KBrO_3$$

Explain why this is a disproportionation reaction.

Answer

The bromine in KOBr is in the +1 oxidation state. In the products, the oxidation number of bromine is –1 in KBr and +5 in $KBrO_3$. The bromine in the single species is oxidised from +1 to +5 and simultaneously reduced from +1 to –1.

Worked example 2

Consider the reaction:

$$NaOCl + NaCl + H_2SO_4 \rightarrow Na_2SO_4 + H_2O + Cl_2$$

Explain why this is not a disproportionation reaction.

Answer

The chlorine in NaOCl is reduced from the +1 oxidation state to the zero state and the chlorine in NaCl is oxidised from the –1 state to the zero state. However, the two chlorine atoms are not in the same *species*, and so it is not a disproportionation reaction.

Worked example 3

Is the decomposition of hydrogen peroxide a disproportionation reaction?

$$2H_2O_2 \rightarrow 2H_2O + O_2$$

Answer

The oxidation number of both oxygen atoms in H_2O_2 is –1. The oxidation number of the oxygen in H_2O is –2 and of the oxygen in O_2 is zero. The oxygen in H_2O_2 has simultaneously been oxidised from –1 to zero and reduced from –1 to –2. Therefore, the decomposition is a disproportionation reaction.

Questions

1 State whether the following changes are oxidation or reduction. Explain your answers in terms of the transfer of electrons.
   a Cr^{2+} to Cr^{3+}
   b HI to I_2
   c Sn^{4+} to Sn^{2+}
   d Cl^- to Cl_2

2 Write ionic half-equations for the changes in Question 1, parts a to d.

3 Write ionic half-equations for the following, which take place in acidic solution:
   a Iodate(v) ions, IO_3^-, to iodine
   b Manganese dioxide, MnO_2, to Mn^{2+} ions
   c VO_2^+ ions to VO^{2+} ions

4 Write overall equations for the following redox reactions:
   a Manganate(VII) ions in acid solution reacting with hydrogen iodide
   b Fe^{3+} ions being reduced to Fe^{2+} ions by Sn^{2+} ions, which are oxidised to Sn^{4+} ions
   c Dichromate(VI) ions in acid solution oxidising nitrogen dioxide, NO_2, to nitrate ions, NO_3^-
   d Iodine oxidising sodium thiosulfate, $Na_2S_2O_3$, to sodium tetrathionate, $Na_2S_4O_6$ — the iodine is reduced to sodium iodide, NaI

5 Deduce the oxidation number of iodine in:
   a I_2O_7
   b IO_3^-
   c KIO_4
   d $Ba(IO_2)_2$

6 Deduce the oxidation number of vanadium in:
   a V^{3+}
   b VO_2^+
   c VO^{2+}

7 Deduce the oxidation number of nitrogen in:
   a NO_2
   b NO_2^-
   c N_2O_5
   d N_2H_4

8 Potassium bromide, KBr, reacts with potassium bromate, $KBrO_3$, in the presence of dilute sulfuric acid according to the equation:

   $$5KBr + KBrO_3 + 3H_2SO_4 \rightarrow 3Br_2 + 3K_2SO_4 + 3H_2O$$

   a Give the oxidation numbers of bromine in KBr, $KBrO_3$ and Br_2.
   b State, with a reason, which substance in the equation is the oxidising agent.

9 Calculate the total change in oxidation number of the carbon atoms when ethanedioate ions, $C_2O_4^{2-}$, are oxidised to carbon dioxide.

10 Write the ionic equation for the reaction of bromine with aqueous sodium hydroxide, to form NaBr, NaOBr and water.

11 Write the equation for the disproportionation reaction of copper(I) ions to copper metal and copper(II) ions.

12 From each of the redox equations below, deduce the two half-equations and identify the oxidising agent and the reducing agent.
   a $Cl_2 + 2NaBr \rightarrow 2NaCl + Br_2$
   b $HSO_3^- + 2Fe^{3+} + H_2O \rightarrow SO_4^{2-} + 2Fe^{2+} + 3H^+$

Chapter 10

The periodic table: group 2

Neon: lights and lasers

Atomic number: 10

Electron configuration: $1s^2\ 2s^2\ 2p_x{}^2\ 2p_y{}^2\ 2p_z{}^2$

Symbol: Ne

Neon is one of the noble gases in group 0 of the periodic table. It does not form any compounds. It is present in very small quantities in air — 88 000 tonnes of liquid air are required to make 1 tonne of neon.

Discharge tubes

When an electric current is passed through a gaseous element, coloured light is emitted. When the high-speed electrons strike a gaseous neon atom, an electron is promoted into a higher orbit. When the electron drops down to a lower state, light of a particular frequency is emitted. Neon produces bright red light of wavelength 633 nm. Neon discharge tubes are used for eye-catching advertising displays. Very little current is required to produce the light, so tiny neon tubes are placed in some electrical sockets; the red glow shows when the socket is live.

Gas lasers

The word laser stands for **l**ight **a**mplification by **s**timulated **e**mission of **r**adiation. A helium/neon laser contains a mixture of the two gases at low pressure. The tube has a fully reflecting mirror at one end and a mirror that allows about 1% of the light to leave at the other. An electric discharge excites the helium atoms, which then transfer their energy to the neon atoms through collision. An electron in a neon atom enters a metastable state. A metastable state is one in which the electron can remain for a brief time. When the electron in one atom spontaneously falls back to a lower state, the light emitted stimulates the electrons in the metastable state of other neon atoms to fall back to the lower state. These emit light *exactly in phase* with the original light. The beam is amplified as it is reflected back and forth by the mirrors at each end. Each photon emitted stimulates the production of another photon and, as the light bounces backwards and forwards, the number of photons increases rapidly. The output is the light that escapes through the partial reflecting mirror at one end.

Neon signs vie for attention in our towns and cities

MEHAU KULYK/SCIENCE PHOTO LIBRARY

The periodic table: group 2

Group 2

The s block consists of the elements in groups 1 and 2. The group 1 elements (lithium, sodium, potassium, rubidium, caesium and francium) are called the **alkali metals**. They form cations with a 1+ charge.

The elements in group 2 are the **alkaline earth metals**. They have a valency of 2, form cations with a charge of 2+ and have an oxidation number of +2 in their compounds.

The elements in group 2 are:
- beryllium, Be
- magnesium, Mg
- calcium, Ca
- strontium, Sr
- barium, Ba
- radium, Ra

Radium is a rare, radioactive metal.

All are metallic elements. The solids consist of a regular arrangement of 2+ ions in a sea of delocalised electrons. The metallic bond is the attraction between the metal ions and the delocalised electrons. Its strength depends on the ionic radius of the 2+ ion and the type of packing. Beryllium and magnesium have close-packed hexagonal structures, calcium and strontium have face-centered cubic structures and barium has a body-centred cubic structure.

The electronegativities of the elements decrease down the group, from 1.5 for beryllium to 0.9 for radium. This means that the elements become more reactive as the atomic number increases. The ionic radius increases from 0.031 nm for the Be^{2+} ion to 0.140 nm for Ra^{2+}. This means that the ions become *less* polarising down the group. Polarisation of an anion causes the value of the lattice energy calculated from a Born–Haber cycle to differ from that calculated based on the purely ionic model. Table 10.1 shows how this difference becomes smaller down the group for the metal chlorides. The difference is an indication of partial covalent character, so it follows that barium chloride is the most ionic of the group 2 chlorides.

	Ionic radius/nm	Lattice energy of chloride/kJ mol⁻¹		Difference/ kJ mol⁻¹
		Born–Haber value	Calculated value	
Magnesium	0.065	−2526	−2326	200
Calcium	0.099	−2258	−2223	35
Strontium	0.11	−2156	−2127	29
Barium	0.14	−2056	−2033	23

Table 10.1 Ionic radii and lattice energy values

Physical properties

- **Melting temperature** — a group 2 metal has a higher melting temperature than the group 1 metal in the same period. This is because each atom loses two electrons to form the metallic bond, which is therefore stronger than the metallic bond in the group 1 metal. Also, the metallic radius of the group 2 element is smaller than that of the group 1 element in the same period. The melting temperature *decreases* down the group. This is because the ionic radius increases, so the strength of the metallic bond between the 2+ ions and the delocalised electrons decreases.
- **Electrical conductivity** — the delocalised electrons in the metal are mobile. Therefore, they can move across an electric potential.

Element	Be	Mg	Ca	Sr	Ba
Atomic number, Z	4	12	20	38	56
Relative atomic mass, A_r	9	24	40	88	137
Electron configuration	[He] $2s^2$	[Ne] $3s^2$	[Ar] $4s^2$	[Kr] $5s^2$	[Xe] $6s^2$
Melting temperature/°C	1278	649	839	769	725
Atomic radius/nm	0.11	0.16	0.20	0.21	0.22
Ionic radius/nm	0.031	0.065	0.099	0.11	0.14
First ionisation energy/kJ mol^{-1}	900	736	590	548	502

Table 10.2 Physical properties of the group 2 elements

Ionisation energy

There is a big jump between the second and the third consecutive ionisation energies of all the group 2 metals. This is because the third electron removed comes from an inner electron shell. Therefore, it is not as well shielded as the outer two electrons and is much closer to the nucleus.

The first ionisation energies decrease down the group (Table 10.3). From magnesium to calcium, the nuclear charge increases by eight but the number of inner shielding electrons also increases by eight. This means that the effective nuclear charge is approximately the same in both. The difference is that the electron being removed in calcium is a $4s$-electron, which is further from the nucleus than the $3s$-electron removed when magnesium is ionised.

	Beryllium	Magnesium	Calcium	Strontium	Barium
First ionisation energy/kJ mol^{-1}	900	736	590	548	502
Nuclear charge, Z	4	12	20	38	56
Electron configuration	[He] $2s^2$	[Ne] $3s^2$	[Ar] $4s^2$	[Kr] $5s^2$	[Xe] $6s^2$
Atomic radius/nm	0.11	0.16	0.20	0.21	0.22
Number of inner shielding electrons	2	10	18	36	54
Approximate effective nuclear charge, Z_{eff}	$4 - 2 = 2$	$12 - 10 = 2$	$20 - 18 = 2$	$38 - 36 = 2$	$56 - 54 = 2$

Table 10.3 First ionisation energies

In the same period, the first ionisation energy of the group 2 metal is larger than that of the group 1 metal. This is because the effective nuclear charge has increased approximately from 1 to 2 and the group 2 atom has a smaller radius than the group 1 atom.

	Sodium	Magnesium
Nuclear charge	11	12
Number of inner shielding electrons	10	10
Approximate effective nuclear charge	11 − 10 = 1	12 − 10 = 2
Atomic radius/nm	0.19	0.16
First ionisation energy/kJ mol^{-1}	494	736

Table 10.4 Comparison of a group 1 and a group 2 element in the same period

Reactions of the group 2 metals

- In all their compounds, the group 2 metals are in the +2 oxidation state.
- Reactivity increases down the group because it becomes easier to remove the two outer electrons.

Reaction with oxygen

All the group 2 metals burn in air. They react with the oxygen and form ionic oxides of formula, MO. For example, with calcium:

$$2Ca + O_2 \rightarrow 2CaO$$

The metal is oxidised. Each atom loses two electrons:
$$Ca \rightarrow Ca^{2+} + 2e^-$$

The metals burn with characteristic colours:

- Magnesium burns with an intense white light.
- Calcium burns with a red flame.
- Strontium burns with a dark red flame.
- Barium burns with a pale green flame.

Flame test data are given in Figure 10.2 (page 212).

Reaction with water

- Beryllium does not react with water.
- Magnesium reacts slowly with cold water to produce an alkaline suspension of magnesium hydroxide and hydrogen gas:
$$Mg + 2H_2O(l) \rightarrow Mg(OH)_2 + H_2$$
When heated in steam, magnesium burns, producing magnesium oxide and hydrogen:
$$Mg + H_2O(g) \rightarrow MgO + H_2$$
- Calcium, strontium and barium react rapidly with cold water to produce alkaline solutions of the metal hydroxide and bubbles of hydrogen gas — for example:
$$Ca(s) + 2H_2O(l) \rightarrow Ca(OH)_2(aq) + H_2(g)$$

Reaction with chlorine

All the group 2 elements react when heated in chlorine — for example:

$$Ca + Cl_2 \rightarrow CaCl_2$$

- Beryllium forms a covalent anhydrous chloride.
- The other group 2 metals form ionic chlorides.
- The chlorides have the formula MCl_2.

Apart from beryllium chloride, the group 2 chlorides dissolve in water, producing solutions that contain hydrated cations of formula $[M(H_2O)_6]^{2+}$ — for example:

$$MgCl_2(s) + aq \rightarrow [Mg(H_2O)_6]^{2+}(aq) + 2Cl^-(aq)$$

The six water molecules are dative-covalently bonded from the lone pair of electrons on the oxygen to the central metal ion. The shape of the hydrated ion is octahedral because the metal ion is surrounded by six bond pairs of electrons and no lone pairs.

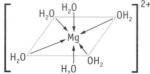

The hexaaquamagnesium ion is deprotonated by solvent water molecules and the solution becomes slightly acidic:

$$[Mg(H_2O)_6]^{2+}(aq) + H_2O(l) \rightarrow [Mg(H_2O)_5(OH)]^+(aq) + H_3O^+(aq)$$

Group 2 oxides and hydroxides

Group 2 oxides and hydroxides are bases. The general formulae are MO and $M(OH)_2$.

Reaction of the oxides with water

- Beryllium oxide does not react with water.
- Magnesium oxide reacts slowly and incompletely to form a slightly alkaline suspension of magnesium hydroxide:

$$MgO + H_2O \rightarrow Mg(OH)_2$$

- Calcium oxide is called quicklime. It reacts very exothermically with water to form an alkaline suspension of calcium hydroxide:

$$CaO + H_2O \rightarrow Ca(OH)_2$$

Solid calcium hydroxide is called slaked lime and solutions of it are called limewater.

- Strontium and barium oxides react with water to form alkaline solutions of the hydroxide:

$$BaO(s) + H_2O(l) \rightarrow Ba^{2+}(aq) + 2OH^-(aq)$$

Reaction with acids

The oxides and hydroxides of the group 2 metals are bases. Therefore, they react with acids to form salts and water.

- Magnesium oxide reacts with dilute sulfuric acid to form magnesium sulfate and water. The equation is:

$$MgO(s) + H_2SO_4(aq) \rightarrow MgSO_4(aq) + H_2O(l)$$

The ionic equation is:

$$MgO(s) + 2H^+(aq) \rightarrow Mg^{2+}(aq) + H_2O(l)$$

- Calcium hydroxide reacts with nitric acid solution to form calcium nitrate and water. The equation is:

$$Ca(OH)_2(s) + 2HNO_3(aq) \rightarrow Ca(NO_3)_2(aq) + 2H_2O(l)$$

The ionic equation is:

$$Ca(OH)_2(s) + 2H^+(aq) \rightarrow Ca^{2+}(aq) + 2H_2O(l)$$

- Beryllium oxide is amphoteric. This means that it reacts both as a base and as an acid:

$$BeO(s) + 2H^+(aq) \rightarrow Be^{2+}(aq) + H_2O(l)$$
$$BeO(s) + 2OH^-(aq) + H_2O(l) \rightarrow Be(OH)_4^{2-}(aq)$$

A solution becomes acidic if H_3O^+ ions are formed.

An amphoteric oxide or hydroxide reacts with an acid to form a simple cation (e.g. Be^{2+}) and with a strong alkali to form an oxy-anion containing four or six –OH groups (e.g. $Be(OH)_4^{2-}$).

$$2NaOH(aq) + H_2SO_4(aq) \rightarrow Na_2SO_4(aq) + 2H_2O(l)$$

Reaction of limewater with carbon dioxide

Carbon dioxide is an acidic oxide. It reacts with the base calcium hydroxide to form the salt, calcium carbonate, and water. Calcium carbonate is insoluble and appears as a milky precipitate. This reaction is the test for carbon dioxide:

$$Ca(OH)_2(aq) + CO_2(g) \rightarrow CaCO_3(s) + H_2O(l)$$

If carbon dioxide is passed into a suspension of calcium carbonate, the milkiness slowly clears as the soluble acid salt, calcium hydrogencarbonate, is formed:

$$CaCO_3(s) + H_2O(l) + CO_2(g) \rightarrow Ca(HCO_3)_2(aq)$$

Calcium hydrogencarbonate is soluble in water, whereas calcium carbonate is insoluble. This is the basis of a test to distinguish between a carbonate and a hydrogencarbonate of a group 1 metal. A solution of the test substance is added to a solution of calcium chloride. If the substance is a carbonate, a white precipitate of calcium carbonate is seen. If the test substance is a hydrogencarbonate, no precipitate forms until the mixture is heated. The hydrogencarbonate ions decompose forming carbonate ions, carbon dioxide and water:

$$2HCO_3^-(aq) \rightarrow CO_3^{2-}(aq) + CO_2(g) + H_2O(l)$$

The carbonate ions then react with the calcium ions to form a precipitate of insoluble calcium carbonate:

$$CO_3^{2-}(aq) + Ca^{2+}(aq) \rightarrow CaCO_3(s)$$

Solubility of group 2 hydroxides

The solubility of group 2 hydroxides *increases* down the group:

- Beryllium hydroxide is insoluble.
- Magnesium hydroxide is very slightly soluble.
- Calcium hydroxide is slightly soluble.
- Strontium and barium hydroxides are more soluble.

A solution of sodium hydrogen-carbonate is slightly alkaline and a solution of sodium carbonate is quite strongly alkaline. When a solution of sodium hydrogen-carbonate is heated, the pH rises.

	$Be(OH)_2$	$Mg(OH)_2$	$Ca(OH)_2$	$Sr(OH)_2$	$Ba(OH)_2$
Solubility/mol dm^{-3}	Insoluble	0.0002	0.015	0.083	0.15
pH of saturated solution	7	10.6	12.5	13.2	13.5

Table 10.5 Solubility of group 2 hydroxides

Solubility of group 2 sulfates

The solubility of the sulfates *decreases* down the group:

- Beryllium and magnesium sulfates are very soluble.
- Calcium sulfate is slightly soluble.
- Strontium sulfate is very slightly soluble.
- Barium sulfate is insoluble.

Barium sulfate is so insoluble that it passes unaffected through the alimentary canal. This has led to its use in hospitals. A paste of barium sulfate is eaten ('barium meal') and, at various times, the patient is X-rayed. The progress of the barium sulfate through the digestive tract is observed and any obstructions seen. Barium is opaque to X-rays because it has a very high nuclear charge.

On adding a solution of sodium sulfate (or any solution containing sulfate ions, such as dilute sulfuric acid) to a solution of calcium, strontium and barium compounds, a white precipitate of an insoluble sulfate is produced. The ionic equation for the reaction of barium chloride solution with sodium sulfate solution is:

$$Ba^{2+}(aq) + SO_4^{2-}(aq) \rightarrow BaSO_4(s)$$

This does not happen with magnesium or beryllium compounds because magnesium sulfate and beryllium sulfate are soluble.

> *e* Barium sulfate is so insoluble that when dilute sulfuric acid is added to a piece of barium metal, the reaction stops almost at once. The initial reaction causes a layer of insoluble barium sulfate to form on the surface of the metal. This layer prevents further acid from reaching the unreacted metal.

Thermal stability of group 1 and group 2 nitrates and carbonates

Thermal stability depends upon the polarising power of the cation, which is determined by its charge density. The larger the charge and the smaller the radius, the greater is the charge density. Compounds containing cations that strongly polarise the anion are more easily decomposed than those with less polarising cations.

This means that, in both groups 1 and 2, the ease of decomposition decreases down the group. This is because the ionic radius increases down the group, reducing the polarising power of the cation.

The ions of the group 2 elements have a charge of 2+, compared with a charge of 1+ for the group 1 elements. The group 2 ions also have a smaller ionic radius than the group 1 ions. Therefore, group 2 cations are considerably more polarising than group 1 cations and so group 2 compounds are less stable to heat and decompose more easily.

Nitrates

Thermal stability of group 1 nitrates

All group 1 nitrates decompose on strong heating. However, lithium nitrate decomposes in a different way from the other nitrates. The products are lithium oxide, nitrogen dioxide and oxygen. The nitrogen dioxide is visible as a brown gas:

$$4LiNO_3 \rightarrow 2Li_2O + 4NO_2 + O_2$$

- The small radius of the lithium ion causes it to have a high charge density, which polarises the O–N bond in the nitrate ion sufficiently to break it and form an O^{2-} ion.
- The other group 1 cations are too large to polarise to this extent. Therefore, on heating, their nitrates do not decompose to give an oxide. Very strong heating causes them to melt, giving off oxygen and leaving a molten nitrite that contains the NO_2^- ion, for example:

$$2NaNO_3 \rightarrow 2NaNO_2 + O_2$$

e Polarisation occurs when a positive ion attracts the electrons of a negative ion, distorting the negative ion. Cations with high charges, or ions with small radii, have a high charge density and so are highly polarising.

Thermal stability of group 2 nitrates

The nitrate ion in a group 2 nitrate is sufficiently polarised by the 2+ cation to result in thermal decomposition by the same pathway as lithium nitrate.

- On heating, group 2 nitrates decompose to the metal oxide, nitrogen dioxide and oxygen. The temperature at which thermal decomposition occurs is lowest for beryllium nitrate and highest for barium nitrate, which requires very strong heating before brown fumes of nitrogen dioxide are seen. The equation for the decomposition of magnesium nitrate is:

$$2Mg(NO_3)_2 \rightarrow 2MgO + 4NO_2 + O_2$$

Carbonates

Thermal stability of group 1 carbonates

Only lithium carbonate decomposes when heated. This is because the Li^+ cation is very small and polarises the O–C bond in the $CO_3{}^{2-}$ ion sufficiently for it to break and form an O^{2-} ion:

$$Li_2CO_3 \rightarrow Li_2O + CO_2$$

The other group 1 cations have larger radii. Therefore, their polarising power is not sufficient to cause decomposition of the *anhydrous* carbonates. When a *hydrated* carbonate, such as hydrated sodium carbonate (washing soda), $Na_2CO_3.10H_2O$, is heated, the water of crystallisation hydrolyses the carbonate and carbon dioxide and steam are given off:

$$Na_2CO_3.10H_2O(s) \rightarrow 2NaOH(s) + CO_2(g) + 9H_2O(g)$$

Thermal stability of group 2 carbonates

All the group 2 carbonates decompose on heating:

- Beryllium carbonate is so unstable that it does not exist at room temperature.
- The equation for the decomposition of, for example, calcium carbonate is:

$$CaCO_3 \rightarrow CaO + CO_2$$

- Barium carbonate requires strong heating before it is decomposed.

Experiments to study thermal decomposition

The relative ease of decomposition of group 2 carbonates or nitrates can be studied by the following experiment:

- Place the same amount of each carbonate (or nitrate) in a series of hard glass test tubes.
- Fix a delivery tube to the test tube and clamp in a stand, as in Figure 10.1.

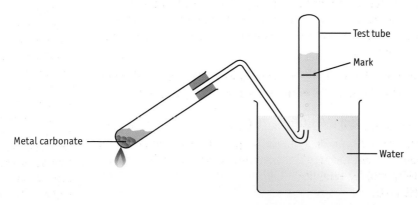

e Remember that the ease of decomposition depends on the charge density of the cation. Ions with a low charge density, such as Ba^{2+} or K^+, form carbonates that are stable to heat. Ions with a high charge density, such as Mg^{2+} or Li^+, form carbonates that are easily decomposed.

Figure 10.1 Heating group 2 carbonates

- Light the Bunsen and measure the time taken for the gas evolved to reach the mark on the test tube in the water bath.
- Repeat with the same amount (in moles) of the remaining carbonates. You must make sure that the Bunsen flame is the same and that the distance between the Bunsen and the hard glass test tube is the same in all experiments.

The group 2 carbonate that, on heating, fills the test tube the fastest is the one that is the easiest to decompose.

As a control, a test tube can be heated. The air in it will expand and some gas will be collected over the water bath.

Group 2 chlorides

The group 2 chlorides have the formula MCl_2.

Preparation
Group 2 chlorides can be prepared by:
- **Direct synthesis** — the metal and chlorine are heated together — for example:
$$Ca + Cl_2 \rightarrow CaCl_2$$
- **Neutralisation** of the oxide by dilute hydrochloric acid — dilute hydrochloric acid is warmed with a group 2 oxide. The metal chloride and water are produced — for example:
$$MgO + 2HCl \rightarrow MgCl_2 + H_2O$$
The chloride can also be produced by neutralising the acid with a carbonate — for example:
$$CaCO_3(s) + 2HCl(aq) \rightarrow CaCl_2(aq) + H_2O(l) + CO_2(g)$$

Properties
Solubility
All the group 2 chlorides are soluble in water.

Bonding
Beryllium chloride, $BeCl_2$, is covalent and is soluble, not only in water but also in organic solvents. There are two pairs of electrons and no lone pairs around the beryllium atom. The two electron pairs repel to give maximum separation and so the gaseous molecule is linear:

$$: \overset{\bullet\bullet}{\underset{\bullet\bullet}{Cl}} \overset{\times}{} Be \overset{\times}{} \overset{\bullet\bullet}{\underset{\bullet\bullet}{Cl}} : \qquad Cl - Be - Cl$$
$$180°$$

The other group 2 chlorides are ionic, although magnesium chloride has some covalent character.

Hydrated chlorides
The tendency to form hydrated chlorides decreases down group 2. Magnesium and calcium chlorides form hydrated solids with six molecules of water of crystallisation, e.g. $MgCl_2.6H_2O$. When heated, these solids decompose because the chloride ions deprotonate water molecules. Hydrogen chloride gas is produced.
$$MgCl_2.6H_2O(s) \rightarrow Mg(OH)_2(s) + 4H_2O(g) + 2HCl(g)$$

ℯ It is a good idea to add state symbols to an equation when a gas or a precipitate is produced.

Summary

Moving down group 2 from beryllium to barium, the trend is for the following properties to *decrease*:

- melting temperature
- ionisation energy
- polarising power of the cation

The trend is for the following properties to *increase*:
- atomic and ionic radii
- reactivity of the metal
- thermal stability of the compounds

Titrations

If the accurate concentration of one solution is known, a titration can be used to determine the concentration of another solution. The method involves having one solution in a burette and pipetting a known volume of the other solution into a conical flask. An indicator is then added to the solution in the conical flask and the solution is titrated from the burette until the indicator changes colour.

At AS, most titrations involve acid versus base. An iodine titration is described in Chapter 11. Other types will be met at A2, but the method and the calculation are the same.

Detailed method

- The burette is rinsed out with distilled water and then with a little of one of the solutions (solution B).
- The pipette is rinsed out with distilled water and then with a little of the other solution (solution P).
- A conical flask is rinsed out with distilled water.
- The burette is filled with solution B. The tap is opened and some of the solution is run out so that the stem below the tap does not contain any air bubbles. The burette is read to 0.05 cm^3 and the volume is recorded.
- Using a pipette filler, the pipette is filled with solution P so that the bottom of the meniscus is on the line.
- The contents of the pipette are discharged into the conical flask. A few drops of indicator are added.
- The solution from the burette solution is added, with constant swirling of the flask, until the indicator changes colour. The burette is read and the difference between the starting volume and the final volume is calculated. This is called the **titre**. This titre is approximate because the burette solution was added quickly and it is, therefore, likely that the end point was overshot.

ANDREW LAMBERT PHOTOGRAPHY/SCIENCE PHOTO LIBRARY

Performing an acid–base titration

e Do not blow out the last drop of liquid from the pipette.

- The conical flask is emptied and washed out thoroughly with water.
- The process is repeated, but this time solution B is added quickly until about 1 cm^3 or 2 cm^3 short of the approximate titre.
- Solution B is then added slowly — finally dropwise — swirling the flask constantly, until the indicator changes to the required colour. At this point, the burette reading is taken and the first accurate titre calculated.
- The titration is repeated until at least two concordant titres are obtained. Concordant means that two titres are within 0.2 cm^3 of each other.
- The concordant titres are then averaged to give the mean (average) titre.

e Always read the volume using the bottom of the meniscus. All burette readings must be recorded to *two* decimal places (or to 0.05 cm^3).

Indicators

The common acid–base indicators are **methyl orange** and **phenolphthalein**. Some indicators and their colours are given in Table 10.6.

Indicator	Colour in acid	Colour in alkali	End-point colour
Methyl orange	Red	Yellow	Orange
Bromophenol blue	Yellow	Blue	Green
Methyl red	Red	Yellow	Orange
Bromothymol blue	Yellow	Blue	Green
Phenolphthalein	Colourless	Purple	Pale pink

Table 10.6 Some indicators and their colours

Calculations

To find the concentration of one of the solutions

The concentration of one solution is known. This solution is called the **standard solution**. The reacting volumes are determined by titration.

The calculation is carried out in three steps.

Step 1: calculate the amount (in moles) of the reagent of known concentration. Use **moles = concentration × volume**

Step 2: calculate the amount (in moles) of the other reagent. Use the stoichiometry of the equation.

Step 3: calculate the concentration of the other solution. Use $\textbf{concentration} = \dfrac{\textbf{moles}}{\textbf{volume}}$

e At AS, the equation for the reaction is normally given in the question. If it is not, you must write it before starting the calculation.

> *Worked example 1*
> In a titration, neutralisation of 25.0 cm^3 of 0.100 mol dm^{-3} hydrochloric acid solution required 26.8 cm^3 of sodium hydroxide solution.
> $$NaOH + HCl \rightarrow NaCl + H_2O$$
> Calculate the concentration of the sodium hydroxide.
>
> **Answer**
> **Step 1:** amount of HCl = concentration × volume
> $$\text{amount} = 0.100 \text{ mol dm}^{-3} \times \frac{25.0 \text{ dm}^3}{1000} = 0.00250 \text{ mol}$$
> **Step 2:** ratio of HCl to NaOH is 1:1
> amount of NaOH = amount (moles) of HCl = 0.00250 mol

The volumes have to be changed from cm^3 to dm^3. This is done by dividing by 1000.

Step 3: concentration of sodium hydroxide solution $= \dfrac{\text{moles}}{\text{volume}}$

$$\text{concentration} = \frac{0.00250 \text{ mol}}{\left(26.8/1000\right) \text{ dm}^3} = 0.0933 \text{ mol dm}^{-3}$$

'Amount' in chemistry is measured in moles.

Worked example 2

25.0 cm^3 of a 0.0504 mol dm^{-3} solution of sulfuric acid was titrated with a solution of sodium hydroxide. The mean titre was 27.3 cm^3. The equation for the reaction is:

$$2NaOH + H_2SO_4 \rightarrow Na_2SO_4 + 2H_2O$$

Calculate the concentration of the sodium hydroxide solution.

Answer

Step 1: amount of H_2SO_4 = concentration × volume

$$\text{amount} = 0.0504 \text{ mol dm}^{-3} \times \frac{25.0 \text{ dm}^3}{1000} = 0.00126 \text{ mol}$$

Care must be taken if the stoichiometric ratio is not 1:1.

Step 2: ratio of NaOH to H_2SO_4 is 2:1

amount of NaOH = 2 × amount of H_2SO_4

$= 2 \times 0.00126$

$= 0.00252$ mol

Step 3: concentration of sodium hydroxide solution $= \dfrac{\text{moles}}{\text{volume}}$

$$\text{concentration} = \frac{0.00252 \text{ mol}}{\left(27.3/1000\right) \text{ dm}^3} = 0.0923 \text{ mol dm}^{-3}$$

Some students like to use the formula

$$\frac{M_1 V_1}{n_1} = \frac{M_2 V_2}{n_2}$$

where n_1 is the number in front of the formula of substance 1 in the equation and n_2 is the number in front of the formula of substance 2 in the equation.

Units of concentration

The most common unit is mol dm^{-3}. However, other units are sometimes used, for example:

- nmol dm^{-3} — the letter 'n' here means that the unit is nanomoles per dm^3. As 1 nmol equals 1×10^{-9} mol, a solution of, for example, 23.4 nmol dm^{-3} is also of concentration of 2.34×10^{-8} mol dm^{-3}.
- g dm^{-3} — this is obtained from a concentration in mol dm^{-3} by multiplying by the molar mass of the solute.
- ppm (parts per million) — this is the mass of solute in grams in one million grams of solution.

Worked example 1

Calculate the concentration in g dm^{-3} of a 0.152 mol dm^{-3} sodium hydroxide solution.

Answer

molar mass of sodium hydroxide = (23.0 + 16.0 + 1.0) = 40.0 g mol^{-1}

concentration = 0.152 mol dm^{-3} × 40.0 g mol^{-1} = 6.08 g dm^{-3}

Worked example 2

What is the concentration in ppm of a solution containing 45 ng (nanograms) of a toxic chemical in 100 cm^3 of water?

Answer

There are 45 ng in 100 g water, which is 4.5×10^{-8} g in 100 g water.

Therefore, there are $\dfrac{4.5 \times 10^{-8} \times 10^{6}}{100} = 4.5 \times 10^{-4}$ g in 1 million grams of water.

The concentration of solute is 0.00045 ppm.

To find the volume needed

In this type of calculation, both concentrations are given in the question. The first two steps are the same as in calculations to find the concentration of a solution. The third step is to use the concentration of the second solution to work out the volume required:

$$\text{volume} = \frac{\text{moles}}{\text{concentration}}$$

Worked example

Calculate the volume of 0.100 mol dm^{-3} hydrochloric acid solution required to neutralise 25.0 cm^3 of 0.0567 mol dm^{-3} sodium carbonate solution. The equation for the reaction is:

$$2HCl + Na_2CO_3 \rightarrow 2NaCl + H_2O + CO_2$$

Answer

Step 1: amount of sodium carbonate = concentration × volume

$$\text{amount} = 0.0567 \text{ mol dm}^{-3} \times \frac{25.0 \text{ dm}^3}{1000} = 0.001418 \text{ mol}$$

Step 2: ratio of HCl to Na$_2$CO$_3$ = 2:1

amount of hydrochloric acid = 2 × 0.001418 = 0.002836 mol

Step 3: volume of hydrochloric acid solution = $\dfrac{\text{moles}}{\text{concentration}}$

$$\text{volume} = \frac{0.002836 \text{ mol}}{0.100 \text{ mol dm}^{-3}} = 0.0284 \text{ dm}^3 = 28.4 \text{ cm}^3$$

Back titrations

A back titration is used when the substance being investigated is either insoluble or, for some other reason, cannot be titrated directly.

The method is as follows:

- Weigh a sample of the substance being investigated.
- Add excess of a standard solution of a substance, usually an acid or a base.
- Either titrate the excess or make up the solution to 250 cm^3 and titrate portions of the diluted solution that contain the excess.

An example is the determination of the purity of a sample of chalk, which is impure calcium carbonate. The procedure is as follows:

- Weigh a sample of impure chalk and place it in a beaker.

- Add 50 cm^3 of 1.00 mol dm^{-3} hydrochloric acid solution. This amount of acid must be enough to react with all the calcium carbonate, with some acid in excess.
- Use a funnel to pour the solution obtained (which contains the excess acid) into a standard 250 cm^3 flask. Wash the beaker and the funnel into the flask and make up to the mark with distilled water. Shake the standard flask thoroughly.
- Titrate 25.0 cm^3 portions of this diluted solution against a solution of sodium hydroxide of known concentration.
- Repeat the titration until two consistent titres have been obtained.

The calculation starts with the sodium hydroxide needed in the titration and comprises three stages:

- mean titre → moles NaOH → moles of excess HCl in 25 cm^3 portion → total moles of excess HCl
- original volume of HCl → moles of HCl taken
- (original moles of HCl) – (total moles of excess HCl) = moles HCl reacted with CaCO$_3$ → moles CaCO$_3$ → mass of CaCO$_3$ in the chalk sample → % purity

℮ If this type of calculation is asked at AS, the question will be structured so that you are guided through the many steps.

Worked example

1.41 g of a sample of chalk, which is mostly calcium carbonate with some inert impurities, was placed in a beaker and 50.0 cm^3 of 1.00 mol dm^{-3} hydrochloric acid solution was slowly added. The equation for the reaction is:

$$CaCO_3 + 2HCl \rightarrow CaCl_2 + H_2O + CO_2$$

When the fizzing had ceased, the contents of the beaker were washed into a standard 250 cm^3 volumetric flask, made up to the mark and thoroughly shaken.

A pipette was used to transfer a 25.0 cm^3 sample into a conical flask. The sample was titrated against 0.100 mol dm^{-3} sodium hydroxide solution. The equation for the reaction is:

$$HCl + NaOH \rightarrow NaCl + H_2O$$

The titration was repeated. The mean titre was 23.6 cm^3.

a Calculate the amount of sodium hydroxide in the mean titre.
b Calculate the total amount of hydrochloric acid in excess.
c Calculate the amount of hydrochloric acid originally taken.
d Use your answers to (b) and (c) to calculate the amount of hydrochloric acid that reacted with the chalk.
e Hence, calculate the amount of calcium carbonate in the chalk sample.
f Calculate the mass of calcium carbonate in the sample and hence the % purity.

Answer

a amount of NaOH in mean titre = concentration × volume

$$\text{amount} = 0.100 \text{ mol dm}^{-3} \times \frac{23.6 \text{ dm}^3}{1000} = 0.00236 \text{ mol}$$

b ratio of HCl to NaOH = 1:1

amount of HCl in 25 cm^3 = 0.00236 mol

amount of HCl in 250 cm^3 = 10 × 0.00236 = 0.0236 mol = amount of excess HCl

c amount of HCl taken = concentration × volume

$$\text{amount} = 1.00 \text{ mol dm}^{-3} \times \frac{50.0 \text{ dm}^3}{1000} = 0.0500 \text{ mol}$$

d amount HCl reacted with CaCO$_3$ = amount taken − amount of excess

amount = 0.0500 mol − 0.0236 mol = 0.0264 mol

e ratio of CaCO$_3$ to HCl = 1:2

amount of CaCO$_3$ in sample = $\frac{1}{2}$ × 0.0264 = 0.0132 mol

f molar mass of CaCO$_3$ = 40.0 + 12.0 + (3 × 16.0) = 100.0 g mol^{-1}

mass of CaCO$_3$ in sample = moles × molar mass

mass of CaCO$_3$ = 0.0132 mol × 100.0 g mol^{-1} = 1.32 g

$$\% \text{ purity} = \frac{1.32 \text{ g}}{1.41 \text{ g}} \times 100 = 93.6\%$$

To find the stoichiometry of a reaction

In questions that ask for the determination of the stoichiometry of a reaction, the concentrations and volumes of both solutions will be given.

The calculation is performed in three steps.

Step 1: calculate the amount (moles) of one reagent.

Step 2: calculate the amount (moles) of the other reagent.

Step 3: find the whole number ratio of these amounts. This is the reactant ratio, which allows the equation to be written.

> ### Worked example
> 25.0 cm^3 of a 0.100 mol dm^{-3} solution of sodium hydroxide required 25.8 cm^3 of a 0.0485 mol dm^{-3} solution of phosphoric(III) acid, H$_3$PO$_3$. Work out the equation for this acid + base reaction.
>
> **Answer**
> **Step 1:** amount of sodium hydroxide = concentration × volume
>
> $$\text{amount} = 0.100 \text{ mol dm}^{-3} \times \frac{25.0 \text{ dm}^3}{1000} = 0.00250 \text{ mol}$$
>
> **Step 2:** amount of phosphoric(III) acid = concentration × volume
>
> $$\text{amount} = 0.0485 \text{ mol dm}^{-3} \times \frac{25.8 \text{ dm}^3}{1000} = 0.001251 \text{ mol}$$
>
> **Step 3:** molar ratio NaOH to H$_3$PO$_3$ = $\dfrac{0.00250}{0.001251}$ = 2:1
>
> 2 mol of NaOH react with 1 mol of H$_3$PO$_3$, so the equation is:
>
> 2NaOH + H$_3$PO$_3$ → Na$_2$HPO$_3$ + 2H$_2$O

Only two of the hydrogen atoms in H$_3$PO$_3$ are replaceable by metals. This is because only two are bonded to oxygen atoms; the third is bonded to the phosphorus atom and is not acidic.

Preparation of a standard solution

A standard solution is one in which the concentration is accurately known. $250 \, cm^3$ of a standard solution is made up as follows:

- From the molar mass and the concentration required, calculate the mass of substance needed.
- Weigh out this mass and pour it into a beaker, washing all the solid into the beaker.
- Dissolve the solid in a small amount of warm water and pour the solution into a standard $250 \, cm^3$ volumetric flask.
- Wash the contents of the beaker into the standard flask and add water up to the mark.
- Shake the flask thoroughly.

The mass is obtained by first calculating the amount needed in moles.

> ### Worked example
> Calculate the mass of sodium carbonate decahydrate, $Na_2CO_3.10H_2O$, needed to make up $250 \, cm^3$ of a $0.0500 \, mol \, dm^{-3}$ solution.
>
> **Answer**
> molar mass of $Na_2CO_3.10H_2O = (2 \times 23.0) + 12.0 + (3 \times 16.0) + (10 \times 18.0)$
> $$= 286.0 \, g \, mol^{-1}$$
> amount (moles) needed = concentration $\times$ volume
> $$amount = 0.0500 \, mol \, dm^{-3} \times \frac{250 \, dm^3}{1000} = 0.0125 \, mol$$
> mass needed = moles $\times$ molar mass
> mass $= 0.0125 \, mol \times 286.0 \, g \, mol^{-1} = 3.58 \, g$

e The solution is made up to a total volume of $250 \, cm^3$, rather than by adding $250 \, cm^3$ of water to the solid. The volume of a solution is not the same as the volume of solvent used to make the solution.

Minimising errors in titrations

Errors because of the apparatus

Burettes and pipettes measure volumes to certain degrees of accuracy. In most cases, the error in each measurement is $\pm 0.05 \, cm^3$. This means that the maximum possible error when calculating the titre is $2 \times 0.05 = 0.1 \, cm^3$. In a titre of $24.00 \, cm^3$, this is a percentage error of $0.1 \times 100/24.00 = 0.42\%$. If the titre had been $12.00 \, cm^3$, the error would be the same, but the percentage error would be $0.1 \times 100/12.00 = 0.83\%$, which is twice as large.

Titrations are planned to give a titre between $20 \, cm^3$ and $30 \, cm^3$ in order to minimise apparatus error.

Errors because of poor technique

Practice and care will reduce errors due to poor technique. Common mistakes include:

- not rinsing out the burette and the pipette with the correct solutions
- not ensuring that there is no air below the tap in the burette
- getting air bubbles in the stem of the pipette
- running in the solution from the burette too quickly and overshooting the end point
- not swirling the flask after each addition of solution from the burette
- not shaking the solution in the standard flask thoroughly

Tests for group 1 and group 2 cations

All compounds of the group 1 elements are soluble in water, as are many compounds of group 2 elements. Therefore, the identity of the cation cannot usually be detected by a precipitation reaction. However, apart from magnesium, they all colour a Bunsen flame.

The flame test

First the compound has to be converted into a chloride because chlorides are more volatile than other types of compound. The procedure is as follows:

- Take a platinum or nichrome wire or a specially marketed flame test rod. Do *not* use a spatula or a wooden splint.
- Check that it is clean by dipping the wire in some concentrated hydrochloric acid on a watch glass and then placing it in the hottest part of a Bunsen flame. The flame should not be coloured. If it is, repeat the treatment until the flame is not coloured.
- Once again, dip the wire in concentrated hydrochloric acid and then into some of the solid under test.
- Place this in the hottest part of the flame and observe the colour of the flame.

The colour of the flame identifies the cation present (Figure 10.2).

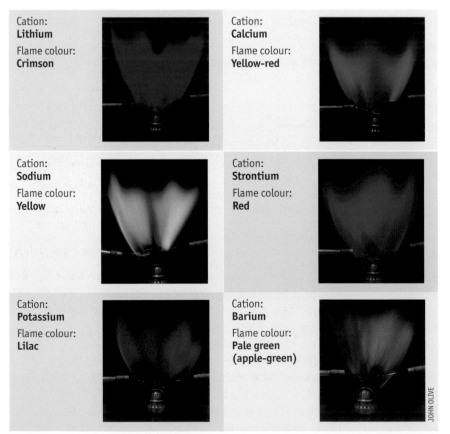

Figure 10.2 Flame test colours

Cation:
Lithium

Flame colour:
Crimson

Cation:
Calcium

Flame colour:
Yellow-red

Cation:
Sodium

Flame colour:
Yellow

Cation:
Strontium

Flame colour:
Red

Cation:
Potassium

Flame colour:
Lilac

Cation:
Barium

Flame colour:
Pale green (apple-green)

JOHN OLIVE

Explanation of flame colour

The heat energy of the flame causes decomposition and promotes an electron in each metal ion into an excited state. This means that the electron is not in the lowest available orbital, but in a higher one. This excited state is not stable, so the electron drops down to the ground state (the lowest available energy level). The energy released is given off in the form of visible light (Figure 10.3). The gaps between the energy levels in different cations are not the same, so the amount of energy, and hence the colour of the light emitted, varies from element to element.

The flame colour is related to the *emission* spectrum. Heat from the Bunsen flame is absorbed, an electron is promoted to a higher energy level, and then light energy is given out as the electron drops down to the ground state. The result is that heat energy is converted to light energy.

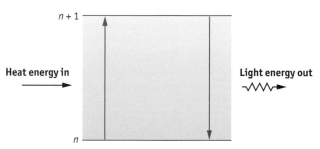

Figure 10.3 *Electronic transitions on heating in a flame*

Questions

1 Define the term 'first ionisation energy'. Write an equation that represents the first ionisation of calcium.

2 Explain why the first ionisation energies decrease down group 2 from beryllium to barium.

3 Explain why the first ionisation energy of calcium is greater than that of potassium.

4 Explain why there is a big jump between the second and third ionisation energies for magnesium.

5 Write an equation for the following reactions:
   a magnesium and oxygen
   b barium and water
   c magnesium and chlorine
   d calcium oxide and nitric acid

6 State and explain the trend in the thermal stability of the group 1 nitrates.

7 $25.0\,cm^3$ of $0.0561\,mol\,dm^{-3}$ sulfuric acid solution was neutralised by $23.45\,cm^3$ of sodium hydroxide solution. The equation for the reaction is:

$$H_2SO_4 + 2NaOH \rightarrow Na_2SO_4 + 2H_2O$$

Calculate the concentration of the sodium hydroxide solution in:
   a $mol\,dm^{-3}$
   b $g\,dm^{-3}$

8 $3.05\,g$ of hydrated ethanedioic acid, $H_2C_2O_4.2H_2O$, was weighed out, dissolved in water and made up to $250\,cm^3$ in a standard flask. Calculate the concentration of the solution in $mol\,dm^{-3}$.

9 Calculate the volume of $0.100\,mol\,dm^{-3}$ hydrochloric acid needed to neutralise $25.0\,cm^3$ of $0.0525\,mol\,dm^{-3}$ sodium carbonate solution. The equation for the reaction is:

$$Na_2CO_3 + 2HCl \rightarrow 2NaCl + H_2O + CO_2$$

10 The skin of peaches can be loosened by standing them in a solution of potassium hydroxide. After a short time, the skins can be removed easily by machine. The skinned peaches are then washed to remove excess potassium hydroxide. The final washings are tested to see how much potassium hydroxide remains. $100\,cm^3$ of the final washings was titrated with $0.0050\,mol\,dm^{-3}$ hydrochloric acid. The mean titre was $7.55\,cm^3$. Calculate the concentration of potassium hydroxide in the final washings in parts per million.

11 The composition by mass of a barium compound was: barium 81.1%; oxygen 18.9%.
   a Calculate the empirical formula of this compound.
   b The molar mass of this compound is $169.3\,g\,mol^{-1}$. Calculate its molecular formula.

12 a State what is observed when strontium nitrate is heated.
   b Write the equation for the thermal decomposition of strontium nitrate.
   c Calculate the volume of gas produced when 2.12 g of strontium nitrate is heated. Assume that, under the conditions of the experiment, 1 mol of gas occupies 24.0 dm^3.

13 Explain why magnesium carbonate decomposes at a lower temperature than calcium carbonate.

14 25.0 cm^3 of barium hydroxide solution were neutralised by 17.6 cm^3 of 0.100 mol dm^{-3} nitric acid solution:

$$Ba(OH)_2 + 2HNO_3 \rightarrow Ba(NO_3)_2 + 2H_2O$$

Calculate the concentration of the barium hydroxide solution.

15 What volume of 0.202 mol dm^{-3} sodium hydroxide reacts completely with 25.0 cm^3 of 0.100 mol dm^{-3} phosphoric(v) acid solution? The equation is:

$$3NaOH + H_3PO_4 \rightarrow Na_3PO_4 + 3H_2O$$

16 25 ibuprofen tablets were reacted with 50.0 cm^3 of 1.00 mol dm^{-3} sodium hydroxide solution.

The reaction is:

$$C_{12}H_{17}COOH + NaOH \rightarrow C_{12}H_{17}COONa + H_2O$$

When the reaction was over, the solution, containing excess sodium hydroxide, was made up to 250 cm^3 with distilled water. 25.0 cm^3 samples were titrated against 0.110 mol dm^{-3} hydrochloric acid solution:

$$NaOH + HCl \rightarrow NaCl + H_2O$$

The mean titre was 23.35 cm^3.
   a Calculate the amount (in moles) of hydrochloric acid used in the titration.
   b Calculate the amount (in moles) of excess sodium hydroxide in a 25.0 cm^3 sample.
   c Calculate the amount (in moles) of total excess sodium hydroxide.
   d Calculate the amount (in moles) of sodium hydroxide originally taken.
   e Use your answers to (c) and (d) to calculate the amount (in moles) of sodium hydroxide that reacted with the 25 ibuprofen tablets.
   f Hence, calculate the amount (in moles) of ibuprofen in 25 tablets.
   g Calculate the mass, in mg, of ibuprofen in one tablet.

Chapter 11

The periodic table: group 7

Sodium: salt, killer and fraudster

Atomic number: 11

Electron configuration: $1s^2\ 2s^2\ 2p^6\ 3s^1$

Symbol: Na

Sodium is an alkali metal, in group 1 of the periodic table.

Salt is sodium chloride — the only food humans eat regularly that is not derived from animal or plant constituents or products. When our tongues come into contact with salt, the brain receives a message that something unnatural is being eaten. The brain orders the nose to investigate and the odour receptors that control our sense of taste are reset to a heightened state of sensitivity. We now perceive even the most bland food as interesting and flavoursome.

However, a recent report from the Department of Health claimed that many foodstuffs are saltier today than 25 years ago. A typical supermarket chicken Caesar sandwich was found to contain 5.9 g of salt — almost the recommended daily amount for an adult. Salt is estimated to contribute to the deaths of around 35 000 Britons a year, mostly as a result of heart attacks and strokes brought on by high blood pressure.

Salt has been used to clean chimneys, weld pipes, glaze pottery and treat toothache. It was recognised as a valuable commodity in China 4000 years ago. The ancient Egyptians used it in mummification. Roman legions were paid monthly in salt, giving us the word 'salary' from the Latin, *sal*. By the Middle Ages, caravans of camels were crossing the Sahara bearing salt as trade.

The discovery that salt could preserve food made the early journeys of exploration possible.

In eighteenth-century France, the *gabelle* (salt tax) fuelled the resentment of the masses and was a link in the chain of events that culminated in the Revolution. In India, the British Raj's harsh control of the pricing and manufacture of salt led to Ghandi's 1930 'Salt March' and hastened the move to independence.

Is salt really harmful? An excess of salt does cause the body to retain fluid, putting extra strain on the heart. However, salt is vital for life; without it we would die. The human body cannot make salt, so we *have* to rely on external sources. Millions of us associate saltiness with taste, which is why there are 5 g of salt in a can of baked beans, 500 mg in a slice of white bread, 750 mg in a bowl of cornflakes and over 3 g in a fast-food meal.

Consumers like the taste and the industry likes the business…which leaves the Department of Health with the job of casting sodium into odium.

(Adapted from an article in the *Sunday Telegraph*, 20 June 2004)

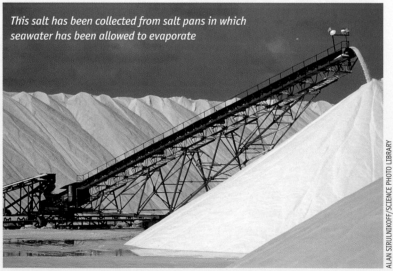

This salt has been collected from salt pans in which seawater has been allowed to evaporate

ALAN SIRULNIKOFF/SCIENCE PHOTO LIBRARY

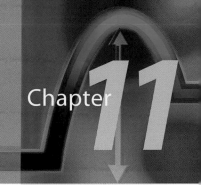

The periodic table: group 7

The halogens and iodine titrations

The elements in group 7 are called the **halogens.** This word is derived from the Greek for salt maker — halogens react with metals to form salts.

The elements are:

- fluorine, F
- chlorine, Cl
- bromine, Br
- iodine, I
- astatine, At

Astatine is a radioactive element with a half-life of 8.3 hours.

Physical properties

At room temperature and pressure, fluorine and chlorine are gases, bromine is a volatile fuming liquid and iodine is a solid that sublimes on heating.

	Fluorine	Chlorine	Bromine	Iodine
Colour	Pale yellow gas	Greenish gas	Brown liquid	Dark grey-black solid which forms a violet vapour
Electron configuration	[He] $2s^2\,2p^5$	[Ne] $3s^2\,3p^5$	[Ar] $3d^{10}\,4s^2\,4p^5$	[Kr] $4d^{10}\,5s^2\,5p^5$
Atomic radius/nm	0.07	0.10	0.11	0.13
Ionic radius/nm	0.14	0.18	0.20	0.22
First ionisation energy/kJ mol^{-1}	1680	1260	1140	1010
First electron affinity/kJ mol^{-1}	−348	−364	−342	−314
Electronegativity	4.0	3.0	2.8	2.5

Table 11.1 Physical properties of the halogens

JOHN OLIVE

e Make sure that you know the colour and physical states of chlorine, bromine and iodine. They are often asked for in Unit Test 2.

First ionisation energy

The first ionisation energy of the halogens decreases down the group. The nuclear charge increases, but the number of inner shielding electrons increases by the same amount. The effective nuclear charge therefore remains approximately the same. The atomic radius increases as new shells are added, so the outermost electrons become less strongly held. This is why, from fluorine to iodine, the first ionisation energy decreases.

First electron affinity

Energy is released when a negative electron is brought from outside the atom towards the positive nucleus and into the outer orbit. The trend is a decrease in electron affinity down the group. This is because the electron is not brought as close to the nucleus in large atoms as it is in smaller atoms. Therefore, less energy is released when a large atom gains an electron than when a smaller atom gains an electron.

Fluorine is an exception. This is because the atom is so small that the repulsion between the in-coming electron and the seven electrons in the second shell reduces the energy liberated by the attraction between the incoming electron and the nucleus.

Solubility and colour of solutions

- Chlorine and bromine are fairly soluble in water. This is because they react reversibly with water to form a mixture of acids — for example:

$$Cl_2 + H_2O \rightarrow HCl + HOCl$$

  Aqueous solutions of chlorine and bromine are called chlorine water and bromine water, respectively. Chlorine water is pale green and bromine water is brown-red.
- Iodine is only very slightly soluble in water. Solutions are pale brown, unless potassium iodide is present, in which case the solution is deep brown-red. This colour is due to the I_3^- ion:

$$I_2(aq) + I^-(aq) \rightarrow I_3^-(aq)$$

- All the halogens are more soluble in inert organic solvents such as carbon tetrachloride or hexane than in water. This means that if hexane is added to an aqueous solution containing a halogen, the coloured halogen is concentrated in the organic layer. A solution of iodine in hexane is violet.

Bonding in compounds of halogens

Ionic bonding

All the halogens form anions with a 1– charge. This is because they all have high electronegativity values and, when a halogen atom accepts an electron, the electron configuration of a noble gas is achieved.

Bonding with group 1 and group 2 metals

Group 1 and group 2 metal halides are ionic — for example, calcium chloride:

$$\left[:\!\overset{\cdot\cdot}{\underset{\cdot\cdot}{Cl}}\!\!\underset{\times}{\times} \right]^{-} \quad \left[\underset{\times\times}{\overset{\times\times}{\times\, Ca\, \times}} \right]^{2+} \quad \left[\overset{\cdot\cdot}{\underset{\cdot\cdot}{\times\, Cl\, :}} \right]^{-}$$

Bonding with aluminium (group 3)

- Aluminium fluoride, AlF_3, is ionic.
- Anhydrous aluminium chloride is covalent, but the hydrated chloride, $AlCl_3.6H_2O$, is ionic. Each aluminium ion is surrounded by six water molecules, with dative covalent bonds from the oxygen to the central Al^{3+} ion.
- Anhydrous aluminium chloride sublimes on heating, producing gaseous molecules of formula Al_2Cl_6. In this dimer, the two $AlCl_3$ units are joined by two dative covalent bonds, each being from a chlorine atom in one $AlCl_3$

molecule to the aluminium atom in the other molecule. In this way, the outer orbit of both aluminium atoms contains an octet of electrons:

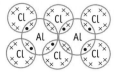

> **e** The aluminium atom in covalent $AlCl_3$ has only six electrons in its outer orbit (its own three plus one from each covalent bond). A lone pair from a chlorine atom in the other $AlCl_3$ molecule forms a dative covalent bond using an empty $3p$ orbital of the aluminium. As a result, the aluminium now has four pairs of electrons in its outer orbit. Therefore, the four chlorine atoms are arranged tetrahedrally around each aluminium. There are no bonds between the two aluminium atoms — drawing a bond here is a very common error in AS answers.

Bonding with d-block metals

The chlorides, bromides and iodides of *d*-block metals are covalent when anhydrous and ionic when hydrated. For example, anhydrous $FeCl_3$ is covalently bonded and is soluble in several organic solvents. The hydrated ion has six molecules of water bonded with dative covalent bonds from the oxygen in the water to the empty $3d$- $4s$- and $4p$-orbitals of the iron(III) ion. The formula of the hydrated ion is $[Fe(H_2O)_6]^{3+}$ and its shape is octahedral:

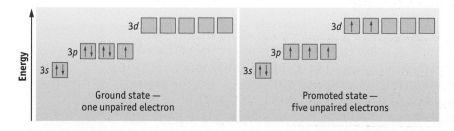

Covalent bonding

The halogens form polar covalent bonds with almost all non-metals.

- Fluorine is the most electronegative of all the elements and so is always in the −1 oxidation state in compounds.
- Chlorine is the second most electronegative of the group 7 elements. In compounds it is in the −1 oxidation state, apart from when it is bonded to fluorine or oxygen.

The halogens, other than fluorine, have empty *d*-orbitals in the valence shell. For example, chlorine has five empty $3d$-orbitals. Therefore, one or more of the $3p$-electrons can be promoted into these energetically similar energy levels. This enables chlorine to form more than one covalent bond (Figure 11.1).

Figure 11.1 Electron promotion before bonding

This only happens when the bond energy released compensates for the energy required to promote the electrons, i.e. when it is bonded to a small electronegative atom such as oxygen:

- Chlorine forms chloric(I) acid (hypochlorous acid), HOCl, and chlorate(I) salts, which contain the ClO^- ion. In these ions, chlorine is in the +1 oxidation state.
- Chlorine also forms chlorate(V) salts, which contain the ClO_3^- ion. In these ions, the chlorine is in the +5 oxidation state. There are five covalent bonds between chlorine and the three oxygen atoms — two double bonds and one single bond.

Bromine and iodine form similar salts in which they are in the +1 and +5 oxidation states.

Reactions of the halogens

- The reactivity of the halogens decreases down the group. (Conversely, the reactivity of many of their compounds *increases* down the group.)
- The halogens are oxidising agents and can remove an electron from many substances.
- The ionic half-equation for chlorine acting as an oxidising agent is:
    $$Cl_2 + 2e^- \rightarrow 2Cl^-$$
- The oxidising power of the halogens decreases down the group.

Reaction with metals

On heating, metals react with halogens to form halides in which the halogen is in the –1 oxidation state. For example, chlorine oxidises iron to iron(III) chloride:
$$2Fe + 3Cl_2 \rightarrow 2FeCl_3$$
Iodine is a less powerful oxidising agent than chlorine. Iodine oxidises iron to iron(II) iodide:
$$Fe + I_2 \rightarrow FeI_2$$

Reaction with phosphorus

- In a limited supply of chlorine, phosphorus trichloride is formed:
    $$2P + 3Cl_2 \rightarrow 2PCl_3$$
    Phosphorus trichloride can then react with more chlorine to form phosphorus pentachloride:
    $$PCl_3 + Cl_2 \rightarrow PCl_5$$
- In excess chlorine, phosphorus pentachloride is formed:
    $$2P + 5Cl_2 \rightarrow 2PCl_5$$
- Iodine reacts with the red allotrope of phosphorus to form phosphorus triiodide:
    $$2P + 3I_2 \rightarrow 2PI_3$$
    If the mixture is damp, this then reacts with water to form hydrogen iodide gas:
    $$PI_3 + 3H_2O \rightarrow 3HI + H_3PO_3$$
    This is how gaseous hydrogen iodide is prepared.

Reactions with solutions of other halides

A halogen will displace a less reactive halogen from one of its salts:

- When chlorine is bubbled into a solution of potassium bromide, the chlorine oxidises the bromide ions. A brown colour of liberated bromine is observed:
$$Cl_2(g) + 2Br^-(aq) \rightarrow Br_2(aq) + 2Cl^-(aq)$$
The chlorine molecule gains two electrons — one from each bromide ion. The process can be described using half-equations:
$$2Br^-(aq) \rightarrow Br_2(aq) + 2e^-$$
$$Cl_2(g) + 2e^- \rightarrow 2Cl^-(aq)$$
- Chlorine displaces iodine from iodides, forming a dark grey precipitate of iodine:
$$Cl_2(g) + 2I^-(aq) \rightarrow I_2(s) + 2Cl^-(aq)$$
- Bromine also displaces iodine from iodides:
$$Br_2(aq) + 2I^-(aq) \rightarrow I_2(s) + 2Br^-(aq)$$

The displacement of bromine by chlorine is used in the extraction of bromine from seawater, which contains a small amount of dissolved bromide ions. Chlorine gas is bubbled into seawater and the bromine produced is removed by blowing air through the solution. On cooling, the bromine vapour condenses and is collected.

Reaction with water

- Chlorine reacts reversibly with water to form an acidic solution containing hydrochloric acid and the weak acid, chloric(I) acid. The solution is acidic because the hydrochloric acid is totally ionised into H^+ and Cl^- ions:
$$Cl_2(aq) + H_2O(l) \rightarrow H^+(aq) + Cl^-(aq) + HOCl(aq)$$

> ℮ This is an example of a disproportionation reaction. The chlorine (Cl_2) starts in the zero oxidation state and is reduced to –1 in the Cl^- ion and oxidised to +1 in chloric(I) acid, HOCl. A single species, chlorine, is simultaneously oxidised and reduced.

- Bromine reacts in a similar way, but the position of equilibrium lies more to the left:
$$Br_2(aq) + H_2O(l) \rightleftharpoons H^+(aq) + Br^-(aq) + HOBr(aq)$$
- Iodine does not react with water.

Reaction with alkali

With alkali at room temperature

Chlorine, bromine and iodine all react with *cold* aqueous sodium hydroxide by disproportionation reactions to produce a mixture of halide and halate(I) salts — for example, with chlorine:
$$Cl_2 + 2NaOH \rightarrow NaCl + NaOCl + H_2O$$
$$0 \qquad\qquad -1 \quad\; +1$$

The chlorine in the zero state is reduced to –1 in NaCl and oxidised to +1 in NaOCl (sodium chlorate(I)).

The ionic equation is:
$$Cl_2(aq) + 2OH^-(aq) \rightarrow Cl^-(aq) + OCl^-(aq) + H_2O(l)$$

℮ Do not say that a violet colour is observed. Iodine is only violet when it is either a gas or is dissolved in an inert organic solvent such as hexane or carbon tetrachloride.

◄ This displacement reaction illustrates the decrease in oxidising power from chlorine to iodine.

Fluorine reacts irreversibly with water to form hydrogen fluoride and oxygen.

If a solution containing chlorate(I) ions is heated, further disproportionation takes place and chlorate(V) ions are formed:

$$3OCl^-(aq) \rightarrow 2Cl^-(aq) + ClO_3^-(aq)$$
$$+1 \qquad\quad -1 \qquad\quad +5$$

The +1 chlorine in OCl^- is reduced to -1 in Cl^- and oxidised to +5 in ClO_3^-.

With hot alkali

This is also a disproportionation reaction. However, the halogen is oxidised to the +5 state in the chlorate(V) ion, ClO_3^-, rather than the +1 state, as is the case with cold alkali:

$$3Cl_2(aq) + 6OH^-(aq) \rightarrow 5Cl^-(aq) + ClO_3^-(aq) + 3H_2O(l)$$
$$0 \qquad\qquad\qquad\qquad\quad -1 \qquad\quad +5$$

> **e** Note that the oxidation number of chlorine rises by 5, so there must be five chlorine atoms being reduced from 0 to −1 to balance the change in oxidation number of the chlorine from 0 to +5. There must be six OH^- ions on the left-hand side of the equation because there are six negative ions on the right and the equation must balance for charge.

Bromine and iodine react similarly. The molecular equation for iodine reacting with hot aqueous sodium hydroxide is:

$$3I_2 + 6NaOH \rightarrow 5NaI + NaIO_3 + 3H_2O$$

The atom economy for forming sodium iodate(V) from iodine is poor, being just 16.7%.

Reactions with reducing agents

Iron(II) compounds are used here as examples of reducing agents.

Chlorine and bromine oxidise iron(II) ions to iron(III) ions. The two half-equations for oxidation by chlorine are:

$$Cl_2(aq) + 2e^- \rightarrow 2Cl^-(aq)$$
$$Fe^{2+}(aq) \rightarrow Fe^{3+}(aq) + e^-$$

The second half-equation has to be multiplied by 2 and then added to the first equation to give the overall equation:

$$Cl_2(aq) + 2Fe^{2+}(aq) \rightarrow 2Cl^-(aq) + 2Fe^{3+}(aq)$$

Iodine is too weak an oxidising agent to oxidise iron(II) to iron(III). The reverse happens. Iodide ions are oxidised by Fe^{3+} ions to iodine:

$$2I^-(aq) + 2Fe^{3+}(aq) \rightarrow I_2(s) + 2Fe^{2+}(aq)$$

> Note that the overall equation balances for charge. The charge on the left-hand side is 4+, as is the charge on the right-hand side: $(2-) + (6+) = 4+$.

The colourless solution containing iodide ions becomes a pale brown solution with a precipitate of iodine. Iodine can be tested for by the addition of starch. An intense blue, almost black, colour is obtained.

Reaction with sodium thiosulfate

The oxidation number, y, of sulfur in sodium thiosulfate ($Na_2S_2O_3$) is given by:

$$2 \times (+1) + 2y + 3 \times (-2) = 0$$
$$2y = 6 - 2 = +4$$

The oxidation number of each sulfur atom $= y = +2$.

With chlorine and bromine

Chlorine and bromine oxidise the sulfur in thiosulfate to the +6 state in sulfate, SO_4^{2-}. The equation with the oxidation numbers underneath is:

$$4Cl_2(aq) + S_2O_3^{2-}(aq) + 5H_2O(l) \rightarrow 8Cl^-(aq) + 2SO_4^{2-}(aq) + 10H^+(aq)$$
$$\quad 0 \qquad 2 \times (+2) \qquad\qquad\quad 8 \times (-1) \quad 2 \times (+6)$$

The total change in oxidation number of sulfur is from +4 to +12 (a change of +8), so eight chlorine atoms or four Cl_2 molecules are needed.

With iodine

Iodine oxidises thiosulfate ions to tetrathionate ions, $S_4O_6^{2-}$, in which the oxidation state of sulfur is 2.5. This is a change of 0.5 per sulfur atom. The equation is

$$I_2(aq) + 2S_2O_3^{2-}(aq) \rightarrow 2I^-(aq) + S_4O_6^{2-}(aq)$$

Iodine titrations

The reaction between iodine and thiosulfate ions can be used to find the concentration of a solution of an oxidising agent. The method is as follows:

- Pipette a known volume of the solution of the oxidising agent into a conical flask. Usually it is necessary to add some dilute acid at this stage, as H^+ ions are often present in their half-equations.
- Add excess potassium iodide solution (or solid potassium iodide) and swirl to mix the solution.
- Titrate the liberated iodine against standard sodium thiosulfate solution until the iodine colour fades to a pale straw colour.
- Add some starch indicator. The remaining unreacted iodine will react reversibly with the starch to form a blue-black complex.
- Continue adding sodium thiosulfate drop by drop until the blue-black colour disappears.
- Repeat the procedure until at least two concordant titres have been obtained.

> **e** This is an important equation for the quantitative estimation of iodine and, therefore, of oxidising agents.

> The starch must not be added too early, as it will then irreversibly form a solid starch-iodine complex.

Worked example

1.00 g of an impure sample of potassium iodate(v), KIO_3, was dissolved in distilled water and the solution made up to 250 cm^3.

25.0 cm^3 portions of this solution were taken and about 25 cm^3 of dilute sulfuric acid and excess potassium iodide were added. Each was titrated with standard 0.100 mol dm^{-3} sodium thiosulfate solution. The mean titre was 23.75 cm^3.

Calculate the purity of the potassium iodate(v).

The equations for the reactions are:

$$KIO_3 + 5KI + 3H_2SO_4 \rightarrow 3I_2 + 3K_2SO_4 + 3H_2O$$
$$I_2 + 2Na_2S_2O_3 \rightarrow 2NaI + Na_2S_4O_6$$

Answer

amount (in moles) of sodium thiosulfate, $Na_2S_2O_3$

$$= 0.100 \, mol \, dm^{-3} \times 0.02375 \, dm^3$$
$$= 2.375 \times 10^{-3} \, mol$$

1 mol KIO_3 produces 3 mol I_2, which require 6 mol $Na_2S_2O_3$

amount (in moles) of potassium iodate(V) in 25 cm^3 = $\dfrac{1}{6} \times 2.375 \times 10^{-3}$

$$= 3.958 \times 10^{-4} \, mol$$

amount (in moles) of potassium iodate(V) in 250 cm^3 = $10 \times 3.958 \times 10^{-4}$

$$= 3.958 \times 10^{-3} \, mol$$

molar mass of KIO_3 = 39.1 + 126.9 + (3 × 16.0) = 214.0 g mol^{-1}

mass of pure KIO_3 in sample = moles × molar mass

$$= 3.958 \times 10^{-3} \, mol \times 214.0 \, g \, mol^{-1} = 0.847 \, g$$

purity of sample = $\dfrac{0.847 \, g}{1.00 \, g} \times 100 = 84.7\%$

Tests for the halogens

Chlorine

Chlorine is a pale green gas. The tests for chlorine are as follows:

- Chlorine *rapidly* bleaches *damp* litmus paper.
- If chlorine gas is passed into a solution of potassium bromide, the colourless solution becomes brown.

A solution of chlorine can be tested in the same way. It is pale green. It rapidly bleaches litmus paper and also turns colourless potassium bromide solution brown.

Bromine

Bromine is a brown fuming liquid that bleaches *damp* litmus *slowly*.

The test for bromine is to add it to excess potassium iodide solution. The colourless solution goes deep red-brown. This is because iodine is liberated, and then reacts with excess I^- ions to form the red-brown I_3^- ion. Excess bromine would give a grey-black precipitate of iodine.

Iodine

Iodine is a grey-black solid. Solutions in aqueous potassium iodide are red-brown and aqueous solutions are pale brown.

The test for iodine is that it turns starch dark blue.

Hydrogen halides

Physical properties

The hydrogen halides are gases at standard temperature and pressure. Their boiling temperatures are given in Table 11.2 and graphically in Figure 11.2.

	HF	HCl	HBr	HI
Boiling temperature/°C	+20	−85	−67	−35

e Many candidates fail to mention that the litmus paper must be damp. Damp blue litmus paper turns red before it is bleached.

Table 11.2 Boiling temperatures of the hydrogen halides

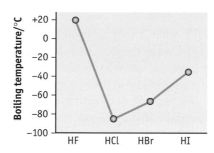

Figure 11.2 Boiling
temperatures of the
hydrogen halides

Hydrogen fluoride, HF, has an anomalous boiling temperature because it forms strong intermolecular hydrogen bonds. The hydrogen atom is δ^+ and the fluorine atom is both very small and is δ^-. Hydrogen chloride, HCl, does not form intermolecular hydrogen bonds because chlorine is less δ^- and the atom is too big.

All the hydrogen halides have instantaneous induced dipole forces (dispersion forces) and dipole–dipole forces between the molecules. These forces are weaker than the hydrogen bonds in hydrogen fluoride. Therefore, less energy is needed to separate the molecules.

Trend for HCl to HI

The strength of the dispersion (London) forces increases from hydrogen chloride to hydrogen iodide, HI, as the number of electrons in the molecule increases. The increase in the strength of the dispersion forces is much greater than the decrease in the strength of the dipole–dipole forces. Therefore, from hydrogen chloride to hydrogen iodide, the total intermolecular force increases and the boiling temperature increases.

Bonding

Anhydrous hydrogen halides are covalently bonded.

Hydrogen atoms have the electron configuration $1s^1$. Therefore, to reach the same configuration as the noble gas helium, one electron is needed. The electron configuration of the halogens is $ns^2\, np^5$. So, a halogen atom needs one electron to reach the configuration of a noble gas. Halogen halide molecules share two electrons, one from the halogen and one from hydrogen. The dot-and-cross diagram for hydrogen chloride is:

$$\text{H} \overset{\times\times}{\underset{\times\times}{\overset{\times}{\underset{\bullet}{\,}} \text{Cl} \overset{\times}{\,}}}$$

Chemical reactions

Hydrogen halides are acids. Hydrogen iodide, HI, is the strongest acid and hydrogen fluoride, HF, the weakest.

Reaction with water

The hydrogen halides are all soluble in water. This is because they react with the water to form ions:

$$HCl(g) + H_2O(l) \rightarrow H_3O^+(aq) + Cl^-(aq)$$

e Make sure that you can explain why hydrogen fluoride has intermolecular hydrogen bonds and the other hydrogen halides do not. Also, candidates often forget to state and explain the trend in the boiling temperatures of the halides from hydrogen chloride to hydrogen iodide. Remember that covalent bonds are *not* broken on boiling, so do not mention them in your answer.

The solutions are acidic because H_3O^+ ions are formed.

The H–halogen covalent bond is broken by the water. This bond breaking requires energy — it is endothermic. The energy is regained by the formation of the dative covalent bond between oxygen in a water molecule and a hydrogen ion, H^+, to form the H_3O^+ ion. Hydrogen iodide is the strongest acid of the hydrogen halides because the H–I bond is the weakest, so the removal of a hydrogen ion from hydrogen iodide by water is the most energetically favourable reaction.

$$HI(g) + H_2O(l) \rightarrow H_3O^+(aq) + I^-(aq)$$
$$HF(g) + H_2O(l) \rightleftharpoons H_3O^+(aq) + F^-(aq)$$

Reaction with ammonia

Ammonia is a base and, therefore, reacts with the hydrogen halides.

When gaseous hydrogen chloride reacts with gaseous ammonia, white fumes of solid ammonium chloride are formed:

$$HCl(g) + NH_3(g) \rightarrow NH_4Cl(s)$$

The same observation is made with the other hydrogen halides. This reaction is used as a test for gaseous HCl, HBr or HI, but it does not distinguish between them.

Hydrogen halides as reducing agents

The hydrogen halides can act as reducing agents. Hydrogen iodide is the strongest reducing agent.

When acting as reducing agents, hydrogen halides are oxidised to the halogen. The general ionic half-equation is:

$$2HX \rightarrow X_2 + 2H^+ + 2e^-$$

- Gaseous hydrogen iodide reduces concentrated sulfuric acid to sulfur and hydrogen sulfide, H_2S. Clouds of violet iodine vapour are given off. The sulfur in sulfuric acid is reduced from the oxidation state of +6 in sulfuric acid to –2 in hydrogen sulfide and 0 in elemental sulfur.
- Gaseous hydrogen bromide is a weaker reducing agent than gaseous hydrogen iodide and so only partially reduces concentrated sulfuric acid to sulfur dioxide, SO_2. It is itself oxidised to brown bromine vapour:

$$2HBr + H_2SO_4 \rightleftharpoons Br_2 + 2H_2O + SO_2$$

  The sulfur in sulfuric acid is reduced from the oxidation state of +6 in sulfuric acid to +4 in sulfur dioxide.
- Hydrogen chloride and hydrogen fluoride are not strong enough reducing agents to reduce concentrated sulfuric acid.

Halide salts

General properties

- Group 1 metals form ionic salts of formula MX, where X represents a halogen — for example, potassium iodide, KI, and sodium chloride, NaCl.
- Group 2 metals form ionic salts of formula MX_2 — for example, calcium bromide, $CaBr_2$.

- The d-block metals form salts of formula MX_2 or MX_3. When hydrated these salts are ionic — for example, copper(II) chloride, $CuCl_2$, and iron(III) chloride, $FeCl_3$. Silver forms halides of formula AgX.

Solubility

All halides are soluble in water, apart from silver halides and lead(II) halides.

If a solution of an ionic chloride such as sodium chloride is added to a solution of silver nitrate, a white precipitate of silver chloride is obtained:

$$Cl^-(aq) + Ag^+(aq) \rightarrow AgCl(s)$$

When a solution of an ionic bromide is added to a solution of silver nitrate, a cream precipitate of silver bromide is formed:

$$Br^-(aq) + Ag^+(aq) \rightarrow AgBr(s)$$

Chemical reactions

Halide ions as reducing agents

The strength of halide salts as reducing agents *increases* from Cl^- to I^-.

The half-equation for the reduction of a halide ion X^- is:

$$2X^- + 2e^- \rightarrow X_2$$

- Chloride ions are very difficult to oxidise. They are only oxidised by very strong oxidising agents such as ClO^- ions or potassium manganate(VII) in dilute sulfuric acid.
- Bromide ions are more easily oxidised. Chlorine, potassium dichromate(VI) in dilute sulfuric acid and potassium manganate(VII) in dilute sulfuric acid all oxidise bromide ions.
- Iodide ions are strong reducing agents and are oxidised by many substances. For example, Fe^{3+} ions are reduced to Fe^{2+} ions by iodide ions in solution. The two ionic half-equations are:

$$2I^-(aq) \rightarrow I_2(s) + 2e^-$$
$$Fe^{3+}(aq) + e^- \rightarrow Fe^{2+}(aq)$$

Thus the overall equation is:

$$2Fe^{3+}(aq) + 2I^-(aq) \rightarrow 2Fe^{2+}(aq) + I_2(s)$$

Reaction of halide salts with concentrated sulfuric acid

When concentrated sulfuric acid is added to solid halides, the hydrogen halide is formed first — for example, with a sodium halide, NaX:

$$NaX + H_2SO_4 \rightarrow NaHSO_4 + HX$$

The hydrogen halide produced might then reduce the concentrated acid.

With chlorides, steamy fumes of hydrogen chloride are produced:

$$NaCl + H_2SO_4 \rightarrow NaHSO_4 + HCl$$

These fumes turn blue litmus red and give white fumes with gaseous ammonia. The hydrogen chloride is not further oxidised by the concentrated sulfuric acid.

With bromides, steamy fumes of hydrogen bromide are produced together with some brown gaseous bromine. Sulfur dioxide is also produced but it is not visible

e The second ionic half-equation has to be multiplied by two, so that the number of electrons on the left-hand side in the reduction half-equation (Fe^{3+} to Fe^{2+}) is the same as the number of electrons on the right-hand side in the oxidation half-equation (I^- to I_2).

because it is a colourless gas. The equations for the reaction with potassium bromide are:

$$KBr + H_2SO_4 \rightarrow HBr + KHSO_4$$
$$2HBr + H_2SO_4 \rightarrow Br_2 + SO_2 + 2H_2O$$

With iodides, some steamy fumes of hydrogen iodide are formed, but the most obvious observation is a cloud of violet iodine vapour, mixed with some yellow sulfur. There is a smell of rotten eggs because hydrogen sulfide is also produced.

The results and observations are given in Table 11.3.

Halide	Reduction of sulfuric acid	Observations
Fluoride	None	Steamy fumes (HF)
Chloride	None	Steamy fumes (HCl)
Bromide	Reduces sulfuric acid to gaseous sulfur dioxide	Steamy fumes (HBr) and brown gas (Br_2)
Iodide	Reduces sulfuric acid to solid sulfur and gaseous hydrogen sulfide	Clouds of violet gas (I_2), smell of bad eggs (H_2S) and yellow solid (S)

Table 11.3 Reaction of halide salts with concentrated sulfuric acid

JOHN OLIVE

Progress of the reaction of potassium bromide with concentrated sulfuric acid

Tests for halides

Dilute nitric acid is added to a solution of the unknown halide, until the solution is *just* acidic. Silver nitrate solution is then added.

Chlorides, bromides and iodides all produce a precipitate of the silver halide. The solubility of the silver halide precipitate in aqueous ammonia varies (Table 11.4).

On addition of dilute nitric acid and silver nitrate solution:

- chlorides (including hydrochloric acid) give a white precipitate, which is soluble in dilute aqueous ammonia
- bromides give a cream precipitate, which is insoluble in dilute aqueous ammonia but soluble in concentrated aqueous ammonia
- iodides give a pale yellow precipitate, which is insoluble in both dilute and concentrated aqueous ammonia

The dilute nitric acid has to be added to prevent the precipitation of other ions (e.g. carbonate) by the silver ions in the silver nitrate.

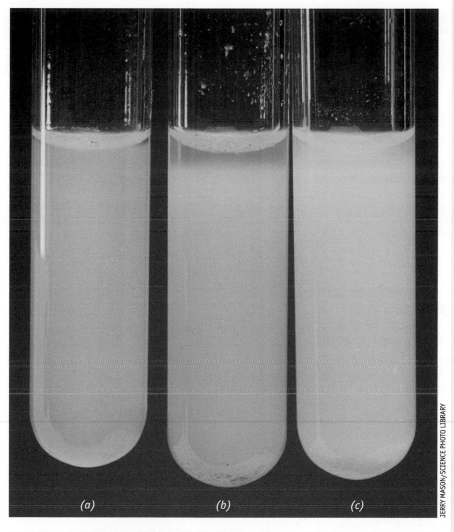

Precipitates of
(a) silver chloride,
(b) silver bromide and
(c) silver iodide

(a)   (b)   (c)

JERRY MASON/SCIENCE PHOTO LIBRARY

Table 11.4 Tests for halides

	Colour of precipitate formed on addition of acidified silver nitrate	Solubility of precipitate in dilute aqueous ammonia	Solubility of precipitate in concentrated aqueous ammonia
Chlorides	White	Soluble	—
Bromides	Cream	Insoluble	Soluble
Iodides	Pale yellow	Insoluble	Insoluble

Action of light on silver halides

Silver halides are slowly decomposed to metallic silver by visible light. For example, with silver chloride the equation is:

$$2AgCl \xrightarrow{\text{light}} 2Ag + Cl_2$$

This reaction causes the blackening of photographic film, enabling black-and-white photographs to be made.

Questions

1 Chlorine reacts at room temperature with aqueous potassium hydroxide.
   a Write the ionic equation for this reaction.
   b State the type of reaction that takes place.

2 Write the equations for the reaction of concentrated sulfuric acid with calcium chloride, and state *one* observation that could be made.

3 What would you observe when concentrated sulfuric acid is added to solid aluminium bromide? Identify the gases produced in this reaction.

4 Describe the chemical test for bromine.

5 Bromine reacts with hydrogen to form hydrogen bromide.
   a Name the type of bonding in gaseous hydrogen bromide.
   b Explain why hydrogen bromide is soluble in water and why the solution is acidic.
   c Write an equation for the reaction of hydrogen bromide solution with sodium carbonate.

6 Potassium iodate, KIO_3, reacts in acid solution with potassium iodide according to the equation:
   $$KIO_3 + 5KI + 6HCl \rightarrow 3I_2 + 6KCl + 3H_2O$$
   a Deduce the oxidation numbers of iodine in KIO_3, KI and I_2.
   b Use your answers from (a) to explain which substance in the reaction is a reducing agent.
   c Describe how you would test the solution to show the presence of iodine.

7 Explain why the first electron affinity of bromine is greater than that of iodine.

8 A white solid is known to be calcium chloride, calcium bromide or calcium iodide. Describe the tests to identify the salt. The answer should include all observations.

9 Write ionic half-equations for the following processes:
   a the reduction of iodine to iodide ions
   b the oxidation of chlorine to ClO^- ions
   c the oxidation of ClO^- ions to ClO_3^- ions
   d the reduction of ClO^- ions to Cl^- ions
   e using your answers to (c) and (d), write the overall equation for the disproportionation reaction of ClO^- ions.

10 Which of chlorine and iodine is the stronger oxidising agent? Describe *two* reactions that show this.

11 a A compound X has the following composition by mass: lithium 7.75%; chlorine 39.25%; oxygen 53.0%. Calculate its empirical formula.

   b The molecular formula of compound X is the same as the empirical formula. When a sample of X was heated, it decomposed into lithium chloride and oxygen gas. The volume of oxygen was found to be $1.77 \, dm^3$. (The molar volume of oxygen under the conditions of the experiment is $24.0 \, dm^3 \, mol^{-1}$.) Write the equation for the decomposition of compound X and calculate the mass of X that decomposed.

12 16.95 g of an iron(III) compound, X, was dissolved in distilled water and the solution made up to $250 \, cm^3$. A pipette was used to measure $25.0 \, cm^3$ portions into conical flasks. Dilute sulfuric acid and excess potassium iodide were added to each flask. The contents were titrated against $0.133 \, mol \, dm^{-3}$ sodium thiosulfate solution using starch as an indicator. The mean titre was $26.45 \, cm^3$.

   The equations for the reactions are:
   $$Fe^{3+} + I^- \rightarrow \tfrac{1}{2}I_2 + Fe^{2+}$$
   $$I_2 + 2S_2O_3^{2-} \rightarrow 2I^- + S_4O_6^{2-}$$
   Calculate:
   a the amount (in moles) of sodium thiosulfate in the mean titre
   b the amount (in moles) of iron(III) ions in $25.0 \, cm^3$ of solution
   c the amount (in moles) of iron(III) ions in $250 \, cm^3$ of solution
   d the mass of iron in the sample
   e the percentage of iron in compound X

13 A substance, Z, containing one cation and one anion gave a pale green flame in a flame test.

   A sample of Z was dissolved in water and the solution was divided into two portions:
   ■ Dilute sulfuric acid was added to one portion. A white precipitate was observed.
   ■ Dilute nitric acid followed by silver nitrate solution was added to the second portion. A pale yellow precipitate was obtained, which was insoluble in concentrated aqueous ammonia.

   Identify the unknown compound Z. Justify your answer.

Chapter *12*

Introduction to kinetics

Magnesium: sacrificial metal and flares

Atomic number: 12

Electron configuration: $1s^2\, 2s^2\, 2p^6\, 3s^2$

Symbol: Mg

Magnesium is in group 2 of the periodic table. Seawater contains $1.35\ \mathrm{g\ dm^{-3}}$ of magnesium ions and is the principal source of the metal.

Corrosion prevention

Iron is the cheapest and most useful metal, but it corrodes in damp air. The cost of rusting to the world's economy is well in excess of a hundred billion pounds a year.

Iron, in contact with oxygenated water, acts as a cathode, where oxygen is reduced to hydroxide ions:

$$O_2(aq) + 2H_2O(l) + 4e^- \rightarrow 4OH^-(aq)$$

The iron nearby acts as an anode and becomes oxidised by the loss of electrons:

$$Fe(s) \rightarrow Fe^{2+}(aq) + 2e^-$$

One way of preventing an iron object from rusting is to place pieces of magnesium at intervals along its surface. Magnesium is more reactive than iron, so is preferentially oxidised. Thus, the magnesium slowly corrodes away and the iron remains free from rust. The pieces of magnesium have to be regularly replaced to continue the protection. The magnesium is acting as a sacrificial anode.

The iron pintle, which holds the rudder onto the stern of a motorboat, is protected from rusting in this way, as are mothballed naval ships, iron storage tanks and underground pipes. In the USA, 12 000 tonnes of magnesium are used every year in the prevention of corrosion.

Flares and fireworks

Magnesium burns with a brilliant white light. A charge of black powder is used to throw the firework contents

A white starburst firework

COREL

or flare hundreds of feet into the air, when pellets of magnesium are ignited. The particle size determines the duration of the flash. Large particles produce the dazzling light for longer. Magnesium is not used in coloured starburst fireworks because the temperature reached in the combustion would cause the decomposition of the molecules that give colour to the flame.

Introduction to kinetics

Introduction

In chemistry, **kinetics** is the study of the speed, or rate, of reactions. Industrial processes must work sufficiently quickly to be economic, so knowledge of kinetics is important in chemical engineering.

Some reactions are extremely rapid. When solutions of silver nitrate and sodium chloride are mixed, a precipitate of silver chloride is produced instantaneously. The combustion of petrol in the engine of a Formula 1 car is very rapid and propels the car at high speeds. Conversely, rusting of iron is slow, as is the gradual deposition of calcium carbonate from dissolved calcium hydrogen carbonate in the formation of stalactites and stalagmites.

A Formula 1 car — 0–60 mph in 3.2 seconds

Speed is measured by the amount a property changes in a given time. The average speed of a car is measured by the distance it travels in a given time (miles per hour or kilometres per hour). The rate of a chemical reaction is determined by the change in concentration of a reactant or product per unit time.

$$\text{rate of reaction} = \frac{\text{change in concentration}}{\text{time for change to happen}}$$

$$= \frac{\Delta[\text{reactant}]}{\Delta \text{time}}$$

$$= \frac{\Delta[\text{product}]}{\Delta \text{time}}$$

tites and mites — in ars

In this definition, $\Delta[\text{reactant}]$ means the change in concentration of the reactant in mol dm^{-3}, and Δtime is the time for that change to occur.

During a reaction, the concentration of the reactant falls. Therefore, the reaction rate falls, until there is no reactant left and the reaction stops. In a graph of the concentration of a reactant against time, the slope of the graph measures the rate of reaction. The slope decreases as the rate decreases and becomes zero (a horizontal line) when all the reactant has been used (Figure 12.1).

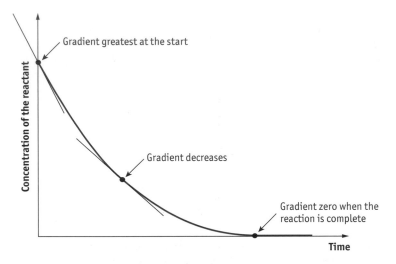

Figure 12.1 *Change in concentration of reactant with time*

It is sometimes easier to measure the amount of product produced over time and plot a graph of the concentration of the product against time (Figure 12.2). As before, the slope measures the rate of reaction and gradually decreases as the reaction slows down. When the slope becomes zero, the rate is zero. This is because all of the reactant has been used.

Figure 12.2 *Change in concentration of product with time*

The steeper the slope, the faster is the reaction.

Collision theory

The explanation of rates of reaction is based upon **collision theory**. Consider the reaction:

A + B → C + D

The first requirement is for molecule A to collide with molecule B. However, this cannot be the *only* requirement, otherwise all reactions would be extremely fast. There are about 10^{27} molecular collisions every second in 1 cm^3 of air in a room, and the frequency of collision is even greater in a liquid.

Collision theory also states that colliding molecules must hit each other with sufficient energy to cause a reaction.

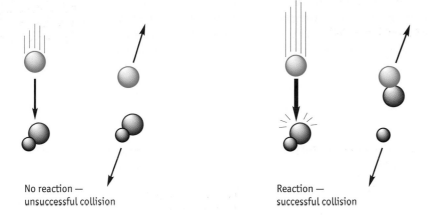

No reaction —
unsuccessful collision

Reaction —
successful collision

Particles (e.g. molecules in a gas or ions in solution) move about randomly at
various speeds. Maxwell and Boltzmann calculated the distribution of speeds of
the molecules in a gas and hence their kinetic energies.

Maxwell–Boltzmann distribution of molecular energies

The Maxwell–Boltzmann distribution of molecular energies is plotted on a graph
of the fraction of molecules with a particular energy against kinetic energy.

Figure 12.3 shows the distribution of molecular energies at two temperatures, T_1
(in blue) and T_2 (in red). T_2 is a higher temperature than T_1, and so the average
kinetic energy of the molecules at T_2 is greater than that at T_1.

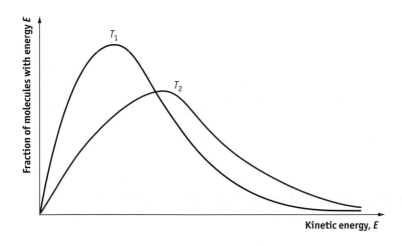

*Figure 12.3
Maxwell–Boltzmann
distribution of
molecular energies at
two temperatures*

Points to note about the graphs in Figure 12.3 are that:
- neither curve is symmetrical
- both curves start at the origin and finish by approaching the *x*-axis
- the area under both curves is the same (the number of molecules is the same)
- the peak of the T_2 (higher temperature) distribution is to the *right* and *lower*
  than the peak of the T_1 distribution

The second requirement of collision theory is that colliding molecules must possess, between them, at least the specific amount of kinetic energy required for a reaction to take place. The minimum energy that colliding molecules must have for the collision to result in reaction is called the **activation energy**.

> The activation energy in a reaction, E_a, is the minimum kinetic energy that the colliding molecules must possess for the collision to be successful and result in the formation of product molecules.

The fraction of collisions that are successful is shown by the Maxwell–Boltzmann distribution in Figure 12.4. The *area* under the curve to the right of the activation energy line is the fraction of molecules that have the necessary energy to react on collision. For a reaction that proceeds steadily at room temperature, only about 1 in 10^{12} molecules possesses this energy.

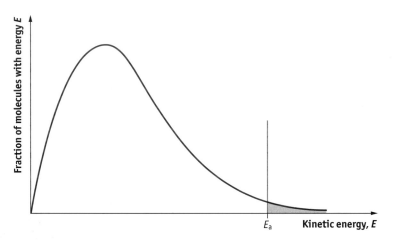

If the activation energy is high, the number of colliding particles that have this energy at room temperature is extremely small, so the reaction does not take place. The system is then said to be **kinetically stable**. The reaction can occur only if the temperature is increased.

If the reactants are heated, enough molecules may gain energy equal to or greater than the activation energy, E_a, for the system to become kinetically unstable. An example is the reaction between hydrogen and oxygen. The mixture is thermo-dynamically very unstable relative to the product water, but the activation energy is so high that the reaction does not occur at room temperature. The mixture is kinetically stable, but if it is heated with a spark, it becomes kinetically unstable and explodes.

The third requirement of collision theory is that the colliding molecules must have the correct orientation on collision. For example, in the reaction between iodomethane, CH_3I, and hydroxide ions, OH^-, the collision has to be between the oxygen in the OH^- ion and the carbon of CH_3I. In addition, the OH^- ion has to approach from the side opposite to the iodine atom (Figure 12.5).

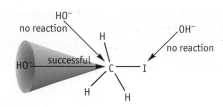

Figure 12.5
Importance of
orientation for a
successful collision

Summary of collision theory

- The molecules must collide. The rate of a reaction depends on the *frequency* of collision.
- A collision will only be successful if the colliding molecules have kinetic energy equal to or greater than the activation energy. A fast reaction has a low activation energy, so a greater *proportion* of collisions will be successful than in a reaction with a higher activation energy.

> **e** A low activation energy means a fast reaction; a high activation energy means a slow reaction or even one that does not occur at room temperature.

- The molecules must collide with the correct orientation. In a reaction involving complex molecules, fewer collisions will have the correct orientation than in a reaction between simpler molecules.

Factors affecting reaction rate

Pressure

For a **homogeneous** reaction of gases, an increase in pressure, at constant temperature, results in an increase in reaction rate.

Pressure can be increased by:
- reducing the volume of the container
- pumping more reactant gas into the container

Either way, the result is the same. There is an increased number of gas particles per cm^3, so the *frequency* of collision is increased. The average kinetic energy of the particles remains the same, so the same proportion of collisions will result in reaction. However, as there are more collisions per second, the rate increases.

Concentration

For reactions in solution, an increase in concentration causes an increase in reaction rate. The frequency of collisions between solute molecules in solutions of concentration 2 mol dm^{-3} is greater than the frequency in solutions of 1 mol dm^{-3} concentration. The average kinetic energy is independent of the concentration, so as the collision frequency is higher, the frequency of *successful* collisions is also higher, so the rate of reaction is faster. Doubling the concentration of one reactant usually causes the rate of reaction to double.

◀ A homogeneous reaction is one that takes place in a single phase — for example, all the species present are gases.

e Do not just say that the number of collisions increases. It is the number per second (or the frequency) that increases.

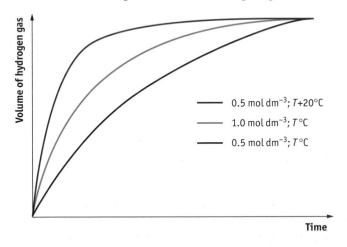

(a)   *(b)*

Marble chips reacting with (a) concentrated HCl and (b) dilute HCl

The effect of increasing the concentration of acid on the rate of reaction between a piece of magnesium ribbon and excess aqueous hydrochloric acid is shown in Figure 12.6. The reaction is:

$$Mg(s) + 2HCl(aq) \rightarrow MgCl_2(aq) + H_2(g)$$

The blue curve is for acid of concentration 0.5 mol dm^{-3}; the green curve is for 1 mol dm^{-3} acid. Note that the green line has a steeper *initial* slope (faster rate) than the blue line, but both lines flatten off at the same volume of hydrogen gas produced. This is because magnesium is the limiting reagent.

Volume of hydrogen gas

—— 0.5 mol dm^{-3}; $T+20°C$
—— 1.0 mol dm^{-3}; $T°C$
—— 0.5 mol dm^{-3}; $T°C$

Time

Figure 12.6 Reaction between magnesium and dilute hydrochloric acid

The red curve in Figure 12.6 shows the effect of increasing the temperature on the reaction between magnesium and 0.5 mol dm^{-3} hydrochloric acid.

Particle size

For **heterogeneous** reactions involving a solid, a larger surface area of the solid results in a faster reaction.

When a solid, such as zinc, reacts with an acid, only collisions between hydrogen ions in the solution and zinc atoms on the *surface* of the solid zinc can result in

A heterogeneous reaction is one in which the reactants are in two phases — for example, a solid and a gas or a solid and a solution.

reaction. If the zinc is powdered, the surface area is increased and hydrogen gas is formed more quickly.

Temperature

An increase in temperature always causes an increase in reaction rate. There are two reasons for this. The more important is that the molecules have a higher average kinetic energy at the higher temperature. This means that a greater fraction of the molecules possess the energy necessary to react on collision (Figure 12.7).

The tinted area under each curve to the right of the activation energy is the fraction of molecules that have sufficient energy to react on collision. The red plus blue area under the higher temperature curve is larger than the blue area under the lower temperature curve. Therefore, at the higher temperature, a greater *proportion* of the collisions result in a reaction. This means that the reaction is faster at the higher temperature.

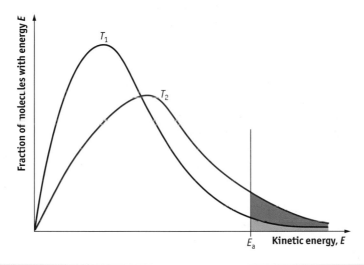

◀ Solid heterogeneous catalysts are manufactured with a large surface area so that there is a greater chance of a collision between a reacting molecule and the catalyst.

◀ An increase of 10°C from room temperature causes the rate of the average reaction approximately to double.

Figure 12.7
Maxwell–Boltzmann diagram showing the effect of temperature

◀ Note that the Maxwell–Boltzmann diagram showing the effect of temperature consists of *two* curves and *one* activation energy.

> **e** You must make it clear in your explanations that the area under the curve to the right of the activation energy is the fraction of molecules that have enough energy for the collision to be successful.

A second, and minor, effect of an increase in temperature is a slight increase in collision frequency. The additional rate caused by this is about 1% for a 10°C rise from room temperature. This effect is swamped by the increased rate due to the greater fraction of molecules with energy equal to or greater than the activation energy and the resulting increase in the fraction of successful collisions.

The effect of raising the temperature on the reaction between a piece of magnesium ribbon and dilute hydrochloric acid is shown in Figure 12.6. The red line shows the volume of hydrogen produced over time with 0.5 mol dm^{-3} acid at a temperature 20°C higher than that of the blue line. The slope of this line is much steeper than the slopes of the other two lines.

Catalyst

Catalysts are specific to reactions and they cause the reaction rate to increase.

> A catalyst speeds up a reaction by providing an *alternative* path with a lower activation energy.

Homogeneous catalysts

A **homogeneous catalyst** is one that is in the same phase as the reactants. An example is the Fe^{2+} ion catalyst in the oxidation of iodide ions, I^-, by persulfate ions, $S_2O_8^{2-}$:

$$S_2O_8^{2-}(aq) + 2I^-(aq) \xrightarrow{\text{Fe}^{2+}(aq)} 2SO_4^{2-}(aq) + I_2(s)$$

◀ Note that the catalyst is written above the arrow in the equation.

A homogeneous catalyst works by reacting with one of the reactants to form an intermediate compound:

$$S_2O_8^{2-}(aq) + 2Fe^{2+}(aq) \rightarrow 2SO_4^{2-}(aq) + 2Fe^{3+}(aq)$$

The intermediate compound then reacts with the other reagent to reform the catalyst:

$$2Fe^{3+}(aq) + 2I^-(aq) \rightarrow I_2(s) + 2Fe^{2+}(aq)$$

The route using the catalyst avoids the necessity of collisions between two negative particles.

Heterogeneous catalysts

A **heterogeneous catalyst** is one that is in a different phase from the reactants. One example is iron when used in the Haber process (page 255):

$$N_2(g) + 3H_2(g) \xrightarrow{\text{Fe}(s)} 2NH_3(g)$$

A heterogeneous catalyst usually promotes a reaction through its active sites. These are places on the surface of the catalyst to which the reactants can bond. The sequence of reactions is:

> gaseous reactants + active sites → adsorbed reactants
> adsorbed reactants → adsorbed products
> adsorbed products → gaseous products + free active sites

There are fewer active sites than there are reactant molecules, so an increase in pressure does not alter the rate of reaction because all the active sites are already occupied.

Another example is platinum when used in the addition of hydrogen to alkenes (page 142):

$$C_2H_4(g) + H_2(g) \xrightarrow{\text{Pt}(s)} C_2H_6(g)$$

How catalysts work

The way that a catalyst speeds up a reaction can be explained using a Maxwell–Boltzmann diagram (Figure 12.8). The presence of a catalyst does not alter the average kinetic energy of the molecules, but the catalysed reaction route has a lower activation energy.

Manganese(IV) oxide is a heterogeneous catalyst. Hydrogen peroxide in the beaker is breaking down into oxygen (seen as white vapour) and water, catalysed by manganese(IV) oxide.

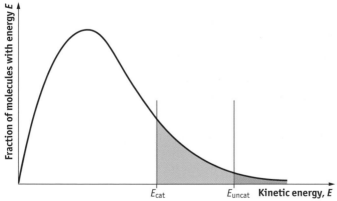

Figure 12.8 Maxwell–Boltzmann diagram for a reaction with and without a catalyst

e Note that this diagram has *one* curve and *two* activation energies. Do not confuse it with the diagram for the effect of temperature, which has two curves and one activation energy.

The tinted area under the curve to the right of each activation energy is the fraction of molecules that have sufficient energy to react on collision. E_{cat} is less than E_{uncat}, so the green plus blue area under the curve to the right of E_{cat} is larger than the blue area to the right of E_{uncat}. This shows that a greater *proportion* of the collisions involving the catalyst result in a reaction. Therefore, the reaction is faster in the presence of the catalyst.

Enzyme catalysts

Enzymes are biological catalysts. They are specific to a particular biochemical reaction and can make the reaction go as much as 10^{10} times faster. Enzymes are large protein molecules that contain one or more **active sites.** The reactant, called the **substrate** in biochemistry, fits into the active site in a similar way to a key fitting into a lock. Only a key with the correct shape fits a lock — the same is the case with a substrate and an active site (Figure 12.9).

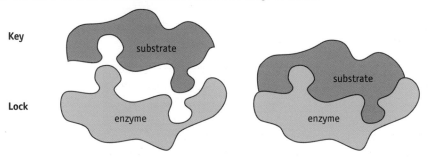

Figure 12.9 The lock-and-key hypothesis of enzyme action

A chemical binding irreversibly into the 'lock' of the enzyme structure can poison the enzyme. For example, nerve gases, such as Sarin, block enzyme activity in one of the reactions that controls the passage of impulses through nerve cells. One enzyme in the reproduction of the HIV virus is HIV-protease. If a poison specific to blocking the active site of this enzyme were to be found, it would prevent the development of HIV into AIDS.

Summary of factors affecting the rate of reaction

- In homogeneous gas-phase reactions, an increase in pressure causes an increased *frequency* of collision and hence an increase in the rate of reaction.
- For reactions in solution, an increase in the concentration of a reactant increases the *frequency* of collisions and so increases the rate of reaction.
- In heterogeneous reactions, an increase in surface area of the solid reactant increases the number of collisions in a given time between the solid and the gaseous or dissolved reactant, and hence increases the rate of reaction.
- A rise in temperature increases the average kinetic energy of the molecules. This results in more of the colliding molecules having energy equal to or greater than the activation energy, E_a. Therefore, a greater *proportion* of the collisions are successful and hence the rate increases.

e These are the explanations that examiners look for.

The rate also increases slightly because of the increase in the frequency of collisions.

- A catalyst speeds up a reaction by providing an alternative path with a lower activation energy. Addition of a catalyst causes more molecules to have energy equal to or greater than E_{cat} than have energy equal to or greater than E_{uncat}. Therefore, the *proportion of successful collisions* increases and hence the rate increases.
- Enzymes are biological catalysts.

Reaction profile diagrams

A **reaction profile diagram** shows the energy levels of the reactants and products of the reaction and of the transition state that the reactants go through. The activation energy, E_a, and the enthalpy change, ΔH_r, are also shown on the diagram.

Consider the reaction:

$$CH_3I + OH^- \rightarrow CH_3OH + I^-$$

As the OH^- ion approaches the carbon atom from the side opposite to the iodine, it is subjected to an increasing force of repulsion. This is caused by the outside of both the oxygen and the carbon atoms being a sphere of negative charge (the outer electrons). If the OH^- ion does not have sufficient energy, it bounces off. However, if the OH^- ion possesses the activation energy, the oxygen approaches close enough to the carbon atom for a covalent bond to start to form. As this bond forms, the C–I bond begins to break. The transition state is the point of highest energy, when the O–C bond has partially formed and the C–I bond has partially broken.

$$\left[H \!-\!\! O \cdots\cdots\overset{\overset{\displaystyle H \quad H}{\diagdown\diagup}}{\underset{\displaystyle H}{C}}\cdots\cdots I \right]^-$$

The energy changes during the course of an exothermic reaction are shown in the reaction profile diagram in Figure 12.10.

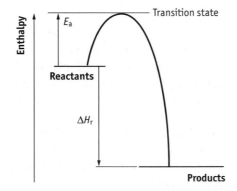

Figure 12.10 Reaction profile of an exothermic reaction

◀ This diagram is similar to an energy level diagram (page 109). The difference is that an energy level diagram does not show the transition state.

A catalysed reaction takes place using a different route from the uncatalysed reaction. An intermediate forms between the reactant and the catalyst, so the reaction takes place in at least two steps. A simplified version of a reaction profile for a catalysed reaction is shown in Figure 12.11.

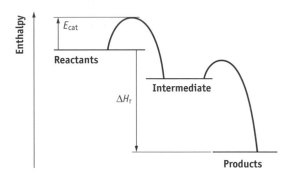

Figure 12.11 Reaction profile of a catalysed reaction

Note that this reaction profile diagram has two humps and that E_{cat} is less than E_a in Figure 12.10. A catalyst has *no* effect on the value of ΔH_r.

Experimental methods

The rate of a reaction can be estimated in a number of ways. A common method is to measure the time, t, for a certain amount of the mixture to react or the time for a certain amount of product to be formed. The assumption is then made that $1/t$ is a measure of the rate. This is not strictly accurate because the rate will have been decreasing from the moment that the reactants were mixed. It is only a fair assumption if the concentrations of the reactants do not change significantly during the time of the experiment.

An example is the reaction between magnesium ribbon and dilute acid. Excess acid is added to a piece of magnesium and the time taken for all the magnesium to disappear is measured. The assumption is then made that the rate is proportional to $1/t$. This is only a fair assumption if the concentration of the acid falls by no more than 10% during the time of the experiment.

Measurable physical changes

There are a number of physical changes that can be used to estimate the rate of a chemical reaction.

Production of a gas

The time taken to produce a certain volume of gas (collected over water or in a gas syringe) can be measured. This is a suitable method for measuring the decomposition of hydrogen peroxide using, for example, either different catalysts or different concentrations of hydrogen peroxide:

$$2H_2O_2(aq) \rightarrow 2H_2O(l) + O_2(g)$$

This method is also suitable for studying the effect of temperature on the rate of reaction between magnesium and hydrochloric acid:

$$Mg(s) + 2HCl(aq) \rightarrow MgCl_2(aq) + H_2(g)$$

Production of a solid

The time taken to produce enough solid to hide a cross on a piece of paper under the reaction apparatus or on the side of the apparatus can be measured.

This method is used to measure the decomposition of sodium thiosulfate by acid. The solutions are mixed and placed in a boiling tube that has a felt-pen cross marked on the side. The time taken for enough sulfur to be produced to hide the cross, when viewed from the other side of the boiling tube, is measured. The experiment can then be repeated, for example, with either different concentrations or at different temperatures.

$$Na_2S_2O_3(aq) + 2HCl(aq) \rightarrow 2NaCl(aq) + SO_2(aq) + H_2O(l) + S(s)$$

Colour change

The change in colour of either a reactant or a product is measured using a colorimeter.

The rate of the reaction between acidified potassium dichromate(VI) solution and an alcohol can be studied by measuring the intensity of the colour of the solution, which changes from orange to green as the dichromate(VI) is reduced by the alcohol.

Questions

1 Explain, using the collision theory, the reaction between two gases A and B.

2 State the effect of increasing the temperature on the rate of:
   a an exothermic reaction
   b an endothermic reaction

3 Draw a diagram that represents the Maxwell–Boltzmann distribution of molecular energies for a gas at two temperatures, T_c and T_h, where T_c is lower than T_h.

   Indicate a suitable value for the activation energy of the reaction and use your diagram to explain the effect of lowering the temperature on the rate of reaction.

4 Hydrogen and iodine react to form hydrogen iodide.

   $$H_2(g) + I_2(g) \rightarrow 2HI(g) \quad \Delta H_r^\circ = 51.8 \text{ kJ mol}^{-1}$$

   State and explain the effect, if any, on the rate of reaction of:
   a halving the volume of the container at constant temperature
   b increasing the pressure by adding more hydrogen at constant volume and temperature

5 Hydrogen peroxide decomposes into water and oxygen:

   $$2H_2O_2(aq) \rightarrow 2H_2O(l) + O_2(g)$$
   $$\Delta H_r^\circ = -196 \text{ kJ mol}^{-1}$$

   a Draw labelled reaction profiles for the reaction with and without a platinum catalyst.
   b Explain the difference between *heterogeneous* and *homogeneous* catalysts.
   c Classify the type of catalyst used in part (a) of this question.
   d Draw a diagram of an apparatus that could be used to measure the rate of this reaction, using platinum and manganese(IV) oxide separately as catalysts. State what measurements you would make and how you would calculate the relative rates of reaction from your data.

6 An experiment was devised to investigate the effect of temperature on the rate of the reaction between magnesium and aqueous hydrochloric acid. 25 cm^3 of a 1.00 mol dm^{-3} solution of hydrochloric acid was added to a beaker. A sample of magnesium of mass 0.06 g was added and a stopwatch started. The watch was stopped

when all the magnesium had reacted. It was assumed that the rate of the reaction is proportional to $1/t$, where t is the time taken for the magnesium to disappear.

$$Mg(s) + 2HCl(aq) \rightarrow MgCl_2(aq) + H_2(g)$$

a Calculate the amounts (in moles) of magnesium and hydrochloric acid taken.

b Calculate the amount of acid used in the reaction.

c Calculate the percentage change in the amount of the acid during the experiment.

d (i) Does this mean that $1/t$ is a fair measure of the rate?

(ii) If not, what change would you make to the volume or concentration of the acid to make it a fairer measure?

7 When petrol is burned in the engine of a car, the combustion is not complete and some carbon monoxide, CO, is produced. Nitrogen monoxide, NO, is another exhaust gas. These gases can be removed by a catalytic converter in the exhaust system.

$$NO(g) + CO(g) \rightarrow CO_2(g) + \tfrac{1}{2}N_2(g)$$

a Draw a Maxwell–Boltzmann distribution of molecular energies at a temperature, T. Indicate the activation energies of the reaction with and without catalyst, and use the diagram to explain how a catalyst speeds up this reaction.

b Explain why petrol that contains lead should not be used in cars fitted with a catalytic converter.

Chapter *13*
Introduction to chemical equilibrium

Aluminium: alloys, aeroplanes and coke cans

Atomic number: 13

Electron configuration: $1s^2\ 2s^2\ 2p^6\ 3s^2\ 3p_x^1$

Symbol: Al

Aluminium is in group 3 of the periodic table. Of all the metals, only beryllium and magnesium have lower densities. Aluminium is the third most common element in the Earth's crust, but only the ore bauxite is a suitable source of the metal. Bauxite is treated with sodium hydroxide solution to separate it from iron oxide and silicon oxide impurities. The now pure aluminium oxide is dissolved in molten cryolite and electrolysed.

Aluminium alloys

Aluminium is much less dense than iron and it is protected by a layer of aluminium oxide, so does not corrode. These properties make it suitable for structures where lack of corrosion and low weight are important. It is too soft to be used by itself and so it is alloyed. An alloy is a substance with metallic properties that is composed of a mixture of two or more elements, at least one of which is a metal.

Aeroplane fuselages

Aeroplane fuselages and wings are made of aluminium alloyed with magnesium, copper and either zinc or manganese. The tensile strength of these alloys is greater than that of steel and, as they are much less dense, their strength-to-weight ratio is very high. Aircraft parts made of aluminium alloy weigh very much less than those made of steel. As the metal does not corrode, its surface remains smooth, which is an essential property for the wings of an aircraft.

Aluminium saucepans

Aluminium saucepans are made from aluminium alloyed with 4% magnesium. The resulting metal is a very good conductor of heat and has a high resistance to corrosion.

Drink cans

Drink cans are also made of an aluminium–magnesium alloy. The electrolytic extraction of aluminium from pure aluminium oxide uses a huge amount of electricity, so recycling aluminium cans makes economic and environmental sense. To obtain 1 mole of aluminium from bauxite requires 297 kJ of electrical energy, compared with only 26 kJ mol^{-1} to melt scrap aluminium. All schools should have recycling centres for drink cans.

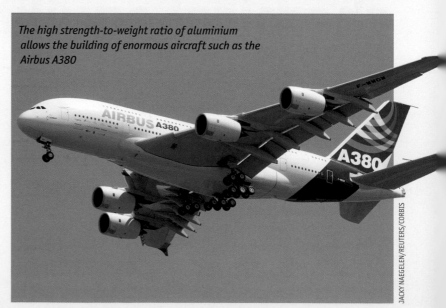

The high strength-to-weight ratio of aluminium allows the building of enormous aircraft such as the Airbus A380

JACKY NAEGELEN/REUTERS/CORBIS

Introduction to chemical equilibrium

Introduction

When a mixture of hydrogen and oxygen in a 2:1 ratio is ignited with a spark, water is produced and no uncombined hydrogen or oxygen is left. This is an example of a complete or **irreversible** reaction.

$$2H_2(g) + O_2(g) \rightarrow 2H_2O(l)$$

Many reactions do not go to completion. If a mixture of hydrogen and iodine in a 1:1 molar ratio is heated to 300°C in a closed vessel, only about 90% of the hydrogen and iodine react.

$$H_2(g) + I_2(g) \rightleftharpoons 2HI(g)$$

> A reversible reaction is one that goes in both directions at the same temperature.

The $\rightleftharpoons$ sign indicates that the reaction does not go to completion and is therefore a reversible reaction.

At 300°C, no matter how long the mixture of hydrogen and iodine is allowed to react, 10% of the reactants will be left uncombined. If some hydrogen iodide is heated to 300°C, it partially decomposes and the composition of the resulting mixture is identical to that produced on starting with equimolar amounts of hydrogen and iodine. When there is no further change in the amounts of reactants and products, the system is said to be in **equilibrium**.

> A system has reached equilibrium when there is no further change in the concentrations of the reactants and products.

When the mixture of hydrogen and iodine is heated to 300°C, the two gases start to react and form hydrogen iodide. As time passes, the concentrations of hydrogen and iodine become lower, so the rate of the reaction decreases. As soon as the reaction starts, some of hydrogen iodide formed begins to decompose to form hydrogen and iodine.

At first, this reaction is very slow, because the concentration of hydrogen iodide is very small. However, as time passes and more and more hydrogen and iodine react, the concentration of hydrogen iodide increases. This means that the rate of the reverse reaction also increases. Eventually, the rates of the forward reaction and the reverse reaction become equal. After this point, there is no further change in concentration. The system is now in equilibrium with the forward and reverse reactions taking place at the same rate. This is called a **dynamic equilibrium**.

Rate of $H_2 + I_2 \rightarrow 2HI$ equals the rate of $2HI \rightarrow H_2 + I_2$.

> In a dynamic equilibrium, the rates of the forward and reverse reaction are equal. Therefore, there is no further change in the concentrations of the reactants and products.

A dynamic equilibrium is illustrated in Figure 13.1.

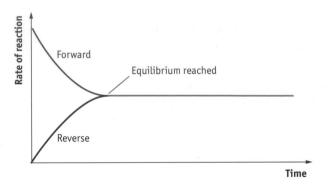

The fact that the reactions do not stop when equilibrium is reached can be proved by using radioactive tracers. Hydrogen iodide, containing a trace of the radio-isotope ^{131}I, is mixed with hydrogen and iodine in *equilibrium* proportions at 300°C and left for several minutes. The amount of each substance does not change, but the radioisotope is found in both hydrogen iodide and iodine molecules. This shows that some hydrogen iodide must have decomposed into iodine and hydrogen. The ^{131}I is distributed between the hydrogen iodide and the iodine in the same ratio as in the overall equilibrium concentrations.

Factors affecting the position of equilibrium

At equilibrium, the ratio of product to reactant concentrations defines the **position of equilibrium**. Using the example of hydrogen iodide, hydrogen and iodine, the position of equilibrium is represented by:

$$\frac{\left[HI\right]}{\left[H_2\right]} = \frac{\left[HI\right]}{\left[I_2\right]} = \frac{9}{1}$$

where [HI] means the concentration of [HI] in mol dm^{-3}.

When the position of equilibrium is expressed as the percentage of a reactant that is converted into product, it is called the **equilibrium yield**.

The actual yield in a process may be less than the equilibrium yield. This occurs in an open system, such as when reactant gases are passed through a catalyst bed. The system may not reach equilibrium because the gases are not in contact with the catalyst for long enough.

Le Chatelier's principle

In 1884, Le Chatelier put forward an idea that can be used to predict (but not to explain) the way in which a system at equilibrium alters as the physical and chemical conditions are changed.

> When the conditions of a system in equilibrium are altered, the position of equilibrium alters in such a way as to try to restore the original conditions.

Effect of temperature

The direction of the change in equilibrium position depends upon whether the reaction is exothermic or endothermic. The direction of change can be predicted using Le Chatelier's principle.

If a system at equilibrium is heated, the temperature rises and the system reacts to remove the heat energy and bring the temperature down. This means that heat energy must be converted into chemical energy. Therefore, the position of equilibrium shifts in the endothermic direction, causing:

- a decrease in the equilibrium yield for an exothermic reaction
- an increase in equilibrium yield for an endothermic reaction

A decrease in temperature moves the equilibrium in the exothermic direction, so that chemical energy is converted to heat energy, thus restoring the temperature.

> An increase in temperature shifts the equilibrium in the endothermic direction.

> ### Worked example 1
>
> Predict the effect on the position of equilibrium of increasing the temperature in the following equilibrium system:
>
> $$N_2O_4(g) \rightleftharpoons 2NO_2(g) \qquad \Delta H = +58\,kJ\,mol^{-1}$$
> colourless   brown
>
> **Answer**
>
> The reaction is endothermic left to right. An increase in temperature will move the position of equilibrium in the endothermic direction, which is to the right, thus causing more of the dinitrogen tetroxide to decompose.

> Dinitrogen tetroxide is a colourless gas; nitrogen dioxide is a brown gas. When dinitrogen tetroxide is heated, the gas darkens as increasing amounts of brown NO_2 are formed from N_2O_4.

JOHN OLIVE

A mixture of N_2O_4 and NO_2 in glass flasks. The flask on the right is at a lower temperature

Worked example 2

Predict the effect on the position of equilibrium of increasing the temperature in the following equilibrium system:

$$N_2(g) + 3H_2(g) \rightleftharpoons 2NH_3 \quad \Delta H_r^{\circ} = -92.4 \text{ kJ mol}^{-1}$$

Answer

The reaction is exothermic from left to right. An increase in temperature will move the position of equilibrium in the endothermic direction, which is to the left, thus reducing the equilibrium yield of ammonia.

ⓔ Reversible reactions go both ways and therefore simply stating that the reaction is exothermic is not sufficient. You must state that it is exothermic *left to right*.

An increase in temperature speeds up the endothermic reaction in the equilibrium more than the exothermic reaction. This is because the activation energy for the reaction in the endothermic direction is greater than that for the reaction in the exothermic direction and reactions with high activation energies are more sensitive to temperature changes than reactions with lower activation energies.

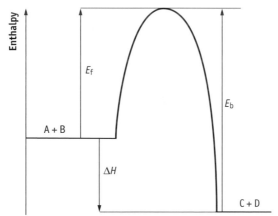

E_f is the activation energy for the reaction $A + B \rightarrow C + D$
E_b is the activation energy for the reaction $C + D \rightarrow CA + B$

Figure 13.2
Energy profile for the exothermic equilibrium reaction:
$A + B \rightleftharpoons C + D$

Effect of pressure

The effect of pressure only applies to equilibrium reactions involving gases. Pressure is caused by the bombardment of the gas molecules on the walls of the container. At a given temperature, the pressure depends only on the number of gas molecules in a given volume.

The pressure of an equilibrium system can be increased by:
- reducing the volume
- adding more gas into the same volume

Both methods result in more molecules per cm^3.

The direction of change of the equilibrium position caused by a change in pressure depends upon the number of gas molecules (or moles of gas) on each side of the equation. If the pressure is increased, Le Chatelier's principle predicts that the system will react in order to try to bring the pressure down again towards the original value. This can only happen if the equilibrium shifts to the side with *fewer* gas molecules.

A decrease in pressure will cause the equilibrium to shift towards the side of the equation with more gas molecules.

◀ An increase in pressure (at constant temperature) will cause the position of equilibrium to shift towards the side of the equation with fewer gas molecules (fewer moles of gas).

Worked example 1

Predict the effect on the position of equilibrium of increasing the pressure in the following equilibrium system:

$$N_2(g) + 3H_2(g) \rightleftharpoons 2NH_3(g) \quad \Delta H_r^\circ = -92.4 \text{ kJ}$$

Answer

There are four gas molecules on the left-hand side of the equation and only two on the right. An increase in pressure on the system at equilibrium will cause the position to shift to the side with fewer gas molecules, which is to the right. High pressures will increase the proportion of nitrogen and hydrogen that is converted into ammonia.

Worked example 2

Predict the effect on the position of equilibrium of increasing the pressure in the following equilibrium system:

$$H_2(g) + I_2(g) \rightleftharpoons 2HI(g) \quad \Delta H_r^\circ = +51.8 \text{ kJ}$$

Answer

There are 2 moles of gas on each side of the equation. Therefore, an increase in pressure will have *no* effect on the position of equilibrium.

Worked example 3

Methane can be trapped in water as the solid methane hydrate, $[CH_4(H_2O)_6]$. The reaction between methane gas and water to form methane hydrate is exothermic:

$$CH_4(g) + 6H_2O(l) \rightleftharpoons [CH_4(H_2O)_6](s)$$

Huge quantities of methane hydrate are found at the bottom of the sea off the Canadian coast, where the pressure is very high.

a Predict what would happen if the pressure on solid methane hydrate were reduced to 1 atm.

b Predict what would happen to the methane hydrate if it were heated.

c Predict what would happen if methane hydrate crystals were brought to the surface of the sea, from the great depths off the Canadian coast.

Answer

a There is 1 mol of gas on the left-hand side of the equation and none on the right. By Le Chatelier's principle, a decrease in pressure will cause the equilibrium to shift to the side with more gas molecules. This means that the methane hydrate will decompose and methane gas will be produced.

b The reaction as written is exothermic. An increase in temperature will cause the equilibrium position to shift in the endothermic direction, which is to the left. Therefore, the methane hydrate will decompose.

c Raising the solid methane hydrate from the cold ocean floor means a decrease in pressure and an increase in temperature. As can be seen from the answers to parts **a** and **b**, this would result in the methane hydrate decomposing into methane gas and water.

Effect of concentration

The effect of concentration applies to systems in which the reactants are dissolved in a solvent, usually water. If the concentration of one of the substances in the equilibrium reaction is increased, Le Chatelier's principle predicts that the position will shift to remove some of that substance. Consider the reaction used as a test for a bromide:

$$AgBr(s) + 2NH_3(aq) \rightleftharpoons [Ag(NH_3)_2]^+(aq) + Br^-(aq)$$

> Concentration is measured by the number of moles of solute per dm^3 of solution.

The position of equilibrium is normally to the left, so very little silver bromide dissolves in dilute aqueous ammonia. However, if concentrated ammonia is added, $[NH_3]$ increases. This drives the position of equilibrium to the right, causing the precipitate of silver bromide to disappear as it reacts to form $[Ag(NH_3)_2]^+$ ions.

If the concentration of one of the species in the equilibrium is decreased, the position of equilibrium shifts to produce more of that species.

Worked example

Dichromate(VI) ions, $Cr_2O_7^{2-}$, are orange and chromate(VI) ions, CrO_4^{2-}, are yellow. In aqueous solution, they are in equilibrium:

$$Cr_2O_7^{2-}(aq) + H_2O(l) \rightleftharpoons 2CrO_4^{2-}(aq) + 2H^+(aq)$$
orange   yellow

Predict and explain the colour change in the system when sodium hydroxide solution is added.

Answer

The alkaline sodium hydroxide reacts with the $H^+(aq)$ ions in the equilibrium mixture, lowering the concentration of $H^+(aq)$ ions. This causes the equilibrium to shift to the right. The solution changes colour from orange to yellow, as chromate(VI) ions are formed from dichromate(VI) ions.

Effect of a catalyst

A catalyst works by providing an alternative route that has a lower activation energy than that of the uncatalysed reaction. Therefore, the rate of reaction with a catalyst is much faster than the rate without the catalyst (page 240). The reaction profile in Figure 13.3 shows the effect of a catalyst on the activation energy of a reaction.

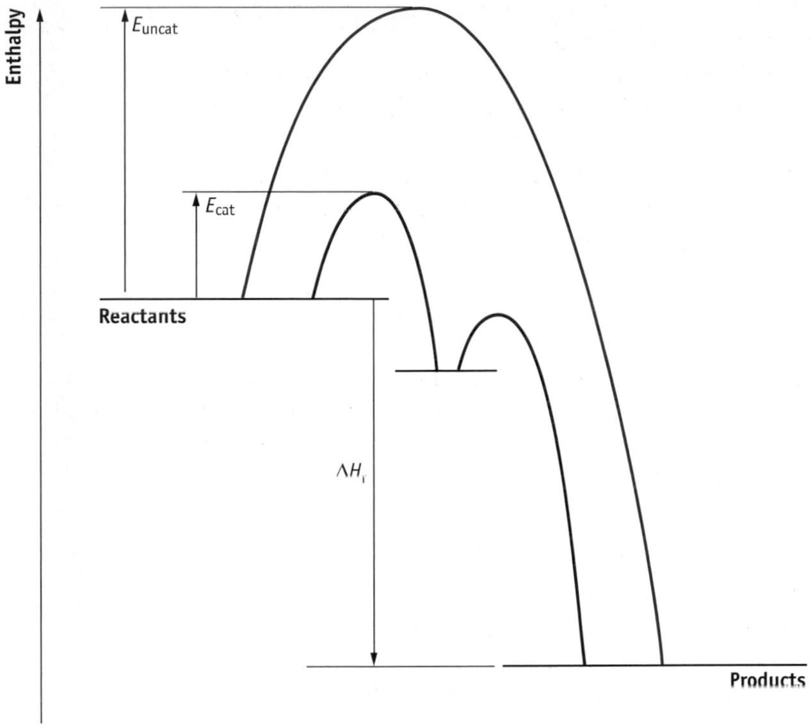

Figure 13.3
Reaction profile for a catalysed and uncatalysed reaction

A catalyst has *no* effect on the position of equilibrium of a reversible reaction. It speeds up the forward and back reactions equally, so equilibrium is reached sooner than without the catalyst.

Catalysts have no effect on the position of equilibrium.

Catalysts are important in industrial processes, many of which are extremely slow at room temperature. Finding a suitable catalyst is therefore very necessary. This is particularly so if the reaction is exothermic, which would result in a low yield at a high temperature.

The Haber process for the manufacture of ammonia is based on the equilibrium:

$$N_2(g) + 3H_2(g) \rightleftharpoons 2NH_3(g) \quad \Delta H^\ominus_r = -92.4 \text{ kJ mol}^{-1}$$

The reaction is very slow at room temperature. However, if it is heated, in the absence of a catalyst, to a temperature at which the rate is economic, the yield is extremely small.

For exothermic reactions such as this:

- a high temperature results in a *low* equilibrium yield, reached *quickly* because the reaction is fast
- a lower temperature results in a *higher* equilibrium yield reached more *slowly* because the reaction rate is slow
- a catalyst is used to allow the reaction to take place rapidly at a lower temperature, so that an acceptable yield is produced at a reasonable rate. This is often called a 'compromise temperature' as it balances an economic yield with an economic rate.

Questions

1 Explain the term 'dynamic equilibrium' with reference to the following equilibrium reactions:
   a $2SO_2(g) + O_2(g) \rightleftharpoons 2SO_3(g)$
   b solid sodium chloride suspended in a saturated solution of salt water, represented by:
   $$NaCl(s) + aq \rightleftharpoons Na^+(aq) + Cl^-(aq)$$

2 Nitrogen and hydrogen react reversibly to form ammonia:
   $$N_2(g) + 3H_2(g) \rightleftharpoons 2NH_3(g)$$
   Under certain conditions, 1 mol of nitrogen and 3 mol of hydrogen were mixed and 20% of the gases reacted. Draw three graphs, on the same axes, showing the number of moles against time, for each of nitrogen, hydrogen and ammonia. Label your graphs.

3 Nickel is purified by passing carbon monoxide gas over the heated metal. A gaseous nickel compound, $[Ni(CO)_4]$, is formed, according to the equilibrium:
   $$Ni(s) + 4CO(g) \rightleftharpoons [Ni(CO)_4](g)$$
   $$\Delta H^\circ_r = -161 \text{ kJ mol}^{-1}$$
   The temperature is then altered and the equilibrium shifts in the reverse direction to produce pure nickel metal.
   Explain whether the temperature has to be increased or decreased to produce this effect.

4 Alcohols react reversibly with organic acids to produce an ester plus water. Predict the effect of increasing the temperature on the position of equilibrium of the esterification reaction between ethanol and ethanoic acid:
   $$C_2H_5OH(l) + CH_3COOH(l) \rightleftharpoons CH_3COOC_2H_5(l) + H_2O(l)$$
   $$\Delta H^\circ_r = 0 \text{ kJ mol}^{-1}$$

5 Gaseous carbon dioxide dissolves in water and forms a solution of carbonic acid, H_2CO_3:
   $$CO_2(g) + H_2O(l) \rightleftharpoons H_2CO_3(aq) \qquad \Delta H^\circ_r = ?$$
   Carbon dioxide is more soluble in cold water than in hot water. Explain whether ΔH for this reaction is exothermic or endothermic.

6 Hydrogen gas is manufactured by passing methane and steam over a heated nickel catalyst:

   $$CH_4(g) + H_2O(g) \rightleftharpoons CO(g) + 3H_2(g)$$
   $$\Delta H^\circ_r = +206 \text{ kJ mol}^{-1}$$
   Explain why a temperature of 750°C and a catalyst are used.

7 The second stage in the manufacture of sulfuric acid is the oxidation of sulfur dioxide in air:
   $$SO_2(g) + \tfrac{1}{2}O_2(g) \rightleftharpoons SO_3(g)$$
   $$\Delta H^\circ_r = -98 \text{ kJ mol}^{-1}$$
   Explain why the conditions used are a temperature of 420°C and a vanadium(V) oxide catalyst.

8 Predict whether the position of equilibrium moves to the left, to the right or is unaltered when the pressure on each of the following systems is increased:
   a $2O_3(g) \rightleftharpoons 3O_2(g)$
   b $C(s) + H_2O(g) \rightleftharpoons CO(g) + H_2(g)$
   c $4NH_3(g) + 5O_2(g) \rightleftharpoons 4NO(g) + 6H_2O(g)$
   d $SO_2(g) + \tfrac{1}{2}O_2(g) \rightleftharpoons SO_3(g)$

9 At 15°C, red-brown iodine monochloride (melting temperature 27°C) reacts with chlorine to form the orange solid iodine trichloride (melting temperature 101°C), according to the equation:
   $$ICl(s) + Cl_2(g) \rightleftharpoons ICl_3(s)$$
   a On heating a mixture at equilibrium, the equilibrium position shifts to the left. What is the sign of ΔH for this reaction? Justify your answer.
   b What would be the effect of increasing the pressure on an equilibrium mixture?
   c What would you observe if a sample of iodine trichloride was heated to 50°C?

10 Lead chloride, $PbCl_2$, is an insoluble solid. When some dilute hydrochloric acid is added, the following equilibrium occurs:
   $$PbCl_2(s) + 2Cl^-(aq) \rightleftharpoons PbCl_4^{2-}(aq)$$
   The position of equilibrium is to the left, so most of the lead chloride is present as a solid. Predict what would happen if concentrated hydrochloric acid were added to solid lead chloride.

Chapter 14

Further organic chemistry

Silicon: glass, seals and semiconductors

Atomic number: 14

Electron configuration: $1s^2\ 2s^2\ 2p^6\ 3s^2\ 3p_x{}^1\ 3p_y{}^1$

Symbol: Si

Silicon is in group 4 of the periodic table. It is the second most abundant element in the Earth's crust. It occurs in sand and all igneous rocks. Sand is silicon dioxide, with impurities that give the sand colour.

A stained glass window

COREL

Glass

Can you conceive of a world without glass? Windows let in light but keep out the wind and the rain. Light bulbs let out the light, but prevent air from reaching the hot filament and burning it. Television and computer screens could not be made without glass.

Glass may feel like a solid, but it is actually a super-cooled liquid. The long chains of silicate ions are not arranged in a regular lattice. This means that glass softens on heating instead of having the sharp melting point characteristic of solids.

The earliest glass objects date from 2500 BC and were found in Mesopotamia. The Romans discovered the art of making flat sheets of glass that could be used for windows.

Glass is made by melting silicon dioxide with sodium carbonate and calcium carbonate. For cut glass, lead carbonate is added, which increases the refractive index of the glass, making it sparkle. Coloured glass is obtained by adding oxides of cobalt (blue), chromium or iron (green), and cadmium or copper (red).

Silicones

Silicones are polymeric compounds consisting of –O–Si–O– chains with side chains of methyl or other organic groups that are hydrophobic and so repel water. Therefore, silicones are used for waterproofing sprays, sealing gutters and filling gaps around baths and showers.

Because silicones are biologically inert, they can be used for cosmetic implants.

Semiconductors

Silicon is a poor conductor of electricity. However, when a small amount of a group 3 or group 5 element is added, its ability to conduct is greatly increased. This process is called doping. If about 1 in 200 000 silicon atoms is replaced by a phosphorus atom, an *n*-type semiconductor is produced, since the phosphorus atom has one more electron than the silicon it replaced. If a boron atom (which has one less outer-shell electron than silicon) replaces the silicon atom, a *p*-type semiconductor is produced.

Further organic chemistry

Alcohols

Alcohols are compounds that contain the C–OH group.

The other atoms attached to the carbon of the C–OH group are either hydrogen or carbon.

Alcohols have the general formula $C_nH_{(2n+1)}OH$. They are named from the parent alkanes, with a number representing the carbon atom in the chain to which the –OH group is joined.

The skeletal formulae, names and structural formulae of some alcohols are shown below:

Butan-1-ol
$CH_3CH_2CH_2CH_2OH$

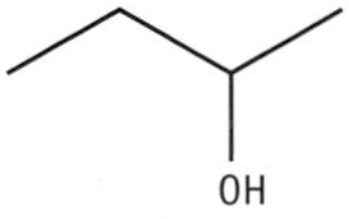

Butan-2-ol
$CH_3CH_2CH(OH)CH_3$

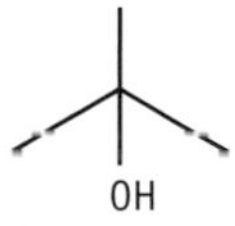

2-methylpropan-2-ol
$(CH_3)_3COH$

Primary alcohols have no more than one carbon atom directly attached to the carbon of the C–OH group. Methanol, CH_3OH, ethanol, CH_3CH_2OH, and propan-1-ol, $CH_3CH_2CH_2OH$, are all primary alcohols. All primary alcohols contain the $–CH_2OH$ group.

Secondary alcohols have two carbon atoms (and hence only one hydrogen atom) directly attached to the carbon of the C–OH group. Propan-2-ol, $CH_3CH(OH)CH_3$, is a secondary alcohol. All secondary alcohols contain the $>CH(OH)$ group.

Tertiary alcohols have three carbon atoms (and hence no hydrogen atoms) directly attached to the carbon of the C–OH group. 2-methylpropan-2-ol is a tertiary alcohol:

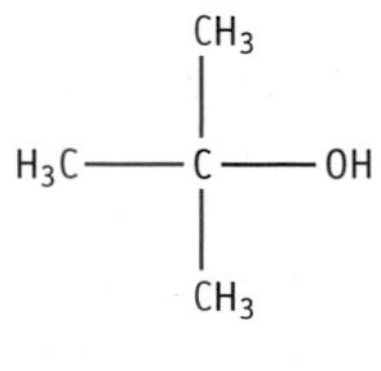

Physical properties

Most alcohols are liquids at room temperature. This is because there is hydrogen bonding between alcohol molecules. The oxygen–hydrogen bond is very polar because of the difference in the electronegativity of the two elements. The δ⁻ oxygen in one alcohol molecule forms a hydrogen bond with the δ⁺ hydrogen atom in another molecule (Figure 14.1). This intermolecular force is stronger and in addition to the instantaneous induced dipole–induced dipole (dispersion) forces such as those acting between alkane molecules and those between the polar halogenoalkane molecules. As the forces are stronger, more energy is required to separate the molecules, and this results in a higher boiling temperature.

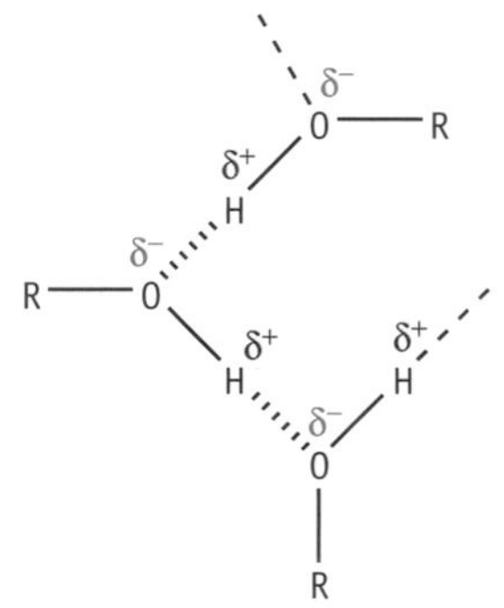

Figure 14.1 Hydrogen bonding between alcohol molecules

The lower members of the homologous series of alcohols are all completely miscible with water. This means that they dissolve in water in all proportions. Higher members are less soluble.

The reason for the solubility is the hydrogen bonding between alcohol and water molecules (Figure 14.2). This is similar in strength to the hydrogen bonding in pure water and in pure alcohol.

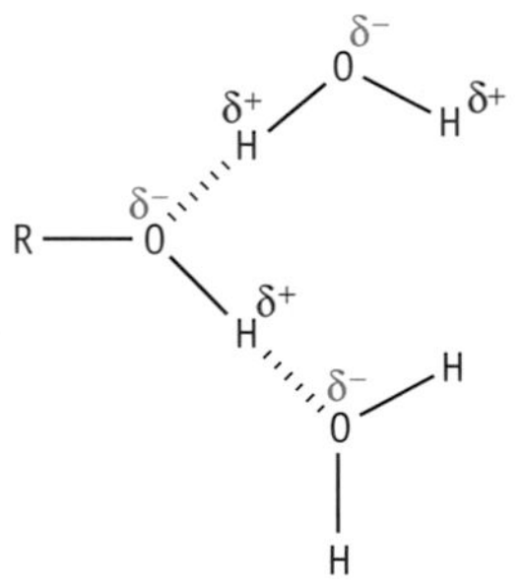

Figure 14.2 Hydrogen bonding between water and alcohol molecules

Manufacture of alcohols

Methanol

Methanol is manufactured by passing hot hydrogen and carbon monoxide gases over a catalyst of zinc and chromium oxides.

$$2H_2(g) + CO(g) \rightarrow CH_3OH(g) \quad \Delta H = -101 \text{ kJ mol}^{-1}$$

Methanol is toxic. It is added to industrial spirits (meths) in order to make it unfit to drink.

Ethanol

Ethanol can be made from carbohydrates, such as grain, by hydrolysis to glucose, followed by fermentation using yeast as a catalyst. The yeast respires anaerobically and converts the glucose into ethanol and carbon dioxide:

$$C_6H_{12}O_6 \rightarrow 2C_2H_5OH + 2CO_2$$

Ethanol can also be made by the catalytic addition of water to ethene, which is obtained from cracking one of the fractions from the primary distillation of crude oil (page 141).

$$H_2C{=}CH_2(g) + H_2O\ (g) \rightarrow CH_3CH_2OH(l)$$

Reactions of alcohols

Combustion

Alcohols, like almost all organic chemicals, burn to form carbon dioxide and water.

Ethanol burns with a clean flame:

$$C_2H_5OH + 3O_2 \rightarrow 2CO_2 + 3H_2O$$

Ethanol can be made by fermenting sugar or grain. Therefore, its use, when mixed with ordinary petrol, helps to reduce the carbon footprint of the fuel (page 305).

Oxidation

- Primary and secondary alcohols are oxidised by a solution of potassium dichromate(VI) and dilute sulfuric acid. The orange potassium dichromate(VI) solution turns green because it is reduced to hydrated Cr^{3+} ions.
- Tertiary alcohols are unaffected by this oxidising agent, which remains orange in colour.

Oxidation of a primary alcohol to an aldehyde

If a **primary** alcohol is heated with a mixture of potassium dichromate(VI) and sulfuric acid and the volatile product is allowed to escape, an aldehyde is produced — for example:

$$CH_3CH_2OH + [O] \rightarrow CH_3CHO + H_2O$$

The aldehyde is not hydrogen bonded and so has a lower boiling temperature than the alcohol. Therefore it boils off from the hot reaction mixture.

A suitable apparatus is shown in Figure 14.3.

[O] can always be used in the equations for oxidation reactions in organic chemistry.

Reagents:	potassium dichromate(VI) and sulfuric acid
Conditions:	add the oxidising agent to a mixture of hot ethanol and acid and distil off the aldehyde as it forms
Product:	ethanal
Reaction type:	oxidation

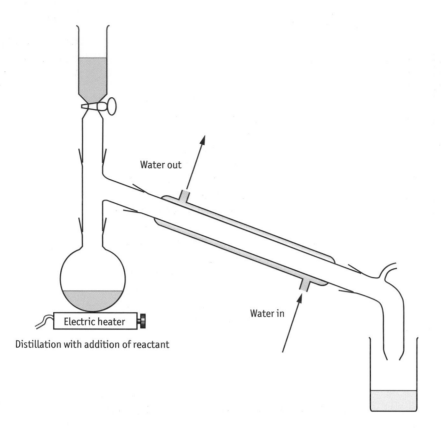

Water out
Water in
Electric heater
Distillation with addition of reactant

Figure 14.3 Apparatus for the oxidation of a primary alcohol to an aldehyde

Oxidation of a primary alcohol to a carboxylic acid

If a primary alcohol is heated under reflux with the oxidising mixture, a carboxylic acid is formed — for example:

$$CH_3CH_2OH + 2[O] \rightarrow CH_3COOH + H_2O$$

The acid can be distilled off after the reaction has gone to completion.

Reagents: potassium dichromate(VI) and sulfuric acid
Conditions: heat the mixture under reflux
Product: ethanoic acid
Reaction type: oxidation

A suitable apparatus is shown in Figure 14.4.

Oxidation of a secondary alcohol to a ketone

If a secondary alcohol is heated under reflux with the oxidising mixture, a ketone is produced — for example:

$$CH_3CH(OH)CH_3 + [O] \rightarrow CH_3COCH_3 + H_2O$$

Reagents: potassium dichromate(VI) and sulfuric acid
Conditions: heat the mixture under reflux
Product: propanone
Reaction type: oxidation

e Make sure that you give the *full* name or formula when a reagent is asked for. Phrases such as 'acidified dichromate' will not score full marks.

A suitable apparatus is shown in Figure 14.4.

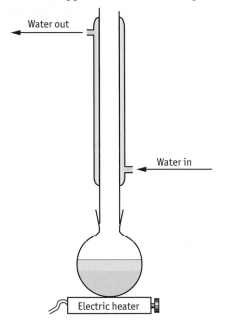

Water out

Water in

Electric heater

Figure 14.4 Reflux apparatus for the production of carboxylic acids and ketones

e If you are asked to draw apparatus, make sure that:
- the apparatus is made up of specific pieces, such as the flask and the condenser, rather than one continuous piece of glassware
- the apparatus is open and not sealed (for heating under reflux, the apparatus is open at the top; for distillation, it is open at the receiving flask)
- you take care when drawing condensers; show the water going in at the bottom and out at the top

Reaction with metallic sodium

Alcohols contain an –OH group and, therefore, react with sodium to form hydrogen gas. The other product is an ionic compound containing a sodium ion, Na^+, and an anion with a negative charge on an oxygen atom.

With ethanol, hydrogen and sodium ethoxide are formed:

$$C_2H_5OH + Na \rightarrow \tfrac{1}{2}H_2 + C_2H_5O^-Na^+$$

If the excess ethanol is evaporated, the sodium ethoxide is seen as a white solid.

If water is added, the ethoxide ion acts as a base and removes a proton from the water, forming an alkaline solution of sodium hydroxide:

$$C_2H_5O^- + H_2O \rightarrow C_2H_5OH + OH^-$$

Note the similarities between this reaction and those of water and of carboxylic acids with sodium:

$$H_2O + Na \rightarrow \tfrac{1}{2}H_2 + Na^+OH^-$$
$$CH_3COOH + Na \rightarrow \tfrac{1}{2}H_2 + CH_3COO^-Na^+$$

The alcohol is acting as an acid. It loses H^+, which takes an electron from the sodium to form hydrogen gas.

Halogenation

In a halogenation reaction, the –OH group of an alcohol molecule is replaced by a halogen atom.

Chlorination

When solid phosphorus pentachloride is added to a dry alcohol, clouds of hydrogen chloride fumes are produced, mixed with the chloroalkane. Phosphorus oxychloride remains in the vessel.

$$CH_3CH_2OH + PCl_5 \rightarrow CH_3CH_2Cl + POCl_3 + HCl$$

Bromination

When an alcohol is heated under reflux with a mixture of potassium bromide and 50% sulfuric acid, hydrogen bromide is first produced, which then reacts with the alcohol to form a bromoalkane:

$$KBr + H_2SO_4 \rightarrow KHSO_4 + HBr$$
$$CH_3CH_2OH + HBr \rightarrow CH_3CH_2Br + H_2O$$

The acid used is only 50% concentrated to prevent the hydrogen bromide that is produced from being oxidised (by sulfuric acid) to bromine (page 226).

Iodination

Warming a mixture of damp red phosphorus and iodine produces phosphorus triiodide, which then reacts with the alcohol to form the iodoalkane and phosphorus(III) acid.

$$2P + 3I_2 \rightarrow 2PI_3$$
$$3CH_3CH_2OH + PI_3 \rightarrow 3CH_3CH_2I + H_3PO_3$$

The moisture is to bring the iodine and phosphorus, both of which are solids, into contact so that they can react.

	Chlorination	Bromination	Iodination
Reagent	PCl_5	KBr and 50% H_2SO_4	I_2 and moist red phosphorus
Conditions	Room temperature	Heat under reflux	Warm
Product	Chloroethane	Bromoethane	Iodoethane
Reaction type	Substitution	Substitution	Substitution

Table 14.1 Halogenation of alcohols

Esterification

The reactions of alcohols with carboxylic acids and with acid chlorides are not required at AS. These reactions are covered at A2.

Summary of reactions of alcohols

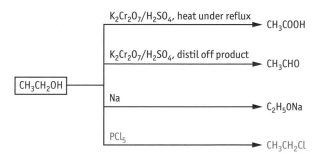

The reactions in red are oxidation reactions.
The reaction in blue is acid + metal.
The reaction in green is substitution (halogenation).

Test for alcohols

Add phosphorus pentachloride to the *dry* substance under test. All alcohols will produce steamy fumes of hydrogen chloride. This is a test for an –OH group, so the same result is obtained with carboxylic acids.

Do not say that *white* fumes or *white smoke* are produced. The steamy fumes of hydrogen chloride produce a white smoke (of ammonium chloride) when a glass rod dipped in concentrated ammonia is held in the fumes.

A common error is to state that a positive result with this test means that the unknown is an alcohol. The correct answer for a positive PCl_5 test is that the unknown substance contains an –OH group.

To distinguish between an acid and an alcohol, test the pH of the substance. Alcohols are neutral, whereas the pH of acids is less than 7.

Further tests to differentiate between types of alcohol

- Tertiary alcohols do not change the colour of a heated solution of potassium dichromate(VI) and sulfuric acid.

Do not forget to give the colour *before* as well as *after*. Primary and secondary alcohols turn hot potassium dichromate in dilute sulfuric acid from orange to green. With tertiary alcohols, the solution stays orange.

- When secondary alcohols containing the $CH_3CH(OH)$ group are warmed with a mixture of iodine and sodium hydroxide solution, a pale yellow precipitate of triiodomethane, CHI_3 (iodoform), is formed. This reaction also takes place with ethanol. This reaction is explained at A2.

Halogenoalkanes

A **halogenoalkane** is a compound in which one or more hydrogen atoms in an alkane has been replaced by halogen atoms and has only single bonds in the molecule.

Halogenoalkanes have the general formula $C_nH_{(2n+1)}X$, where X is a halogen atom. They are named from the parent alkanes, with a number representing the carbon atom in the chain to which the halogen is joined.

The skeletal formulae, names and structural formulae of some halogenoalkanes are shown below:

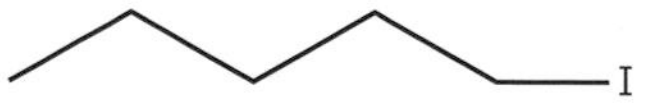

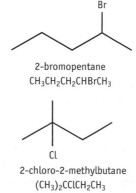

1-iodopentane
$CH_3CH_2CH_2CH_2CH_2I$

2-bromopentane
$CH_3CH_2CH_2CHBrCH_3$

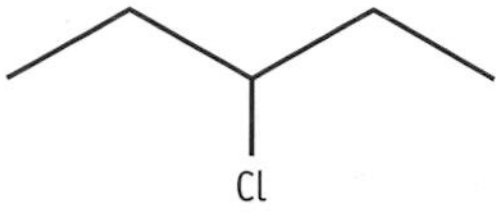

3-chloropentane
$CH_3CH_2CHClCH_2CH_3$

2-chloro-2-methylbutane
$(CH_3)_2CClCH_2CH_3$

Primary halogenoalkanes have no more than one carbon atom directly attached to the carbon in the C–halogen group. Some examples are shown below:

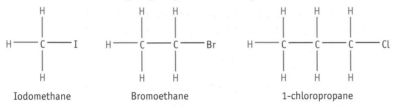

Iodomethane   Bromoethane   1-chloropropane

Secondary halogenoalkanes have two carbon atoms (and hence only one hydrogen atom) directly attached to the carbon in the C–halogen group — for example, 2-chloropropane:

$$
\begin{array}{c}
\text{H} \quad\quad \text{H} \quad\quad \text{H} \\
| \quad\quad\quad | \quad\quad\quad | \\
\text{H} - \text{C} - \text{C} - \text{C} - \text{H} \\
| \quad\quad\quad | \quad\quad\quad | \\
\text{H} \quad\quad \text{Cl} \quad\quad \text{H}
\end{array}
$$

Tertiary halogenoalkanes have three carbon atoms (and hence no hydrogen atoms) directly attached to the carbon in the C–halogen group — for example, 2-chloro-2-methylpropane:

$$
\begin{array}{c}
\text{H} \\
| \\
\text{H} - \text{C} - \text{H} \\
\text{H} \quad\quad | \quad\quad \text{H} \\
| \quad\quad\quad | \quad\quad\quad | \\
\text{H} - \text{C} - \text{C} - \text{C} - \text{H} \\
| \quad\quad\quad | \quad\quad\quad | \\
\text{H} \quad\quad \text{Cl} \quad\quad \text{H}
\end{array}
$$

Halogens are electronegative elements. The carbon–halogen bond in halogeno-alkanes is polarised, making the carbon atom δ^+ and the halogen atom δ^-:

$$
\begin{array}{c}
| \quad\quad \delta^+ \quad \delta^- \\
- \text{C} - \text{Cl} \\
|
\end{array}
$$

Physical properties

Chloromethane, bromomethane and chloroethane are gases at room temperature. Iodomethane and higher members of the homologous series are liquids. The boiling temperatures of halogenoalkanes are higher than those of alkanes, primarily because halogenoalkanes contain more electrons and so have stronger instantaneous induced dipole–induced dipole forces between the molecules. Also, the molecules are polar, so there are permanent dipole–dipole forces between the molecules, which strengthen the intermolecular forces and so increase the boiling temperature.

Preparation of halogenoalkanes

Chloroalkanes

The simplest method of preparation of a chloroalkane is to add phosphorus pentachloride to the alcohol. For example, to make 2-chloropropane, add phosphorus pentachloride to propan-2-ol:

$$CH_3CH(OH)CH_3 + PCl_5 \rightarrow CH_3CHClCH_3 + POCl_3 + HCl$$

An alternative method is to add concentrated sulfuric acid to a mixture of propan-2-ol and potassium chloride. The sulfuric acid protonates the alcohol and releases hydrogen chloride from the potassium chloride:

$$KCl + H_2SO_4 \rightarrow HCl + KHSO_4$$
$$CH_3CH(OH)CH_3 + HCl \rightarrow CH_3CHClCH_3 + H_2O$$

This method is less efficient because some dehydration of the alcohol to propene also occurs.

Bromoalkanes

Potassium bromide cannot be used with concentrated sulfuric acid because much of the hydrogen bromide produced initially would be oxidised to bromine by the concentrated acid (page 226).

An alternative method is to use 50% sulfuric acid. This is slowly added, with cooling, to the alcohol. The mixture is then added to solid potassium bromide and heated. The bromoalkane (boiling temperature 59°C) is more volatile than the alcohol (boiling temperature 82°C). Therefore, it distils off and is condensed and collected.

$$KBr + H_2SO_4 \rightarrow HBr + KHSO_4$$
$$CH_3CH(OH)CH_3 + HBr \rightarrow CH_3CHBrCH_3 + H_2O$$

Iodoalkanes

Even 50% sulfuric acid oxidises hydrogen iodide completely. To prepare an iodoalkane, the alcohol is warmed with a mixture of damp red phosphorus and iodine. The iodine reacts with the phosphorus to form phosphorus triiodide, which then reacts with the alcohol to form the iodoalkane:

$$2P + 3I_2 \rightarrow 2PI_3$$
$$3C_2H_5OH + PI_3 \rightarrow 3C_2H_5I + H_3PO_3$$

Even though they are polar molecules, halogenoalkanes are insoluble in water. This is because the molecules do not contain δ^+ hydrogen atoms, so they cannot hydrogen-bond with water. They are soluble in a variety of organic solvents, such as ethanol and ethoxyethane (ether), $C_2H_5OC_2H_5$.

Substitution reactions

A substitution reaction is a reaction in which an atom or group is replaced by another atom or group.

There are always two reactants and two products in a substitution reaction.

The halogen atom in a halogenoalkane molecule can be replaced by an –OH, –CN or –NH$_2$ group. These reactions are examples of **nucleophilic substitution**, because the attacking group is a **nucleophile**. In these reactions, the reagent forms a bond to the carbon using its lone pair of electrons and the halide ion is released.

> A nucleophile is a species with a lone pair of electrons that is used to form a covalent bond with a δ^+ atom in another molecule.

The nucleophile attacks the δ^+ carbon atom in the halogenoalkane.

Reaction with aqueous alkali

When a halogenoalkane is heated under reflux with an *aqueous* solution of an alkali, such as sodium hydroxide or potassium hydroxide, the halogen is replaced by an –OH group and an alcohol is produced — for example:

$$CH_3CH_2CH_2Br + NaOH \rightarrow CH_3CH_2CH_2OH + NaBr$$

The ionic equation is:

$$CH_3CH_2CH_2Br + OH^-(aq) \rightarrow CH_3CH_2CH_2OH + Br^-(aq)$$

Reagent: sodium (or potassium) hydroxide
Conditions: heat under reflux in *aqueous* solution
Product: propan-1-ol
Reaction type: nucleophilic substitution (also called hydrolysis)

> ⓔ This reaction is sometimes carried out with the halogenoalkane dissolved in alcohol and the alkali dissolved in water. This is because the organic compound is not water soluble, and so little contact would occur between it and the hydroxide ions. If a solution of the alkali in *ethanol* is used, a different reaction occurs, particularly with secondary and tertiary halogenoalkanes (pages 270–71).

A lone pair of electrons on the oxygen atom in the OH^- ion attacks the δ^+ carbon in the carbon–bromine bond and forms a new carbon–oxygen σ-bond. The bromine gains the electrons from the carbon–bromine σ-bond, which breaks, forming a Br^- ion.

The OH^- ion is the nucleophile.

The rate of this reaction varies considerably according to the nature of the halogenoalkane. The rate of reaction is in the order:

iodo- > bromo- > chloro- > fluoro-

The reason for this is the difference in bond enthalpies of the C–halogen bonds (Table 14.2).

Bond	Average bond enthalpy/kJ mol^{-1}
C–F	+484
C–Cl	+338
C–Br	+276
C–I	+238

Table 14.2 Bond enthalpies

A weaker carbon–halogen bond means that there is a smaller activation energy (page 236) for a reaction involving breaking that bond. This results in a faster reaction. The reaction is fastest with tertiary and slowest with primary halogenoalkanes.

The explanation for this is complex. A –CH_3 group (and to a lesser extent a –CH_2 group) has a tendency to push electrons in the σ-bond away from itself. This facilitates the release of the halide ion because there is a general shift of electrons away from the –CH_3 group towards the halogen atom.

Mechanism of nucleophilic substitution

Halogenoalkanes react with nucleophiles such as OH^-, H_2O, NH_3 and CN^- in substitution reactions. This is because the carbon atom joined to the halogen is slightly δ^+ and is, therefore, attacked by nucleophiles.

There are two distinct mechanisms by which halogenoalkanes react. Which mechanism is followed depends on whether they are primary, secondary or tertiary halogenoalkanes.

Primary halogenoalkanes

These react in an S_N2 **reaction**.

The mechanism is called S_N2 because it is a substitution (S), is nucleophilic (N) and there are two species involved at the beginning of the mechanism.

The mechanism is a single step that goes through a **transition state.**

An example of an S_N2 mechanism involving a transition state is the reaction between hydroxide ions and bromoethane:

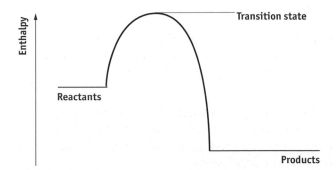

The red curly arrow shows the movement of a lone pair of electrons from the oxygen to the carbon as a covalent bond forms. The green arrow represents the electrons in the C–Br σ-bond moving to the bromine atom as the bond breaks.

The transition state occurs when the new O–C bond is half-formed and the C–Br bond is half-broken.

The reaction profile diagram for this type of reaction is shown in Figure 14.5.

> A transition state is not a species that can be isolated. It changes immediately into the product.

> *e* Do not forget to include the negative charge on the transition state.

Figure 14.5 Reaction profile for an S_N2 reaction involving a transition state

The energy released in the formation of the O–C bond is enough to provide the energy to break the C–halogen bond. The weaker the C–halogen bond, the faster the rate of the reaction. Therefore, since the C–Cl bond is the strongest and the C–I bond is the weakest, the rate order is:

C–I > C–Br > C–Cl

Tertiary halogenoalkanes

Tertiary halogenoalkanes react by S_N1 mechanisms.

The mechanism is called S_N1 because it is a substitution (S), is nucleophilic (N) and there is only one species involved at the beginning of the mechanism.

This type of reaction takes place in two steps.

Step 1: the carbon–halogen bond breaks, a carbocation is formed and a halide ion is released. This is the slower rate-determining step.

Step 2: the carbocation is attacked by the nucleophile in a faster reaction.

The rate of the first step is slower than that of the second step. Therefore, the activation energy for the first step is greater than that of step 2. This is shown in Figure 14.6.

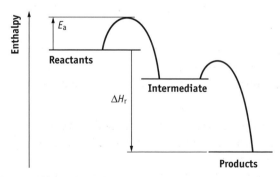

Figure 14.6 Reaction profile for an S_N1 reaction involving an intermediate

Secondary halogenoalkanes
- The rate of reactions with S_N2 mechanisms decreases in the order primary > secondary > tertiary halogenoalkane.
- The rate of reactions with S_N1 mechanisms increases in the order primary < secondary < tertiary halogenoalkane.

These relationships are shown graphically in Figure 14.7.

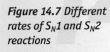

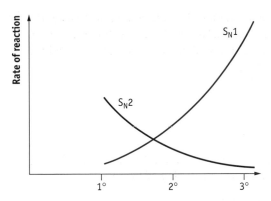

Figure 14.7 *Different rates of S_N1 and S_N2 reactions*

Figure 14.7 shows that primary halogenoalkanes react almost entirely by an S_N2 mechanism and that tertiary halogenoalkanes react by an S_N1 mechanism.

Secondary halogenoalkanes react by both mechanisms. However, S_N1 is faster and is, therefore, the dominant mechanism.

The overall rate is fastest with a tertiary halogenoalkane and slowest with a primary. For example, 2-chloro-2-methylpropane, $(CH_3)_3CCl$, produces an instant precipitate of silver chloride with aqueous silver nitrate, whereas 1-chloropropane gives a precipitate only after heating for a long period.

Reaction with water containing silver nitrate solution

The halogenoalkane reacts with the water to form an alcohol and a solution that contains the halide as an anion:

$$R{-}X(aq) + H_2O(l) \rightarrow R{-}OH(aq) + H^+(aq) + X^-(aq)$$
where X represents a halogen

This halide ion then reacts with silver ions from the silver nitrate to form a precipitate of silver halide:

$$X^-(aq) + Ag^+(aq) \rightarrow AgX(s)$$

The rates of the reaction of the halogenoalkanes can be compared by noting the time taken for the precipitates to appear.

The reaction involves nucleophilic attack on the carbon atom joined to the halogen by the lone pair of electrons on the oxygen in water. Since water is not as good an electrophile as hydroxide ions, the reaction is much slower than that with aqueous sodium hydroxide.

Equal volumes of each halogenoalkane are placed in test tubes, which are then put in a water bath at 60°C. Some silver nitrate solution is then added and the time taken for a precipitate to appear is noted.

- Primary halogenoalkanes hardly react with water. The C–Cl bond is too strong for any noticeable precipitate to be observed. 1-bromoalkanes give a precipitate very slowly; 1-iodoalkanes give a precipitate slightly more quickly.
- Secondary halogenoalkanes react slowly to form a secondary alcohol and a precipitate is soon observed. Precipitate formation is quicker with secondary iodoalkanes than with secondary bromoalkanes, which in turn form precipitates more quickly than secondary chloroalkanes.

- Tertiary halogenoalkanes react rapidly to produce a tertiary alcohol. A precipitate is seen as soon as the silver nitrate is added.

The carbon–halogen bond has been broken by water, so the reaction is called **hydrolysis.**

Reaction with ammonia

The reaction between a halogenoalkane and ammonia produces an **amine**, which is a molecule containing the $-NH_2$, $>N-H$ or $>N-$ group.

Ammonia is a gas that is soluble in water. However, a solution cannot be heated under reflux because ammonia gas would be liberated. This would escape because it would not be condensed by the reflux condenser. Therefore, the halogenoalkane and the ammonia solution must be heated in a sealed container. Alternatively, a concentrated ammonia solution can be used and the mixture left at room temperature for a long time.

A simplified equation for the reaction between 1-chloropropane and ammonia is:

$$CH_3CH_2CH_2Cl + 2NH_3 \rightarrow CH_3CH_2CH_2NH_2 + NH_4Cl$$

Reagent: excess ammonia
Conditions: concentrated solution of ammonia at room temperature or heated in a sealed tube
Product: 1-aminopropane
Reaction type: nucleophilic substitution

The lone pair of electrons on the nitrogen atom attacks the δ^+ carbon atom. Ammonia is therefore the nucleophile. As this happens, the C–halogen bond breaks as in the S_N2 reaction on page 269. Note that the intermediate is neutral, because the attacking species is neutral.

The final step is the loss of H^+, which is removed by a second NH_3 molecule:

The product, 1-aminopropane, can react with more 1-chloropropane to produce the secondary amine, $(CH_3CH_2CH_2)_2NH$, and some tertiary amine, $(CH_3CH_2CH_2)_3N$.

> **e** Under these alkaline conditions, the salt NH_4Cl is produced rather than the acidic HCl. Stating that HCl is a product of this reaction is a common error at AS.

Reaction with aromatic compounds

Halogenoalkanes react with aromatic compounds such as benzene. These reactions are covered at A2 and so are not dealt with here.

Elimination reactions

An elimination reaction is a reaction in which two atoms, or one atom and one small group, are removed from adjacent carbon atoms, forming an unsaturated compound.

When a halogenoalkane is heated under reflux with a concentrated solution of potassium hydroxide in *ethanol*, the halogen atom is removed from the carbon atom to which it is attached and a hydrogen atom is removed from an adjacent carbon atom. The organic product is an alkene or a mixture of alkenes.

With either 1-chloropropane or 2-chloropropane, the only elimination product is propene:

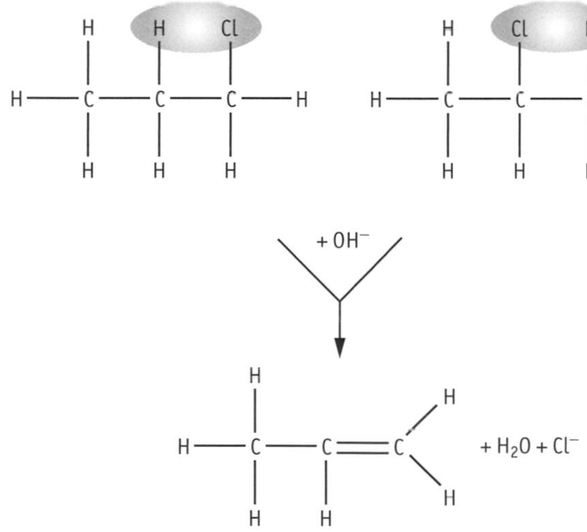

JOHN OLIVE

Heating under reflux

Reagent: concentrated potassium hydroxide
Conditions: heat under reflux in solution in *ethanol*
Product: propene
Reaction type: elimination

1-chlorobutane produces but-1-ene:

$$CH_3CH_2CH_2CH_2Cl + OH^- \rightarrow CH_3CH_2CH{=}CH_2 + H_2O + Cl^-$$

However, 2-chlorobutane produces a mixture of but-1-ene and both geometric isomers of but-2-ene.

◀ The conditions determine whether elimination or substitution takes place. For substitution, the alkali is aqueous; for elimination, it is dissolved in ethanol.

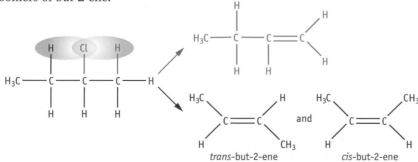

trans-but-2-ene   and   *cis*-but-2-ene

In practice, the situation is more complex because, as well as elimination reactions, substitution reactions take place. Elimination is favoured with tertiary halogenoalkanes and substitution with primary alkanes. The reaction of bromoethane with ethanolic potassium hydroxide produces only about 1% ethene, whereas the reaction using 2-bromopropane gives about 80% propene. Tertiary halogenoalkanes give almost 100% alkenes as the product under these conditions.

Summary of reactions of halogenoalkanes

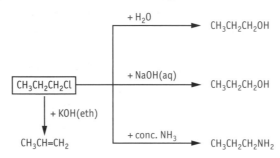

The reactions in red are nucleophilic substitution reactions.
The reaction in blue is elimination.

Test for halogenoalkanes

- Add a few drops of the halogenoalkane to an aqueous solution of sodium hydroxide and warm for several minutes.
- Cool.
- Add dilute nitric acid until the solution is just acidic to litmus.
- Add silver nitrate solution.

Result
- Chloroalkanes give a white precipitate that is soluble in dilute ammonia solution.
- Bromoalkanes give a cream precipitate that is insoluble in dilute ammonia solution but soluble in concentrated ammonia.
- Iodoalkanes give a pale yellow precipitate that is insoluble in dilute and concentrated ammonia.

The sodium hydroxide solution hydrolyses the halogenoalkane and liberates the halide ion. The solution has to be made acidic to prevent the precipitation of silver oxide when the silver nitrate is added. The precipitates are silver chloride (soluble in dilute ammonia), silver bromide (soluble only in concentrated ammonia) and silver iodide (insoluble in concentrated ammonia).

Uses of halogenoalkanes

Solvents
Compounds with several halogen atoms are used as solvents to remove grease from metals before electroplating. An example is 1,1,2-trichloroethane.

Refrigerants

Fully substituted compounds, such as the CFCs freon 12, CF_2Cl_2, and freon 113, $CF_2ClCFCl_2$, are used as coolants in refrigerators and in air conditioning. They are chemically inert, non-toxic and non-flammable. When released, their inert nature causes an environmental problem. They diffuse unchanged up into the stratosphere, where they are decomposed by ultraviolet radiation, producing chlorine radicals. The chlorine radicals break down ozone in a chain reaction. It has been estimated that a single chlorine atom will cause the decomposition of 100 000 ozone molecules. During the Antarctic winter, the chlorine radicals build up on ice crystals in the air, and when the ice is melted by the spring sun, massive depletion of the polar ozone occurs, causing a hole in the ozone layer in the southern hemisphere.

New refrigerant coolants, which contain hydrogen atoms as well as chlorine and fluorine, are being developed. An example is HCFC-22, which has the formula $CHClF_2$. HCFCs are more reactive and are broken down at lower altitudes, avoiding harm to the ozone layer. However, they are potent greenhouse gases.

Herbicides and pesticides

Herbicides kill plants. Two examples are 2,4-D and 2,4,5-T, which have the following structures:

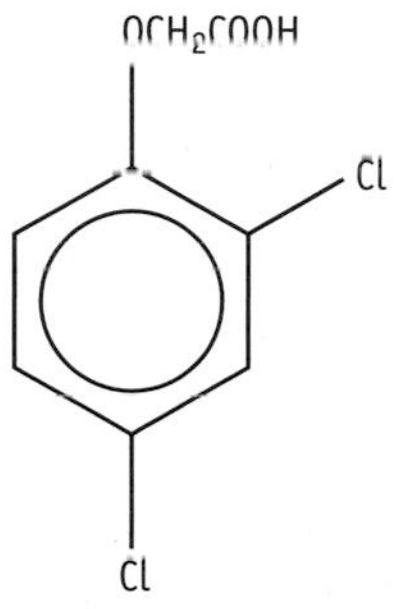

2,4-D 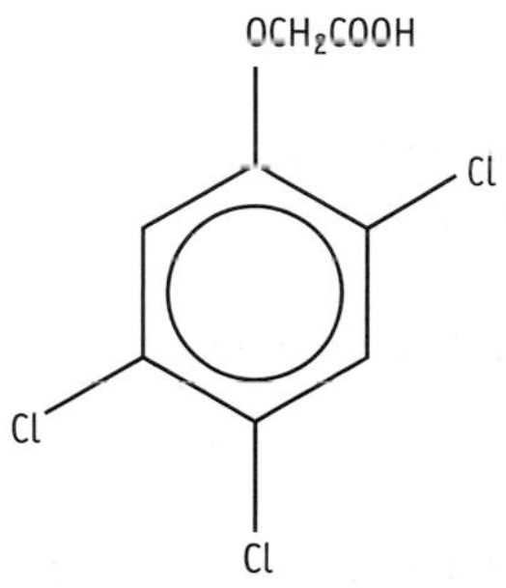   2,4,5-T

Pesticides kill insects. One of the most effective was DDT (dichlorodiphenyltrichloroethane). Unfortunately, it was found that its concentration built up in the fat of animals high in the food chain, causing problems with reproduction. Therefore, its use has been banned in many countries. DDT eradicated malaria from some countries and helped to control it in others. Since its ban, deaths from malaria in some tropical countries have risen.

The problem associated with the use of halogen-containing herbicides and pesticides is that they are inert to chemical and biological attack. They remain in the environment for a very long period. In the 1970s, the US military sprayed the forests of Vietnam with the defoliant Agent Orange, a mixture of 2,4-D and 2,4,5-T, and its adverse effects are still being seen. These chemicals are inert mainly because of the strength of the carbon–chlorine bond, which is particularly strong if bonded to benzene rings, as in DDT.

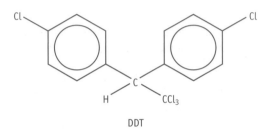

DDT

Summary of mechanisms

Types of reaction

Addition

In an addition reaction two reactants form a single product. An example is the addition of hydrogen to an alkene:

$$CH_2=CH_2 + H_2 \rightarrow CH_3-CH_3$$

Substitution

In a substitution reaction, one atom or group of atoms replaces another atom or group of atoms in a molecule. An example is the reaction between hydroxide ions and bromoethane:

$$CH_3CH_2Br + OH^- \rightarrow CH_3CH_2OH + Br^-$$

Elimination

In an elimination reaction, two atoms are removed from adjacent carbon atoms in a molecule of an organic substance. An example is the removal of a hydrogen and a chlorine atom from 2-chloropropane to form propene:

$$CH_3CHClCH_3 + OH^- \rightarrow CH_3CH=CH_2 + Cl^- + H_2O$$

Polymerisation

Polymerisation occurs when thousands of monomer molecules join together to form a long polymeric chain. An example is the polymerisation of ethene to poly(ethene):

$$nCH_2=CH_2 \rightarrow (CH_2-CH_2)_n$$

Oxidation

In organic chemistry, oxidation is best defined as the addition of oxygen or the removal of hydrogen.

> The oxidation agent is usually written as [O] in oxidation equations.

Secondary alcohols are oxidised to ketones. Two hydrogen atoms are removed from the alcohol. Therefore, it is oxidised:

$$CH_3CH(OH)CH_3 + [O] \rightarrow CH_3COCH_3 + H_2O$$

Hydrolysis

Hydrolysis occurs when water, often with a catalyst of acid or alkali, splits an organic molecule into two parts. The –OH group joins one part and the hydrogen atom joins the other part. An example is the hydrolysis of 2-iodopropane:

$$CH_3CHICH_3 + HOH \rightarrow CH_3CH(OH)CH_3 + HI$$

Type of reagent

Free radical

A free **radical** is an atom or group of atoms with an unpaired electron that it uses to form a covalent bond. A free radical is formed when a bond breaks homolytically, with one electron going to each atom in the bond. When chlorine is exposed to ultraviolet light, the Cl–Cl bond breaks homolytically and two chlorine radicals are produced. A fishhook arrow shows the movement of each electron:

$$Cl \text{—} Cl \longrightarrow Cl\bullet + Cl\bullet$$

Nucleophile

A nucleophile is a species with a lone pair of electrons that it uses to form a covalent bond. Nucleophiles can be:

■ negative ions — for example, OH^-, I^- and CN^-
■ neutral molecules containing oxygen or nitrogen atoms — for example, water and ammonia

Nucleophiles attack δ^+ centres in molecules.

Electrophile

An electrophile attacks places of high electron density — for example, the C=C group in alkenes. The electrophile forms a covalent bond with a pair of electrons from the electron-rich site. Electrophiles can be:

■ positive ions — for example, H^+
■ δ^+ atoms in molecules — for example, the δ^+ hydrogen atom in HBr
■ neutral molecules that can be polarised — for example, Br–Br.

Understanding organic reactions

Organic reactions can be learnt or they can be understood. The first thing to do is to look at the molecule and ask:

■ Is it polar with a δ^+ site? If so, it is likely to be attacked by nucleophiles. A halogenoalkane has a δ^+ carbon joined to the halogen and so undergoes substitution reactions with nucleophiles, resulting in the loss of a halide anion. As ions are formed during this type of reaction, it is usually carried out under aqueous conditions.
■ Does it have an electron-rich site such as a non-polar π-bond? If so, it is likely to be attacked by an electrophile and the π-electrons will form a new bond with the electrophile. Alkenes are attacked by electrophiles.
■ Is the molecule non-polar, without a site of high electron density? If so, it will react with free radicals. Light or peroxides are usually needed to produce the free radicals, so a non-aqueous solvent (or gaseous-phase reaction) and ultraviolet light will probably be necessary. Alkanes react in this way.

Quantitative organic chemistry

Formulae

The **molecular formula** of a compound shows the number of atoms of each element in one molecule of the substance.

e A bond in a molecule breaks heterolytically when it is attacked by a nucleophile. Heterolytic fission of a bond results in both bonding electrons going to the same atom. For example, when the OH^- nucleophile attacks chloroethane, both electrons in the C–Cl bond go to the chlorine, forming a Cl^- ion.

For example, the molecular formula of ethanoic acid is $C_2H_4O_2$ and that of glucose is $C_6H_{12}O_6$.

The **empirical formula** of a substance is the simplest whole number ratio of the atoms of each element in the substance.

For example, the empirical formula for both ethanoic acid and glucose is CH_2O.

The **structural formula** is an unambiguous structure that shows how the atoms in the molecule are arranged.

For example:

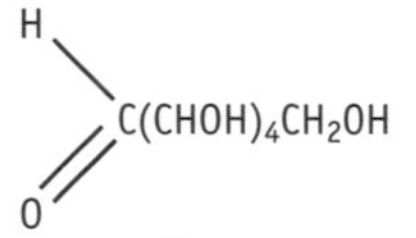

Glucose

The *full* structural formula of a compound shows every atom and every bond:

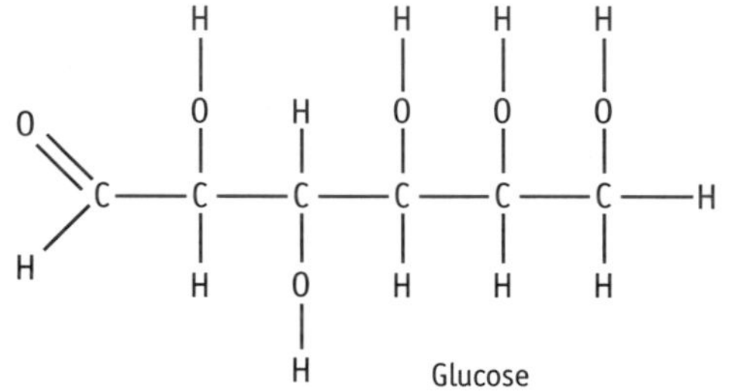

Glucose

The full structural formula is sometimes known as the displayed formula.

Glucose also exists as a six-membered ring compound:

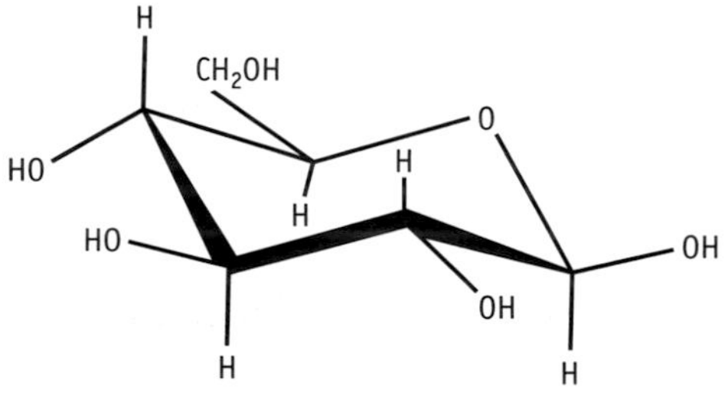

Empirical formula

This can be worked out from the percentage by mass of each element in the compound. The calculation is done in three stages.

Step 1: divide each percentage by the relative atomic mass of the element to give the number of moles of each element in 100 g of the compound.

Ⓔ Make sure you use atomic masses, *not* atomic numbers, in this step.

Step 2: these values, which are the relative number of moles of each element, are then divided by the smallest value. Give the results to one decimal place.

Step 3: if the numbers are not whole numbers, multiply by 2 to see if that converts them into integers. If not, try multiplying by 3 and so on, until all the values are whole numbers.

Ⓔ The calculation is best done in the form of a table.

The integers obtained in the final step are the simplest ratio of the numbers of each element present in the compound. The empirical formulae can now be written.

Worked example 1

A compound contains 62.1% carbon, 10.3% hydrogen and 27.6% oxygen by mass. Calculate its empirical formula.

Answer

Element	%	Divide by r.a.m.	Divide by the smallest	Ratio
Carbon	62.1	62.1/12 = 5.175	5.175/1.725 = 3.0	3
Hydrogen	10.3	10.3/1 = 10.3	10.3/1.725 = 5.97 → 6.0	6
Oxygen	27.6	27.6/16 = 1.725	1.725/1.725 = 1.0	1

The empirical formula is C_3H_6O.

Worked example 2

A compound contains 31.9% carbon, 5.3% hydrogen and 62.8% chlorine by mass. Calculate its empirical formula.

Answer

Element	%	Divide by r.a.m	Divide by the smallest	Multiply by 2 to find the whole number ratio
Carbon	31.9	31.9/12 = 2.65	2.65/1.77 = 1.5	3
Hydrogen	5.3	5.3/1 = 5.3	5.3/1.77 = 3.0	6
Chlorine	62.3	62.8/35.5 = 1.77	1.77/1.77 = 1.0	2

The empirical formula is $C_3H_6Cl_2$.

Molecular formula

If the molar mass is known, the molecular formula can be calculated from the empirical formula. This is done in three stages.

Step 1: calculate the empirical mass.

Step 2: divide the molar mass by the empirical mass. The answer is a whole number.

Step 3: multiply the number of atoms of each element by the integer obtained in step 2 to give the actual numbers of atoms of each element in the molecule.

Worked example

A substance X has the empirical formula, CH. The molar mass of X is 78 g mol^{-1}. Calculate the molecular formula.

Answer

Step 1: empirical mass = 12 + 1 = 13

Step 2: $\dfrac{\text{molar mass}}{\text{empirical mass}} = \dfrac{78}{13} = 6$

Step 3: molecular formula = C_6H_6

e Make sure that you show your working when calculating the molecular formula from the empirical formula.

Reaction yield

Theoretical yield

The theoretical yield is the mass of product that would be formed if the reaction went to 100% completion, with no side reactions.

It is calculated in three steps:

Step 1: calculate the amount of reactant in moles.

Step 2: use the reaction stoichiometry to calculate the amount of product in moles.

Step 3: Convert moles of product to mass of product.

> ### Worked example
>
> Calculate the theoretical yield when 1.23 g of buta-1,3-diene reacts with excess bromine according to the equation:
>
> $$CH_2=CH-CH=CH_2 + 2Br_2 \rightarrow CH_2BrCHBrCHBrCH_2Br$$
>
> **Answer**
>
> **Step 1:** molar mass of buta-1,3-diene = 54 g mol^{-1}
>
> $$\text{amount of buta-1,3-diene} = \frac{1.23\ g}{54\ g\ mol^{-1}} = 0.0228\ mol$$
>
> **Step 2:** ratio of product to reactant = 1:1
>
> amount of product = 0.0228 mol
>
> **Step 3:** molar mass of product = 54 + (4 × 80)
>
> $$= 374\ g\ mol^{-1}$$
>
> theoretical yield = amount of product (moles) × molar mass
>
> $$= 0.0228\ mol \times 374\ g\ mol^{-1}$$
>
> $$= 8.53\ g$$

Percentage yield

The percentage yield is defined as:

$$\% \text{ yield} = \frac{\text{actual yield}}{\text{theoretical yield}} \times 100$$

The actual yield is the measured mass of the product obtained in the experiment.

In questions, the actual yield is usually given and the percentage yield calculated from this. However, the calculation can be done the other way round — the actual yield can be calculated from the percentage yield.

> ### Worked example 1
>
> When 3.21 g of propan-2-ol was heated under reflux with 50% sulfuric acid and potassium bromide, 4.92 g of 2-bromopropane was produced. Calculate the percentage yield.
>
> $$CH_3CH(OH)CH_3 + HBr \rightarrow CH_3CHBrCH_3 + H_2O$$
>
> **Answer**
>
> $$\text{amount of propan-2-ol} = \frac{3.21\ g}{60\ g\ mol^{-1}}$$
>
> $$= 0.0535\ mol$$

ratio of product to reactant = 1:1
amount of 2-bromopropane = 0.0535 mol
theoretical yield = 0.0535 × 123 g mol^{-1}
$$= 6.58 \text{ g}$$
% yield = $\dfrac{4.92 \text{ g} \times 100}{6.58 \text{ g}}$
$$= 74.8\%$$

Worked example 2

The dehydration of a secondary alcohol, such as propan-2-ol, should result in a 75% yield. Calculate the volume of propene gas, $CH_3CH=CH_2$, that should be obtained when 2.46 g of propan-2-ol, $CH_3CH(OH)CH_3$, is dehydrated.

$$CH_3CH(OH)CH_3 \rightarrow CH_3CH=CH_2 + H_2O$$

(Under the conditions of the experiment, the molar volume of a gas = 24 dm^3 mol^{-1}.)

Answer

amount of propan-2-ol = $\dfrac{2.46 \text{ g}}{60 \text{ g mol}^{-1}}$ = 0.0410 mol

ratio of propene to propan-2-ol = 1:1

theoretical yield of propene = 0.0410 mol
$$= 0.041 \text{ mol} \times 24 \text{ dm}^3 \text{ mol}^{-1} = 0.984 \text{ dm}^3$$

actual yield = 75% of theoretical yield
$$= 0.984 \text{ dm}^3 \times \dfrac{75}{100}$$
$$= 0.74 \text{ dm}^3 = 740 \text{ cm}^3$$

If a reaction takes place in two steps, each with a 60% yield, the percentage yield for the overall reaction is:

60% of 60% = $\dfrac{60}{100} \times \dfrac{60}{100} \times 100 = 36\%$

Questions

1 State which of the following is the most polar molecule and which is the least polar:
 a chloromethane, bromomethane or iodomethane
 b fluoromethane or methanol

2 Explain why fluoromethane is insoluble in water, whereas methanol is totally soluble.

3 Explain why ethanol has a higher boiling point than methanol.

4 Name and give the formula of the organic product of the reaction of 1-bromo-2-methylpropane with:
 a an aqueous solution of potassium hydroxide

 b a concentrated solution of sodium hydroxide in ethanol
 c excess concentrated ammonia

5 Explain why 2-iodopropane is hydrolysed by aqueous sodium hydroxide at a faster rate than 2-chloropropane.

6 Give an example of a halogenated polymer. State one use of this polymer and one environmental problem related to its disposal.

7 Give the structural formula of:
 a a primary alcohol of molecular formula $C_5H_{12}O$, which has a branched carbon chain

b a primary alcohol of molecular formula $C_5H_{12}O$, which has an unbranched carbon chain

c a secondary alcohol of molecular formula $C_5H_{12}O$

d a tertiary alcohol of molecular formula $C_5H_{12}O$

8 You are given three unlabelled samples of organic liquids. You know that one is hexene, one is hexanol and the other is hexanoic acid. Describe how you would identify each substance.

9 Identify the organic substance remaining when each of the following is heated under reflux with a solution of potassium dichromate(VI) in sulfuric acid.

a $CH_3CH(CH_3)CH_2OH$

b $C_2H_5C(OH)(CH_3)_2$

c $CH_3CH_2CH(OH)CH_2CH_3$

10 Describe the mechanism of the reaction between 2-bromobutane and aqueous hydroxide ions.

11 Suggest the mechanism for the reaction between 2-chloropropane and cyanide ions $^-:C{\equiv}N$ in aqueous solution.

12 Explain why it is necessary to have some ethanol present for the reaction between dilute aqueous sodium hydroxide and 2-iodopentane.

13 A chloroalkane contains 37.2% carbon and 55.0% chlorine by mass. Calculate its empirical formula.

14 A carboxylic acid contains 50% carbon, 5.6% hydrogen and 44.4% oxygen by mass. It turns brown bromine water colourless.

a Calculate the empirical formula of the acid.

b Suggest the simplest structural formula for this acid.

15 A compound X contains 53.3% carbon, 11.1% hydrogen and 35.6% oxygen by mass.

a Calculate the empirical formula of X.

b The molar mass of X is 90 g mol^{-1}. Deduce the molecular formula of X.

c When excess phosphorus pentachloride was added, 9.0 g of dry X gave off 4.8 dm^3 of hydrogen chloride gas. Suggest one structural formula for X, giving your reasons.

(Under the conditions of the experiment, 1 mol of gas occupies 24 dm^3.)

16 5.67 g of ethanol was added to a hot solution of potassium dichromate(VI) and sulfuric acid. The ethanal produced was distilled off and 4.88 g of pure ethanal was obtained. Calculate the percentage yield.

17 2-iodopropane can be made from 1-chloro-propane by the two stage synthesis:

$$CH_3CH_2CH_2Cl \rightarrow CH_3CH{=}CH_2 \rightarrow CH_3CHICH_3$$

If stage 1 has a 30% yield and stage 2 has a 90% yield, calculate the mass of 2-iodopropane made from 78.5 g of 1-chloropropane.

18 100 g of the fat glyceryl tristearate was boiled with excess sodium hydroxide and the soap, sodium stearate, salted out. The yield of soap was 87.1 g. The reaction is:

$C_{17}H_{35}COOCH_2$
|
$C_{17}H_{35}COOCH$ + 3NaOH $\longrightarrow$ 3$C_{17}H_{35}COONa$
| + $CH_2(OH)CH(OH)CH_2OH$
$C_{17}H_{35}COOCH_2$

Calculate the percentage yield.

Chapter 15

Mass spectra, infrared spectra and the greenhouse effect

Phosphorus: fuel for life, insecticides and nerve gas

Atomic number: 15

Electron configuration: $1s^2\ 2s^2\ 2p^6,\ 3s^2\ 3p_x{}^1\ 3p_y{}^1\ 3p_z{}^1$

Symbol: P

Phosphorus is in group 5 of the periodic table. It is mined as phosphate ores. Phosphorus occurs in many biological molecules.

Adenosine triphosphate, ATP

The synthesis of proteins is endothermic, but the energy is supplied by the exothermic hydrolysis of adenosine triphosphate, ATP:

$$\text{Adenine-ribose}-C-O-\overset{O}{\overset{\|}{P}}-O-\overset{O}{\overset{\|}{P}}-O-\overset{O}{\overset{\|}{P}}-O^-(aq)+H_2O$$
$$\underset{O^-}{|}\quad\underset{O^-}{|}\quad\underset{O^-}{|}$$

ATP

$$\text{Adenine-ribose}-O-\overset{O}{\overset{\|}{P}}-O-\overset{O}{\overset{\|}{P}}-O^-(aq)+HPO_4^{2-}(aq)+H^+(aq)$$
$$\underset{O^-}{|}\quad\underset{O^-}{|}$$

ADP

Some biology books talk about the 'high-energy bond' in ATP. This is wrong. The breaking of the O–P bond is endothermic. The energy released comes from the hydration of the H^+ and HPO_4^{2-} ions formed. The energy

Aftermath of the Tokyo Sarin attack

TOPFOTO

released in this reaction is 2 kJ mol^{-1} more than that required to form the dipeptide alanylglycine from alanine and glycine.

ADP is then converted back to ATP using the energy from the metabolism of glucose:

$$C_6H_{12}O_6(aq) + 6O_2(g) \rightarrow 6CO_2(g) + 6H_2O(l)$$
$$\Delta H = -2800\ \text{kJ mol}^{-1}$$

Insecticides and nerve gas

Organophosphate insecticides are esters of phosphoric acid. Parathion is a typical example:

$$C_2H_5O-\overset{S}{\overset{\|}{P}}-O-\bigcirc-NO_2$$
$$\underset{C_2H_5O}{|}$$

Parathion

Electrical switching centres (synapses) occur in the nervous systems of vertebrates and insects. A nerve impulse arriving at a particular synapse stimulates the release of acetylcholine, which travels across the synapse and triggers an impulse in the next nerve cell. The enzyme cholinesterase breaks down acetylcholine. Cholinesterase is inhibited by organophosphates; acetylcholine is not broken down and this results in electrical impulses occurring continuously across the synapse. The result is uncontrolled, rapid twitching of muscles, paralysed breathing, convulsions and eventually death.

Organophosphates are used as crop sprays, in sheep dips and in nerve gases. Iraq used the organophosphate Sarin in the Iran–Iraq war, killing thousands of people. A terrorist group in Japan used Sarin in an attack in an underground station in Tokyo, killing six people and harming over a thousand.

In case of a nerve gas attack, soldiers in the two Gulf wars carried self-injecting syringes of atropine, which is an antidote to organophosphates.

Many farmers have suffered nervous system problems as a result of using organophosphates in sheep dips.

Mass spectra, infra-red spectra and the greenhouse effect

Mass spectra

A diagram and an explanation of how a mass spectrometer works can be found on pages 18–20. Some examples of its use are given on pages 20–21.

Valuable information about the molecular structure of an organic molecule can be obtained from a **mass spectrum**. The first thing to look for in a mass spectrum is the peak due to the **molecular ion**. This is normally the peak with the highest m/e value. However, sometimes, there is a peak with an m/e value 1 more than that of the molecular ion peak — the $M+1$ peak. This is caused by the presence of the carbon-13 isotope. The carbon-13 content of natural carbon is 1.1%, so a molecule containing six carbon atoms has an $M+1$ peak about 7% of the height of the molecular ion peak.

e In some mass spectra the molecular ion peak is missing. This is caused by the high energy of the ionisation process, which is sufficient to split all the molecular ions into fragments.

> ### Worked example
> **a** An organic compound has the empirical formula C_2H_4O. The largest m/e value in its mass spectrum is 88. Calculate the molecular formula of the compound.
> **b** When phosphorus pentachloride is added to this compound, 1 mol of HCl per mole of the organic compound is liberated. Suggest possible formulae for the compound.
>
> ### Answer
> **a** The mass of $C_2H_4O = (2 \times 12) + 4 + 16 = 44$. This is half of 88, so the molecular formula is $C_4H_8O_2$.
> **b** The HCl is produced in a 1:1 ratio, so there can be only one –OH group in the molecule. Some possible substances are:
> - $CH_3CH_2CH_2COOH$
> - $(CH_3)_2CHCOOH$
> - $CH_3COCH_2CH_2OH$
> Further study of the mass spectrum can help to identify the substance.

Fragmentation of the molecular ion

Various functional groups cause molecules to break apart in characteristic ways. Ketones tend to break where the C=O group bonds onto one of the alkyl groups — for example, butanone, $CH_3CH_2COCH_3$, gives a molecular ion peak at $m/e = 72$ and peaks at m/e values of 57 (loss of CH_3), 43 (loss of C_2H_5), 15 (caused by CH_3^+) and 29 (caused by $C_2H_5^+$). A typical fragmentation would be:

$$CH_3CH_2COCH_3^+ \rightarrow CH_3\bullet + CH_3CH_2CO^+$$

e Note that a radical and an ion are produced in fragmentation.

The fragments give clues to the groups that are present in the molecule. A CH_3 group has a mass of 15 units, so an ion of mass $(M - 15)$ indicates that the substance has a CH_3 group:

$$(RCH_3)^+ \rightarrow R^+ + CH_3\bullet$$

The masses of some common fragments that are often lost are given in Table 15.1.

m/e value	Group
15	CH_3
29	C_2H_5 or CHO
31	CH_2OH
45	COOH
77	C_6H_5

Table 15.1 Common fragments lost

The mass spectrum of the substance in the worked example above (highest m/e value 88) also had peaks at m/e values 43 and 59. These fragments enable identification of the substance.

The three suggested formulae are:
- $CH_3CH_2CH_2COOH$
- $(CH_3)_2CHCOOH$
- $CH_3COCH_2CH_2OH$

The m/e value of 43 is 45 units less than 88 and is caused by the loss of a COOH group. This discounts the third structure. The m/e value of 59 is 29 units less than 88 and is caused by the loss of C_2H_5. This discounts the second structure.

Thus, only the first structure — $CH_3CH_2CH_2COOH$ — is possible.

Propan-1-ol and propan-2-ol can be oxidised to propanal and propanone respectively. The mass spectra of the oxidation products are shown below in Figures 15.1 and 15.2.

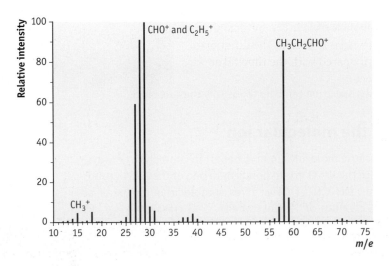

Figure 15.1 Mass spectrum of propanal

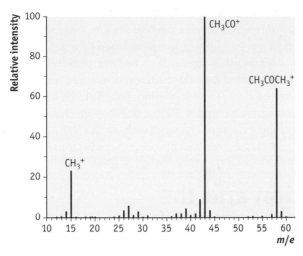

Figure 15.2 Mass spectrum of propanone

Note the small $(m+1)$ peaks in both spectra. These are caused by a ^{13}C atom in the molecule.

The molecular ion peaks at $m/e = 58$ can be seen clearly in both spectra. In the spectrum for propanal there is a large peak at $m/e = 29$, which is due to a combination of the $C_2H_5^+$ and CHO^+ peaks:

m/e can also be written as m/z

$$C_2H_5CHO^+ \rightarrow C_2H_5^+ + CHO\bullet \quad \text{and}$$
$$C_2H_5CHO^+ \rightarrow C_2H_5\bullet + CHO^+$$

The spectrum of propanone shows peaks due to fragments at $m/e = 15$ (CH_3^+) and $m/e = 43$ (CH_3CO^+):

$$CH_3COCH_3^+ \rightarrow CH_3^+ + CH_3CO\bullet$$
$$CH_3COCH_3^+ \rightarrow CH_3\bullet + CH_3CO^+$$

Modern techniques

Time-of-flight mass spectrometers

The sample is bombarded by electrons and becomes ionised. The positive ions are accelerated across a potential difference, as in a traditional mass spectrometer. However, instead of being deflected by a magnetic field, the time taken for each ion to pass to the detector is measured. The kinetic energy ($\frac{1}{2}mv^2$) of all the ions is the same, but their speed depends upon their mass. Heavier ions move more slowly and take longer to reach the detector. The electron beam producing the ions is pulsed; the duration of each pulse must be short compared with the time taken for the ion to move to the detector.

Researcher loading potential cancer drugs into a liquid chromatography mass spectrometer

Mass spectrometers in the pharmaceutical industry

When potential drugs are developed, their identity must be confirmed. The reaction mixture is first separated into pure components by high-performance liquid chromatography (HPLC). The theory is the same as that of the chromatography studied at GCSE but, in HPLC, the solution mixture is forced under pressure through a column containing a solid adsorbent material. The different components are separated quickly and effectively. Samples are then analysed by a specialised mass spectrometer.

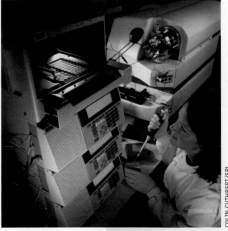

COLIN CUTHBERT/SPL

Most pharmaceuticals are not thermally stable. Therefore, the usual method of bombarding their vapour would cause considerable decomposition. Instead, molecules containing a base, such as an $-NH_2$ group, are protonated by adding methanoic acid. This causes the potential drug molecule to become positively charged. A solution of the charged molecule is sprayed into the spectrometer and the solvent is rapidly evaporated using a laser. The positive ions pass through the spectrometer and their molar masses are determined. Before being sprayed into the spectrometer, molecules containing an acidic group (such as COOH) are treated with ammonia, which deprotonates them and produces negative ions.

Infrared absorption and the greenhouse effect

Electromagnetic radiation

Electromagnetic radiation consists of an oscillating electric and magnetic field of a wide range of frequencies (Table 15.2). In a vacuum, light travels at a speed of 3.00×10^8 m s^{-1} (670 000 000 miles per hour).

Frequency is measured in hertz, Hz (1 Hz = 1 s^{-1}). The frequency determines the colour and the energy of the light, which is calculated by the expression formulated by Max Planck:

$$E = h\nu$$

where h is Planck's constant and ν is the frequency.

The speed of light, its frequency and the wavelength are combined in the equation:

$$c = \lambda\nu$$

where c is speed of light and λ is the wavelength.

The frequency can also be expressed as a wavenumber:

$$\text{wavenumber} = 1/\lambda$$

where the units of λ are cm.

Type of radiation	Frequency/MHz	Wavenumber/cm^{-1}	Energy/J per photon
X-rays	$> 10^{11}$	$> 3 \times 10^6$	$> 7 \times 10^{-17}$
Ultraviolet	1×10^9	3×10^4	7×10^{-19}
Visible — blue	6×10^8	2.1×10^4	4.2×10^{-19}
Visible — yellow	5×10^8	1.6×10^4	3.2×10^{-19}
Visible — red	4×10^8	1.4×10^4	2.8×10^{-19}
Infrared	$< 3 \times 10^8$	$< 10\,000$	$< 2 \times 10^{-19}$
Microwaves	$< 3 \times 10^4$	< 1	$< 2 \times 10^{-22}$
Radiowaves	< 100	$< 3 \times 10^{-3}$	$< 7 \times 10^{-26}$

Table 15.2 Frequency, wavenumber and energy of different types of electromagnetic radiation

When electromagnetic radiation is passed through a diffraction grating, it is split up into a spectrum according to the frequency of the radiation. Visible light is split by passing it through a prism. Blue light is refracted (bent) more than red light.

A rainbow is caused by droplets of rain splitting up white light into its component colours.

Infrared radiation is a form of electromagnetic radiation that has a lower frequency, or longer wavelength, than that of visible light. It is more convenient to use wavenumber, rather than hertz, for frequency, because this is a more manageable number.

$$\text{wavenumber} = \frac{\text{frequency}}{\text{speed of light}}$$

where the units of wavenumber are cm^{-1}.

Infrared light has a wavenumber below $10\,000\ cm^{-1}$, whereas yellow light has a wavenumber of $17\,000\ cm^{-1}$. An infrared line at $1720\ cm^{-1}$ has a frequency of 5×10^{13} Hz and a wavelength of $6000\ nm$.

Absorption of energy

A molecule can absorb energy in a number of ways:

- **Translational energy** — a molecule gains kinetic energy and moves faster. Heating a substance causes it to gain kinetic energy.
- **Rotational energy** — a molecule rotates about its centre of mass. Absorption of microwaves causes an increase in rotational energy.
- **Vibrational energy** — the atoms in a molecule vibrate in different ways. The bonds can stretch or bend. Absorption of infrared energy causes an increase in vibrational energy.
- **Electron excitement** — an electron in an atom or bond is promoted to a higher energy level. Absorption of visible or ultraviolet light results in electron excitement. In some cases, the electron is excited to such a high level that a covalent bond breaks. For example, the absorption of ultraviolet or visible light by chlorine results in the Cl–Cl bond breaking homolytically to form two chlorine radicals:

$$Cl_2 \xrightarrow{\text{light}} 2Cl\bullet$$

Infrared absorption

For a molecule to absorb in the infrared, the vibration must involve a change in the **dipole moment** of the molecule. Molecules with non-polar bonds, such as oxygen and nitrogen, do not absorb infrared radiation. This is why they are not **greenhouse gases**.

Non-polar molecules

Non-polar molecules that have polar bonds absorb in the infrared. These include:

- carbon dioxide
- methane

Carbon dioxide

Oxygen is more electronegative than carbon, so the carbon atom in carbon dioxide is δ^+ and the oxygen atom is δ^-. However, the molecule is linear, so the dipoles cancel out and the molecule is non-polar. There are three ways that the molecule can vibrate:

Asymmetrical stretching

Symmetrical stretching

Bending

The symmetrical stretching does not cause a change in dipole moment and so does not absorb infrared radiation. The other two modes cause a change in dipole moment from zero and are said to be 'infrared active'. Carbon dioxide, therefore, absorbs infrared radiation and is a greenhouse gas (page 291). The infrared spectrum of carbon dioxide is shown in Figure 15.3.

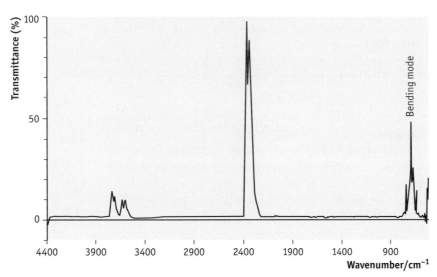

Figure 15.3
Infrared spectrum
of carbon dioxide

Methane

The C–H bond is polar — the electronegativity of carbon is 2.5 and that of hydrogen is 2.1. However, the methane molecule is tetrahedral, so the dipoles cancel and the molecule has no dipole moment. Any asymmetrical bending or stretching causes the molecule to have a dipole moment. Therefore, methane absorbs infrared radiation.

Polar molecules

All polar molecules are infrared active.

Water

The oxygen atom in a molecule of water is δ^- and has two lone pairs of electrons. This results in the molecule being non-linear and, therefore, polar. It absorbs infrared radiation in all its bending and stretching modes. It is the major greenhouse gas in the atmosphere.

The unit of dipole moment is the debye, symbol D

Oxides of nitrogen

The electronegativity of nitrogen is 3.0 and that of oxygen is 3.5. Therefore, all bonds between nitrogen and oxygen are polar. Nitric oxide, NO, is a diatomic polar molecule and has a dipole moment of 0.16 D. Nitrogen dioxide, NO_2, is a V-shaped molecule with a dipole moment of 0.40 D. Therefore, both oxides absorb infrared radiation and are greenhouse gases.

The greenhouse effect

Solar radiation falling on Earth has wavelengths mostly between 400 nm and 700 nm, which is the visible region of the electromagnetic spectrum. The atmosphere absorbs some of the ultraviolet light but the radiation in the visible region passes through the gaseous molecules in the air. Some of this radiation is reflected back into space by the white snow and ice fields of the polar regions and high mountains. The remainder is absorbed by the clouds or by the Earth itself.

The radiation emitted back into space by the Earth is of longer wavelength, being in the infrared and far infrared regions. This outgoing radiation has to pass through clouds and the atmosphere. The clouds absorb much of it, which is why cloudy nights are warmer than clear nights. Some of the molecules in the atmosphere absorb certain frequencies of this infrared radiation. These substances are called **greenhouse gases**. Their molecules absorb infrared radiation and go into a higher vibrational energy level. The molecules then return to the ground state, either by spontaneously emitting radiation or through collision. Collision converts the vibrational energy into heat. In this way, some of the radiation emitted by the Earth is re-radiated back to Earth or is trapped as heat and does not escape into space.

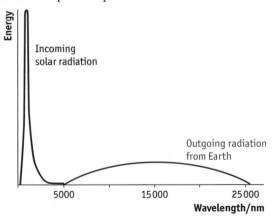

Figure 15.4 Incoming and outgoing radiation

Water vapour is the most important greenhouse gas. It is estimated that nearly 95% of the radiation absorbed by the atmosphere is absorbed by water vapour. This explains why nights in the desert are so cold — there is little water vapour in the air to absorb the heat energy being radiated out.

There is huge concern about the role of carbon dioxide as a greenhouse gas. Levels of carbon dioxide have been rising steadily from the Industrial Revolution until the present time. What is more worrying is that the rate of this rise is increasing. It has been estimated that over 2000 years ago, the carbon dioxide level was about

280 ppm. In 1977, this had risen to 320 ppm. Today, it is over 380 ppm. Every year the mass of carbon dioxide produced by burning fossil fuels increases by over 1 billion tonnes.

Other gases that play a significant role in absorbing infrared radiation are methane, HCFCs and CFCs, as well as oxides of nitrogen. All, apart from the halogenated hydrocarbons, occur naturally in air. However, it is the *increase due to human activity* that is concerning.

It is difficult to estimate the effect that this increase has on global temperatures, compared with natural events such as the Sun's activity and changes in the Earth's orbit. Past ice ages and mini ice ages were caused naturally. In fact, the surface of the Earth has been fashioned by these dramatic climate changes, as glaciers swept back and forth over the land. Two centuries ago the River Thames froze over each winter. Go back several hundred years and there was a time when there was no ice at the North Pole.

Natural and anthropogenic sources of greenhouse gases

There is a natural balance between production and release of carbon dioxide. Animals and plants convert carbohydrates into carbon dioxide (respiration) and plants convert carbon dioxide into carbohydrates (photosynthesis). The sea dissolves carbon dioxide, which is used by marine plants in photosynthesis. Carbon dioxide is released in large quantities by volcanic activity, including hydrothermal events deep on the ocean floor. Shellfish store large amounts of carbon dioxide as calcium carbonate, which then falls to the ocean floor when they die. In recent geological time, the level of carbon dioxide in the atmosphere and dissolved in the sea has remained fairly constant, with only minor changes due to natural climate change and volcanic activity.

In the geological past, during the Carboniferous period, carbon dioxide was locked up as oil, natural gas and coal — the fossil fuels. When burnt, fossil fuels release carbon dioxide that had been removed from the atmosphere by photo-synthesis over 250 million years ago. With the advent of the Industrial Revolution, the combustion of fossil fuels increased dramatically, causing the atmospheric and ocean levels of carbon dioxide to rise. This increase through human actions is called **anthropogenic activity** (Figure 15.5).

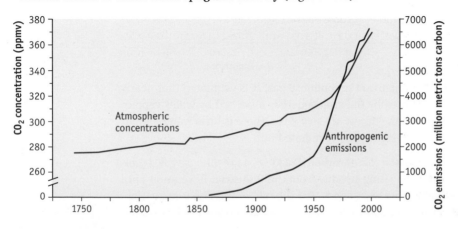

Figure 15.5 *Trends in atmospheric carbon dioxide*

(Source: Oak Ridge National Laboratory, Carbon Dioxide Information Analysis Center, http://cdiac.esd.oml.gov/)

Methane is produced naturally and anthropogenically. Anaerobic decomposition of vegetable matter produces methane, so marshes are a natural source. Agriculture is another source. Cows and sheep produce methane in their digestive systems and it is also released from flooded rice fields. The decay of organic material in landfill sites also produces significant amounts of methane.

Oxides of nitrogen are greenhouse gases. Nitric oxide, NO, is produced naturally by lightning and by soil bacteria. The main anthropogenic sources of nitrogen oxides are the decay of nitrogenous fertilisers and the combustion of fuel in car and aeroplane engines. At the high temperature produced during the combustion, the nitrogen and the oxygen in the air react to form nitric oxide:

$$N_2 + O_2 \rightarrow 2NO$$

Most nitric oxide is removed from car engines by catalytic converters, but in jet engines it is released into the atmosphere.

CFCs and HCFCs are potent greenhouse gases. They are only produced anthropogenically, as is sulfur hexafluoride, which is used in the manufacture of magnesium. Although CFCs are no longer being produced, some remain in the upper atmosphere. HCFCs, for example $CHClF_2$, are produced as refrigerants and can escape into the environment.

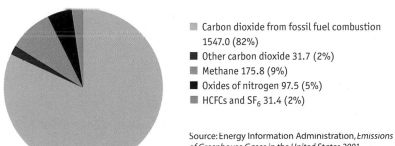

- Carbon dioxide from fossil fuel combustion 1547.0 (82%)
- Other carbon dioxide 31.7 (2%)
- Methane 175.8 (9%)
- Oxides of nitrogen 97.5 (5%)
- HCFCs and SF_6 31.4 (2%)

Source: Energy Information Administration, *Emissions of Greenhouse Gases in the United States 2001* (Washington, DC, 2002)

Figure 15.6 Atmospheric greenhouse gas emissions (million tonne of carbon equivalent)

It is not just the quantity of each greenhouse gas that has to be considered. Its effectiveness in absorbing infrared radiation and the time that the gas spends in the atmosphere before being reacted or deposited on the Earth's surface have to be taken into account. Methane absorbs more radiation per mole than does carbon dioxide. Oxides of nitrogen and HCFCs are even more powerful absorbers. The combination of these two factors is measured by the 'global warming potential' (GWP) of each gas. This is based on the ability that the gas has to absorb infrared radiation and the extent to which it is removed from the atmosphere. Carbon dioxide is used as the standard and has been assigned a GWP rating of 1. The GWPs of the common greenhouse gases are listed in Table 15.3.

If the global warming potentials and the concentrations of the gases are known, their effect on the environment can be calculated. The contribution of these gases to global warming is shown in Figure 15.7.

Table 15.3 GWPs of the common greenhouse gases

Greenhouse gas	GWP
Carbon dioxide, CO_2	1
Methane, CH_4	20
Nitrogen oxides, NO_x	300
HCFC-22, $CHClF_2$	1 300
CFC-12, CCl_2F_2	6 500
Sulfur hexafluoride, SF_6	22 000

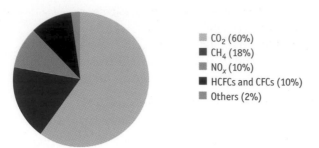

Figure 15.7
Contributions to the
greenhouse effect

CO$_2$ (60%)
CH$_4$ (18%)
NO$_x$ (10%)
HCFCs and CFCs (10%)
Others (2%)

Figure 15.6 shows that total carbon dioxide emissions account for 84% of greenhouse gas emissions. However, the effect of carbon dioxide is only 60% because of its lower GWP (Figure 15.7).

Three notes of caution:

■ The figures for GWP (from the International Panel on Climate Change) are quoted to an accuracy of ±35%.
■ All the numbers quoted above ignore the most important greenhouse gas — water vapour, which contributes 95% to the total greenhouse effect. If water vapour is factored into the calculations the data shown in Table 15.4 are obtained.
■ Only 3% of the carbon dioxide in the atmosphere is of anthropogenic origin. This means that anthropogenic carbon dioxide contributes just 0.1% to the total greenhouse effect.

Greenhouse gas	Contribution to greenhouse effect (%)
Water vapour	94.6
Carbon dioxide	3.8
Methane	1.1
Nitrogen oxides	0.5
Others	< 0.1

Table 15.4 Effect of greenhouse gases, including water vapour

Infrared spectroscopy

An infrared spectrometer measures the amounts of radiation absorbed. It does this by splitting the incident beam of radiation into two. One beam passes through a cell containing a reference sample (usually solvent) and then goes to the detector; the other goes through the dissolved sample to the detector. The difference between the two amounts of radiation arriving at the detector is the amount absorbed by the chemical. The angle of the diffraction grating is slowly altered so that different frequencies sweep through the apparatus.

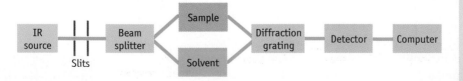

Figure 15.8
Representation of an infrared spectrometer

Different bonds in a covalent molecule absorb radiation of different frequencies, which are normally measured as wavenumbers. An infrared spectrum usually has a range from 4000 cm^{-1} to 600 cm^{-1}.

The C=O bond absorbs at around 1700 cm^{-1}, but the actual value depends on the other atoms attached to the C=O group. This is shown in Table 15.5.

Other types of bond absorb at different frequencies, but the actual value again depends on the neighbouring atoms and groups.

Type of compound	Wavenumber/cm^{-1}
Aliphatic aldehyde	1720–1740
Aliphatic ketone	1700–1730
Aromatic aldehyde	1690–1715
Aromatic ketone	1680–1700
Carboxylic acid	1700–1725
Ester	1735–1750
Acid chloride	1815–1825
Amide	1640–1680

Table 15.5 Absorption frequencies of different C=O groups

Bond	Functional group	Wavenumber/cm^{-1}
O–H	Alcohols (hydrogen bonded)*	3200–3600
	Alcohols (not hydrogen bonded)	3600–3700
	Carboxylic acids	2500–3300
N–H	Amines (hydrogen bonded)*	3300–3500
C–H	Alkanes	2850–3000
	Alkenes and arenes	3000–3100
C–C	Alkanes	1360–1490
C=O	See Table 14.3	
C=C	Aromatic	1450–1650
	Alkenes	1650–1700
C–Cl	Chloroalkanes	800–600
C–Br	Bromoalkanes	600–500
C–I	Iodoalkanes	about 500

*These peaks are very broad because of intermolecular hydrogen bonding.

Table 15.6 Absorption frequencies of some common groups

The region below about 1300 cm^{-1} is known as the fingerprint region. It shows a complex series of peaks that depends on the exact compound being analysed. Just as human fingerprints can be matched by computer to give a unique identification, so computer analysis of the fingerprint region can be used to identify a pure unknown organic substance.

Some infrared spectra are shown below.

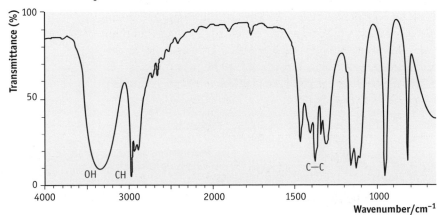

Figure 15.9 Infrared spectrum of propan-2-ol

In the infrared spectrum of propan-2-ol shown in Figure 15.9:
- the broad peak at around 3330 cm^{-1} is due to hydrogen-bonded O–H
- the peak at 2970 cm^{-1} is due to the C–H bond
- the peak at 1380 cm^{-1} is due to the C–C bond

Worked example 1

Examine the infrared spectrum of methylbenzene and assign the peaks at 3000 cm^{-1}, 2900 cm^{-1} and 1500 cm^{-1}.

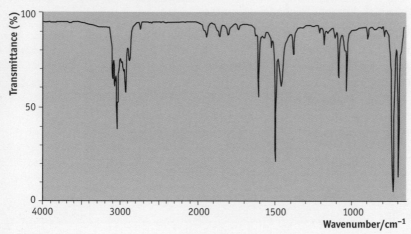

Answer

The peak at 3000 cm^{-1} is due to the stretching of the aromatic C–H bond.
The peak at 2900 cm^{-1} is due to the stretching of the C–H bond in the CH$_3$ group.
The peak at 1500 cm^{-1} is due to the aromatic C=C bond.

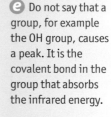 Do not say that a group, for example the OH group, causes a peak. It is the covalent bond in the group that absorbs the infrared energy.

Worked example 2

Examine the infrared spectrum of ethanoic acid.

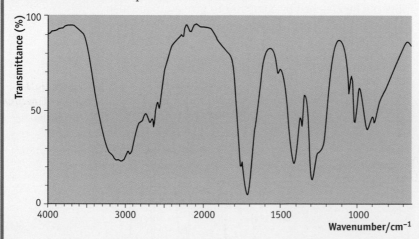

a Assign the peaks at 3000 cm^{-1}, 1700 cm^{-1} and 1400 cm^{-1}.
b Why is the peak at 3000 cm^{-1} very broad?

Answer

a The peak at 3000 cm^{-1} is due to the O–H bond in the carboxylic acid group.
The peak at 1700 cm^{-1} is due to the C=O bond in the carboxylic acid group.
The peak at 1400 cm^{-1} is due to the C–C bond.

b The peak at 3000 cm^{-1} is very broad because of the effects of intermolecular hydrogen bonding.

e The absorption by C–H bonds is often masked by the broad peak due to the O–H bond at similar frequencies.

Success of reaction

Infrared spectroscopy can be used to show whether a reaction has taken place. The spectrum of the final product is compared with that of the original reactant. For example, if an alcohol is being oxidised to an aldehyde or a ketone, the peak between 3200 and 3600 cm^{-1} caused by the stretching of the O–H bond should not be found in the spectrum of the product, but a peak between 1700 and 1740 cm^{-1} due to the C=O bond should be clearly visible.

If butan-2-ol is heated with an oxidising agent, butanone should be produced. By looking at the spectra before and after the experiment and comparing them with a library of spectra (www.aist.go.jp or www.rod.beavon.clara.net), it can be seen whether or not the oxidation took place.

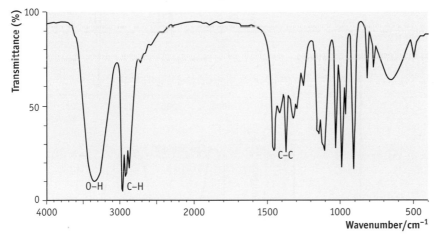

Figure 15.10
Infrared spectrum of butan-2-ol

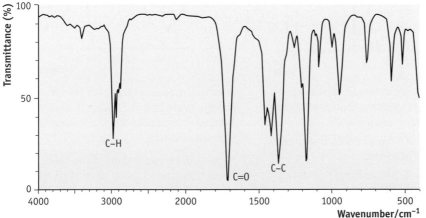

Figure 15.11
Infrared spectrum of butanone

1 Consider a mass spectrometer. State:
   a how the positive ions are produced
   b how the ions are accelerated
   c how the ions are deflected

2 A compound of molar mass 74 g mol^{-1} has a molecular formula of $C_3H_6O_2$. It has been identified as either compound X, formula $CH_3CH(OH)CHO$, or compound Y, formula CH_2OHCH_2CHO. The mass spectrum has peaks at m/e values of 74, 29 and 31 but none at 15 or 59. Identify the compound. Justify your answer.

3 Bromine has two isotopes, ^{79}Br and ^{81}Br, in approximately equal proportions. Sketch the mass spectrum of bromine in the range m/e 155 to 165.

4 Ethanal, CH_3CHO, has a peak at $m/e = 29$ in its mass spectrum.
   a Identify the species responsible for this peak.
   b Write the equation for its formation.

5 The mass spectrum below is that of propanone.

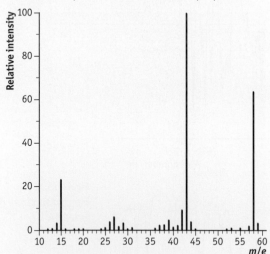

Identify the peaks at m/e values of 58, 43 and 15 and write equations to show the formation of the species responsible for these peaks.

6 Calculate the energy of 1 mol of photons of green light of frequency 5.7×10^{14} Hz.
   (Avogadro constant = 6.02×10^{23} mol^{-1}; Planck's constant, $h = 6.63 \times 10^{-34}$ J s)

7 Calculate the frequency required to break a chlorine molecule homolytically.
   (Cl–Cl bond enthalpy = +242 kJ mol^{-1}; Avogadro constant = 6.02×10^{23} mol^{-1}; Planck's constant, $h = 6.63 \times 10^{-34}$ J s)

8 Explain why water vapour is a powerful greenhouse gas.

9 a Explain the difference between anthropogenic and natural causes.
   b List *two* anthropogenic sources of oxides of nitrogen and *one* natural source.

10 Examine the infrared spectra A and B below. One spectrum is that of propanal; the other is that of propanoic acid.

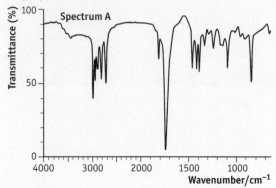

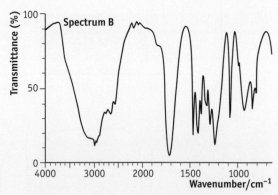

Identify the peaks in A at 1715 cm^{-1} and 1421 cm^{-1} and the peaks in B at 2986 cm^{-1}, 1716 cm^{-1} and 1416 cm^{-1}. Hence, decide which spectrum is that of propanoic acid.

11 Suggest the wavenumbers of the peaks you would see in the infrared spectrum of ethyl ethanoate, $CH_3COOCH_2CH_3$.

Chapter *16*

Green chemistry

Sulfur: drugs and racing tyres

Atomic number: 16

Electron configuration: $1s^2\ 2s^2\ 2p^6\ 3s^2\ 3p_x^2\ 3p_y^1\ 3p_z^1$

Symbol: S

Molecular formula: S_8

Sulfur is in group 6 of the periodic table. It is found near volcanoes and in rocks in underground deposits.

Sulfonamides

Sulfonamides, also known as sulfa drugs, are a group of chemotherapeutic agents that are effective in treating a number of infectious diseases. They were the first and only effective antibiotics available before penicillin and their use was continued during the early years of the Second World War; they are credited with saving the lives of tens of thousands of patients, including Franklin Roosevelt and Winston Churchill. During the war, American soldiers were issued a first-aid kit containing sulfa powder and were told to sprinkle it on any open wound. It prevented wounds becoming infected and turning gangrenous.

The simplest of the sulfa drugs is sulfanilamide (4-aminobenzenesulfonamide):

$$H_2N-\bigcirc-\overset{\overset{\displaystyle O}{\|}}{\underset{\underset{\displaystyle O}{\|}}{S}}-NH_2$$

The drug works by preventing bacteria from making folic acid, which is essential for nucleic acid production. Mammals do not make folic acid in their cells — it has to be part of the diet — so they are not affected by the drug.

Rubber tyres

Rubber is made either from the latex of rubber trees or from oil-based chemicals containing two alkene double bonds. The simplest monomer is isoprene, which is the monomer of poly(isoprene). In rubber latex, the methyl groups in the polymer chain are all *cis-* to each other. Latex has few direct commercial uses as it softens on heating and has low tensile strength.

$$n\quad CH_2=\overset{\overset{\displaystyle CH_3}{|}}{C}-\overset{\overset{\displaystyle CH_3}{|}}{C}=CH_2$$
$$\downarrow$$

Goodyear invented a process that made rubber harder and more elastic. The natural rubber was heated with sulfur. This process is called **vulcanisation**.

In untreated rubber, there is one double bond per repeat unit and it is this that enables rubber to be vulcanised. The S_8 rings break down into chains that vary in length from two to eight atoms. One end of a chain attacks a double bond in the polymer; the other end bonds to a double bond of another polymer. In this way, the polymer chains are cross-linked.

Short sulfur cross-links, with just one or two sulfur atoms in the link, confer good heat resistance. Cross-links with higher number of sulfur atoms give the rubber good dynamic properties, but less heat resistance. Dynamic properties are important for flexing movements such as the movement of a side-wall of a car tyre.

Talent can win races, but so can choice of tyres

Green chemistry

Carbon footprint and carbon neutrality

When an organic fuel is burnt, carbon dioxide is produced. As discussed in Chapter 15, the level of this gas in the atmosphere is rising at an increasing rate. Even if Europe and North America reduce their carbon emissions, the new powerhouse economies of China and India will increase theirs. Scientific debate on this issue must address the total amount of carbon dioxide produced.

Carbon footprint

Carbon footprint of a fuel

This is the *total* mass of carbon dioxide produced by a fuel from its production to its subsequent combustion. For a fossil fuel, it is the amount released on combustion *plus* the amount used in the extraction of the oil, coal or gas, its transportation, refining and distribution.

The carbon footprint of a biofuel is more complicated. If the fuel is made from crops such as sugar cane or plants that produce vegetable oils, the amount of carbon dioxide absorbed during growth should be subtracted from the total mass of carbon dioxide produced on combustion, during growth of the crop (respiration), its conversion to usable fuel and its distribution. This means that although a biofuel releases a significant net amount of carbon dioxide, the carbon footprint is much less than that of a fossil fuel.

No fuel has a zero-carbon footprint. Even hydrogen has a considerable footprint because it is made from methane — a fossil fuel — in an energy-demanding process.

Carbon footprint of an organisation or activity

This is the net amount of carbon dioxide emitted through the use of fossil or biofuels:

- For a business or factory, it is the net amount released as part of its everyday operations.
- For individuals or households, it is the net amount released as part of their daily lives.
- For a product bought at a supermarket, it is the net carbon dioxide emission in all processes involved in its production and transportation to and from the market.

The units of carbon footprint are tonnes of carbon dioxide per event.

Carbon neutrality

No activity is truly carbon neutral, except perhaps going for a walk. Somewhere in the chain from manufacture, to distribution and finally to use, fuels will have been used.

The term 'carbon neutral' is often used wrongly to denote no net carbon dioxide produced at the point of use. Thus, biofuels and hydrogen are sometimes described as carbon neutral, ignoring all the carbon dioxide released before they are used. Similarly, a wind turbine is manufactured using high-energy processes.

Fuels

The properties necessary for a useful fuel are that it should:
- be abundant or easily manufactured
- ignite easily
- have high energy output
- be easily stored and transferred to the point of use
- be non-toxic
- cause minimal environmental damage

An offshore oil extraction platform

INGRAM

Fuels can be divided into two types, according to the source of the fuel.

Fossil fuels and fuels derived from fossil fuels

A fossil fuel is a mixture of hydrocarbons laid down over millions of years through the anaerobic decay of animal and vegetable matter.

Fossil-fuel power-stations burn natural gas, heating oil or coal. Data in Table 16.1 show that, in terms of producing the most carbon dioxide per 1000 kJ of energy released, coal is the worst fossil fuel and natural gas is the best, producing less than half the carbon dioxide emissions of coal. As carbon dioxide is a major greenhouse gas, the change from coal- or oil-based power-stations to natural gas markedly reduces the extent of greenhouse gas production. This change has helped the UK towards meeting the target of the Kyoto agreement. Much more needs to be done to reduce carbon dioxide emissions and, therefore, lower the human causes of global warming (page 292).

Most fossil fuels contain sulfur compounds as impurities.

Methane
- Methane is a gas that cannot be liquefied unless the temperature is reduced to −83°C.
- It is found as natural gas in many deposits throughout the world.
- It burns cleanly, producing very few pollutants and less carbon dioxide per unit of energy than any other fuel except hydrogen.
- It can be piped directly into houses in urban areas.
- It is not suitable as a fuel for vehicles because it would require very large, heavy cylinders to contain enough gas, which would have to be under high pressure.

Table 16.1
Comparison of fuels

Fuel	Physical state	Energy released/ kJ cm^{-3}	Energy released/ kJ g^{-1}	CO_2 released per 1000 kJ energy released/ moles MJ^{-1}
Hydrogen, H_2	Gas	0.012	143	0
Methane (natural gas), CH_4	Gas	0.037	56	1.12
Butane (LPG), C_4H_{10}	Liquefied gas	30	50	1.39
Gasoline (petrol), C_8H_{18}	Liquid	34	48	1.46
Heating (domestic) oil, $C_{12}H_{26}$	Liquid	36	48	1.48
Fuel oil, $C_{20}H_{42}$	Liquid	35	44	1.51
Methanol, CH_3OH	Liquid	18	23	1.38
Ethanol, C_2H_5OH	Liquid	23	30	1.46*
Coal	Solid	73	33	2.50

* Ethanol for fuel is manufactured from grain, which grows via photosynthesis. This converts carbon dioxide into sugars and starch. Therefore, the *net* production of carbon dioxide is zero.

Hydrogen

Hydrogen is not found naturally. It is manufactured from natural gas by passing it, mixed with steam, over a heated catalyst:

$$CH_4 + H_2O \rightarrow CO + 3H_2$$

Hydrogen can also be produced by the electrolysis of water containing acid or alkali. However, the electricity for this process has to be generated by some energy-requiring means. Therefore, hydrogen is currently a fuel derived from fossil fuels. Thus, it has a significant carbon footprint. Potentially, the means of generating energy could be solar, wind or wave power, but the economics are, as yet, unfavourable. Even so, this would only result in hydrogen being carbon neutral if the renewable energy used in its production was not at the expense of generating industrial and domestic electricity by these means.

Hydrogen cannot be stored easily because it can only be liquefied below −240°C. It is dangerously explosive and so is not used as a domestic fuel. Storage problems prevent its use as a fuel in motor vehicles. Hydrogen has to be stored in the car either as a liquid below −240°C or at extremely high pressure as a gas. If cryogenic (very cold) hydrogen is used, it will steadily evaporate, risking an explosion in the

A hydrogen-powered car

TOPFOTO

car or garage. If high-pressure hydrogen is the preferred storage method, then the weight of the strong cylinders is a major handicap. There are also considerable refilling problems with both methods. However, where energy per unit mass is critical, as in the booster engines of a space rocket, liquid hydrogen is by far the best fuel.

Butane–propane mixtures

Butane is found with methane in natural gas and is also produced when oil is cracked. It is often mixed with propane and used as a domestic fuel in rural areas and for camping. It is transported in pressurised refillable cylinders that contain a mixture of liquefied butane and propane.

As a liquid under pressure, butane is also used as a fuel for motor vehicles. The fuel is called LPG (liquefied petroleum gas) or Autogas. Cars have to be specially adapted to use this fuel and specialist equipment has to be provided to refill these vehicles at the filling station.

Butane produces less carbon dioxide than petrol and leaves fewer unburnt hydrocarbons, so it is less polluting.

Octane and other liquid hydrocarbons

Octane and other liquid hydrocarbons are obtained from crude oil. Generally, they are the most convenient fuels. They are easy to transport and transfer to the user, as they are liquids, which are not dangerously volatile.

On combustion, they produce a high yield of energy per gram or cm^3. However, they produce much more carbon dioxide than would be produced by burning methane. If air supply is limited, as in a car engine, some carbon monoxide is produced and unburnt hydrocarbons are emitted. The former is toxic and the latter produce smog. The carbon monoxide can be removed by a catalytic converter, but the high pressure in the converter reduces the efficiency of changing the chemical energy of the fuel into the kinetic energy of the vehicle.

Alcohols

Methanol can be manufactured from coal or from methane. Both these fossil fuels react with steam in an endothermic reaction to form carbon monoxide and hydrogen. These two gases can then be made to react in the presence of a catalyst to produce methanol:

$$CO + 2H_2 \rightarrow CH_3OH$$

Methanol is highly toxic, but is an excellent fuel, burning cleanly and producing less carbon dioxide per 1000 kJ of energy produced than liquid fossil fuels. It is more expensive than methane or octane.

Biofuels

A biofuel is a fuel made from renewable organic materials, such as sugar, grain or vegetable oils.

Biofuels have a carbon footprint because fertilisers and fuel are used when the plants are grown and energy is required to manufacture the fuel and transport it to its point of use.

Bioethanol

Ethanol can be made by the fermentation of carbohydrates.

First, starch is broken down into glucose by heating with an acid catalyst or by enzymes from sprouting barley:

$$(C_6H_{10}O_5)_n + nH_2O \rightarrow nC_6H_{12}O_6$$

Sucrose is converted into a mixture of glucose and its isomer fructose, by the enzymes in yeast:

$$C_{12}H_{22}O_{11} + H_2O \rightarrow C_6H_{12}O_6 + C_6H_{12}O_6$$

Other enzymes ferment both glucose and fructose to produce a dilute solution of ethanol:

$$C_6H_{12}O_6 \rightarrow 2C_2H_5OH + 2CO_2$$

The **atom economy**, in terms of carbon, is a maximum of 67% and the ethanol has to be obtained from the mixture by fractional distillation, which is an energy-demanding process. The result is a mixture of 96% pure ethanol with 4% water.

The energy output of ethanol is $23 \, kJ \, cm^{-3}$, which is 63% that of petrol, so a car fuelled on bioethanol will travel for a shorter distance per litre of fuel than it would with petrol. Much of the petrol sold in the UK contains up to 10% ethanol. In some countries, such as Brazil, the percentage is higher.

New alternatives to petrol

Recent research at the University of Wisconsin in the USA has developed a method of making a better biofuel from sugar. A two-stage process is used to convert fructose into 2,5-dimethylfuran (DMF), C_6H_8O:

or using skeletal formulae

DMF has a higher energy density than ethanol. Its energy output is $33 \, kJ \, cm^{-3}$, which is almost identical to that of petrol. Unlike ethanol, it does not absorb water and it does not require an energy-intensive distillation process because, as it is insoluble in water, it can be separated easily from an aqueous solution during manufacture. In terms of carbon, the atom economy is over 75%, which is an improvement on bioethanol.

Energy density is the energy released in combustion per unit volume of the fuel. Its units are $kJ \, cm^{-3}$ or $MJ \, dm^{-3}$.

Other researchers have found a way of making DMF from glucose, although the yield is lower.

Since it is potentially better than bioethanol, DMF could be a biofuel of the future.

Biodiesel

Vegetable oils are esters of propane-1,2,3-triol and unsaturated fatty acids containing 18 carbon atoms and at least one carbon–carbon double bond. Without further treatment, they are unsuitable as fuels. A vegetable oil is converted into biodiesel by mixing it with methanol and an acid catalyst and a **transesterification** reaction takes place. This produces the methyl ester of the unsaturated acids and free propane-1,2,3-triol. A typical methyl ester in biodiesel is $C_{17}H_{31}COOCH_3$. These methyl esters have similar combustion properties to diesel produced from petroleum (petrodiesel). They can be used either pure or mixed with petrodiesel. The amount of biodiesel is given by the 'B-number' of the fuel. A B20 diesel contains 20% biodiesel.

> Transesterification is a reaction between an ester and an alcohol that produces a different ester and a different alcohol.

Many crops can be used as sources of vegetable oil. Yields from some crops are shown in Table 16.2. However, it has been estimated that a farmer uses $80 \, dm^3 \, ha^{-1}$ of fuel growing such crops, so biodiesel still leaves a carbon footprint.

Table 16.2 Yields of oil from various crops

	Yield/kg ha^{-1}
Corn (maize)	140
Soya	375
Sunflower	800
Oilseed rape	1 000
Algae	40 000

Certain algae are grown that produce a particularly high yield of oil that is suitable for conversion to biodiesel. This process is still at the experimental stage, but the plan is to grow the algae in areas of high sunshine in ponds at wastewater treatment plants or at sewage works.

Another source of vegetable oil for conversion to biodiesel is waste vegetable oil from restaurants and fish-and-chip shops. McDonald's is planning to run its fleet of delivery lorries on biodiesel made from its own waste frying oil.

The main problem with growing biofuels is that they use land that could otherwise be used for food production, or they require the destruction of natural forests in order to provide the land for growing the oil-bearing trees, such as oil palms. One way around this problem is to grow the plant *Jatropha*. This will grow on very poor soils and in dry areas and so utilises land that is otherwise non-productive. The rail line in India between Mumbai and Delhi is planted with *Jatropha* and the train itself runs on 15–20% biodiesel. In Africa, cultivation of *Jatropha* is being promoted and it is grown successfully in countries such as Mali. Seed yields under cultivation have produced over $500 \, dm^3$ of biofuel per hectare.

Jatropha *bushes growing as a hedge in southern India*

Summary

Fuels made from crops are **renewable resources**. As fossil fuels are used up and as the problems of climate change increase, renewable fuels will play an important role in energy production. However, the fossil fuels used in the production of the crops, and the manufacture of the fuel from the crops, must be taken into account when considering their impact on atmospheric carbon dioxide levels.

The environmental and social consequences of growing crops for fuel must also be considered. If forests are cut down and burnt to provide land on which to grow fuel crops such as oil palms, the effect on the environment will be negative. This has already occurred in the Philippines. An increase in biodiesel production will lead to the deforestation of tens of millions of acres, resulting in loss of habitat that could endanger many plant and animal species.

Another factor is that land converted to growing fuel crops, rather than food, will push up the price of grain and sugar. This will affect the poorest in society and could result in starvation in the least-developed countries.

The greener chemical industry

The social and economic costs of pollution are understood by the chemical industry. Air and rivers can no longer be polluted without incurring sanctions. The industry is also aware of its public image.

A number of measures are being taken by the chemical industry to reduce harmful effects on the environment.

Use of less hazardous chemicals

The chemical industry produces many herbicides, insecticides and fungicides, which are essential for high-volume agriculture. However, they pose threats to the environment. DDT was the main insecticide, until it was realised that it became concentrated in the food chain with harmful effects on some higher animal species. New insecticides are being developed that are based on natural chemicals produced by plants as defence mechanisms. These compounds do not last as long in the environment as traditional insecticides and, therefore, pose less risk to higher animal species and humans.

Renewable resources

Water, forest and biomass are examples of renewable resources. However, the rate of use must be slower than the rate of replenishment. This is of a particular concern with forests.

Solar energy can be used to heat domestic water and can also be converted to electricity using photovoltaic cells. These are expensive and are not very efficient at turning sunlight into electricity. It has been estimated that it would cost over £20 000 to buy enough photovoltaic cells to provide all the energy for a single house.

Wind power is a more suitable renewable energy resource. Worldwide, only 1% of electricity is generated this way, but in Denmark the amount is over 20%. One problem with wind power is that it has to be backed up by fossil fuel power stations when there is not enough wind. Since it takes time for a conventional power station to be fired up and generate electricity, these back-up power stations have to be kept running continuously. Another problem is that wind farms are normally a long way from population centres and a significant amount of electrical energy is lost as heat during transmission along the national grid.

Some waste can also be a renewable resource. Aluminium drink cans are easily turned back into pure aluminium. This preserves aluminium ore deposits and requires less energy than the electrolytic extraction of the metal from purified ore. This reduces the carbon footprint of the recycled aluminium by about 90%. Paper is also recycled, which prevents forest from being felled.

Biomass could be a major resource in the future. It can be used to make bio-fuels. Research is being carried out into its potential use in polymer production. The bacterium *Alcaligenes eutrophus* makes 3-hydroxypropanoic acid, CH_2OHCH_2COOH, by the fermentation of natural sugars. It can be polymerised to a polymer called Biopol. Thus, Biopol is produced naturally by renewable agricultural resources and, most importantly, it is fully biodegradable. It is stable when stored in air, even in humid conditions. Degradation to carbon dioxide and water occurs only when the polymer is exposed to microorganisms found naturally in soil, sewage, river beds and landfill sites. The initial commercial applications for Biopol are as packaging for products such as shampoos and cosmetics and in the manufacture of biodegradable cutlery. Biopol is more expensive than polymers made from fossil fuels, but research is being carried out

Solar, wind and wave power and geothermal energy are perpetual resources that can be used as renewable energy resources.

into genetically engineering the gene from the bacterium and transferring it into cells of plants such as potatoes.

The use of plant products to make biofuels is described earlier in this chapter (page 304).

Higher atom economies

The **atom economy** is the percentage of the atoms in the reactants that is present in the desired product.

Ethanoic acid is a major feedstock in the chemical industry. It used to be made by the high-temperature catalytic oxidation of alkanes containing three, four or five carbon atoms. The atom economy of this process is low because there are side reactions and, therefore, unwanted products. A new method of production has been developed. Methanol and carbon monoxide can be made from methane and water by reactions with high atom economies. They combine to form ethanoic acid in a reaction with 100% atom economy:

$$CO + CH_3OH \rightarrow CH_3COOH$$

The secret lies in the catalyst, which is an iridium complex that produces ethanoic acid in a 99% yield.

Efficient use of energy

Heat is expensive and high temperatures are often essential in chemical processes. Therefore, good thermal insulation and heat recovery are necessary.

Hydrogen is made from methane and steam in an endothermic reaction at a high temperature. It is essential to insulate the reaction vessel to make the process economic.

In the manufacture of sulfuric acid, sulfur dioxide and oxygen react together over a vanadium(V) oxide catalyst in an exothermic reaction. The resulting hot gas mixture is cooled and returned into the catalyst chamber to obtain a higher conversion. This is repeated before absorption into concentrated sulfuric acid. This heat energy is used to produce steam, which can generate enough electricity to run the plant.

In Malawi, in southern Africa, an ethanol production plant has been designed that has high-energy efficiency. Sugar cane is crushed and the soluble sugar is extracted. It is fermented to ethanol and the ethanol is fractionally distilled. The energy for this process, and for pumping irrigation water from the nearby lake, is produced entirely by burning the woody residues of the sugar cane. Therefore, there is no carbon footprint in the manufacture of the ethanol.

Pharmaceutical companies use microwaves to heat reaction mixtures. This is efficient, because there are almost no heat losses and the solutions are warmed up rapidly, thus shortening reaction times. The reduced carbon footprint and the faster production benefit both the environment and the pharmaceutical company.

Waste management

Waste is a huge environmental problem. One solution is to use less packaging, so there is less to throw away. Another is to recycle as much waste as possible.

Plastics such as poly(ethene) and PVC are not biodegradable because they are derived from fossil fuels and no microorganism exists that can break them down. Plastics are insoluble and do not easily degrade. They are bulky, so they take up a lot of space in landfill sites. They can be incinerated, but this has to be done carefully to avoid producing harmful materials, such as dioxins. When incinerated, PVC also produces hydrogen chloride. The best solution is to recycle the plastics. However, this is expensive because the plastics have to be sorted first. Much of the plastic waste in the UK is transported in container ships to China, where some of the waste is sorted and recycled but much of it is burnt in an uncontrolled way.

Another strategy is to make plastics that are easily broken down in the environment. These are sometimes wrongly called 'biodegradable plastics'. Disintegration of the plastic object is initiated either by the action of light or by organisms in the landfill site. The latter is achieved by adding starch granules to the plastic before it is made into the plastic object. The microorganisms in the landfill break down the starch, which turns the plastic into a fine powder, thus taking up less space. Such plastics are **biodispersible** — organisms cause the plastic to crumble and become dispersed in the environment.

Metals are more easily recycled. Iron can be separated from other rubbish by using a magnet. It is then mixed with iron ore and put in a blast furnace. Recycled aluminium cans are melted down and recast, saving a huge amount of energy compared with the high-temperature electrolysis of alumina. The extraction of the costly metals used in catalytic converters is economically viable and, at the same time, preserves non-renewable resources.

To reduce pollution to acceptable levels, gases released through chimneys and water that is returned to the environment by factories are monitored for pollutants.

> ◀ Biodegradable means that bacteria cause the decomposition of the organic material to carbon dioxide or methane and water.

The ozone layer

Stratospheric ozone and oxygen protect the Earth from harmful ultraviolet (UV) radiation. Ozone absorbs ultraviolet radiation of wavelengths in the range 200–300 nm:

$$O_3 \xrightarrow{\text{UV}} O_2 + O\bullet$$

Oxygen absorbs ultraviolet light of wavelengths in the range 180–240 nm:

$$O_2 \xrightarrow{\text{UV}} O\bullet + O\bullet$$

> ◀ The breaking of a bond by light is called photolysis

The oxygen radicals react with further diatomic oxygen to form ozone in an exothermic reaction.

$$O\bullet + O_2 \rightarrow O_3 \qquad \Delta H = -600\,kJ\,mol^{-1}$$

An equilibrium is set up and the concentration of ozone should be a constant value. However, human activity interferes with this equilibrium. The most harmful gases are CFCs and oxides of nitrogen — for example, nitric oxide, NO.

Effect of CFCs on the ozone layer

CFCs are decomposed by ultraviolet light to form chlorine radicals:

$$CF_2Cl_2 \overset{UV}{\rightarrow} CF_2Cl\bullet + Cl\bullet$$

The chlorine radicals undergo a series of chain reactions, with each step in the chain causing the destruction of ozone. There are few chain-breaking processes at the low pressure in the stratosphere, so one chlorine radical can be responsible for the destruction of thousands of ozone molecules.

$$Cl\bullet + O_3 \rightarrow ClO\bullet + O_2$$

The ClO radicals react with oxygen radicals produced by the photolysis of ozone, forming chlorine radicals:

$$O_3 \rightarrow O_2 + O\bullet$$
$$ClO\bullet + O\bullet \rightarrow Cl\bullet + O_2$$

Addition of these three reactions gives the overall equation:

$$2O_3 \rightarrow 3O_2$$

Alternatives to CFCs

CFCs are no longer manufactured in the western world. They have been replaced by compounds that contain at least one hydrogen atom, as well as chlorine and fluorine. The generic name for these compounds is HCFCs — for example, HCFC-22, which has the formula $CHClF_2$. The presence of the hydrogen atom allows the molecule to be oxidised. The resultant chlorine atoms are produced at lower altitudes, where they are removed by water and returned to Earth. HCFCs do not destroy the ozone layer, but they are potent greenhouse gases (page 293).

Effect of nitric oxide on the ozone layer

Oxides of nitrogen are produced by the reaction of nitrogen with oxygen. This happens naturally in lightning strikes and also occurs in car and jet engines. Rain washes out most of the nitric oxides that are produced near the Earth's surface. However, jet aircraft fly close to the stratosphere. The nitric oxide produced by their engines rises up to the ozone layer, where reactions occur that are similar to those between CFCs and ozone.

Nitric oxide removes an oxygen atom from an ozone molecule:

$$NO + O_3 \rightarrow NO_2 + O_2$$

The nitrogen dioxide reacts with oxygen radicals that are generated by the natural photolysis of ozone.

e A nitrogen atom has seven electrons and an oxygen atom has eight electrons. Thus, nitric oxide is a free radical. There is an odd number of electrons in the molecule, so one electron must be unpaired.

$$O_3 \rightarrow O_2 + O\bullet$$
$$NO_2 + O\bullet \rightarrow O_2 + NO$$

Nitric oxide is re-formed and so can destroy more ozone.

Addition of these three reactions gives the same overall equation as with chlorine radicals:

$$2O_3 \rightarrow 3O_2$$

The Antarctic hole in the ozone layer

It is known that a 'hole' in the ozone layer appears in early spring in the southern hemisphere. During the Antarctic winter, ice clouds form in air trapped over the South Pole. These ice clouds store chlorine as chlorine nitrate. When the clouds are evaporated by the first sunshine of spring, the chlorine nitrate decomposes into chlorine, which is then photolysed by visible light into chlorine radicals. The sudden surge in concentration of chlorine radicals causes a rapid depletion of ozone — the 'ozone hole'. Gradually, the chlorine radicals are removed and the ozone concentration builds up, only to be destroyed again the following spring.

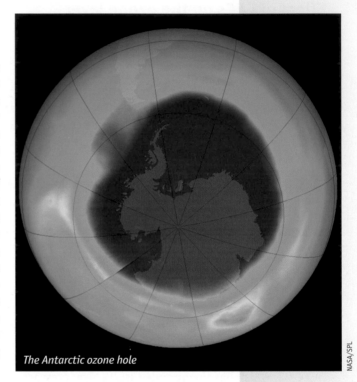

The Antarctic ozone hole

NASA/SPL

Questions

1 Discuss the advantages and disadvantages of hydrogen as a fuel for motor vehicles.

2 Why is bioethanol not carbon neutral?

3 What is transesterification?

4 What are the advantages and disadvantages of using algae as a source of biodiesel?

5 Explain why wind power is not 'carbon zero'.

6 Give two environmental reasons for recycling used aluminium cans.

7 Define the term 'atom economy'.

8 Why are plastics made from fossil fuels not biodegradable?

9 State two problems with burning waste plastics such as PVC.

10 What is the difference between biodegradable and biodispersible plastic?

11 Define the term 'free radical'.

12 What is meant by a chain reaction?

13 Why have CFCs been replaced by HCFCs?

Chapter 17

Laboratory chemistry

Chlorine: poison gas and Agent Orange

Atomic number: 17

Electron configuration: $1s^2\ 2s^2\ 2p^6\ 3s^2\ 3p_x^2\ 3p_y^2\ 3p_z^1$

Symbol: Cl

Molecular formula: Cl_2

Relative atomic mass: 35.453

Poison gas

In the First World War, both sides used chlorine as a poison gas during trench warfare in northern France. Thousands of soldiers were killed by, or suffered lung damage from, this gas. The Germans filled shells with liquid chlorine under pressure and then fired them at the Allied lines. The British released the gas from cylinders in their own trenches, hoping that it would be blown towards the German trenches by the prevailing wind. Gas masks filled with activated charcoal were a defence against chlorine gas attacks.

Gas masks were carried by soldiers in the First World War

Agent Orange

Agent Orange was used by the Americans in the Vietnam war. The plan was to defoliate the forest that was giving cover to the North Vietnamese.

Agent Orange gains its name from the orange-striped barrels in which it was shipped. It is a mixture of two herbicides — 2,4-dichlorophenoxyethanoic acid (2,4-D) and 2,4,5-trichlorophenoxyethanoic acid (2,4,5-T). When sprayed on broad-leaved plants these induce rapid, uncontrolled growth, which results eventually in defoliation. When sprayed on fields of crops such as wheat, maize or grass, they selectively kill the broad-leaved plants (the weeds) leaving the crop relatively unaffected.

2,4-D

2,4,5-T

At the time that Agent Orange was sold to the US government for use in Vietnam, internal memos by its manufacturers show that it was known that a dioxin, 2,3,7,8-tetrachlorodibenzodioxin (TCDD), is produced as a by-product of the manufacture of 2,4,5-T and was, therefore, present in herbicides that contained 2,4,5-T. TCDD is carcinogenic to humans.

TCDD

2,4-D does not contain dioxin and is one of the most-used herbicides today. Many gardeners use it to kill the weeds in their lawns.

Laboratory chemistry

This chapter is designed to help with the internal assessment of practical work. It will also be useful in answering question in Units 1 and 2 that are based on practical work — for example, enthalpy experiments and titrations.

Tests

Tests for gases

The tests used to identify the common laboratory gases have to be learnt. It is also important to be able to draw conclusions from the identification of these gases. For example, if dilute sulfuric acid is added to a solid and a gas is produced that turns limewater milky, the conclusions are that:
- the gas is carbon dioxide
- the unknown solid is a carbonate or a hydrogencarbonate

Hydrogen
Test: ignite the gas
Observation: burns with a squeaky pop

Hydrogen is produced by:
- the reaction between an acid and a reactive metal
- the reaction between an alcohol and sodium
- the reaction between water and either a group 1 metal, or calcium, strontium or barium

Oxygen
Test: place a glowing wooden splint in the gas
Observation: splint catches fire

Oxygen is produced by:
- heating a group 1 nitrate (apart from $LiNO_3$)
- heating other nitrates, but NO_2 is also present
- the catalytic decomposition of hydrogen peroxide, H_2O_2

Carbon dioxide
Test: pass gas into limewater
Observation: limewater goes cloudy/milky

Carbon dioxide is produced by:
- the reaction between an acid and a carbonate or hydrogencarbonate
- heating a carbonate (apart from sodium, potassium or barium carbonates)
- heating a group 1 hydrogencarbonate

Ammonia
Test: place damp red litmus paper in the gas
Observation: litmus paper goes blue

Ammonia is produced by:
- heating aqueous sodium hydroxide with an ammonium salt
- adding aqueous sodium hydroxide and aluminium powder to a nitrate

Nitrogen dioxide
Test: observe colour
Observation: gas is brown

Nitrogen dioxide is produced by:
- heating a group 2 nitrate or lithium nitrate

Chlorine
Test: place damp litmus paper in the gas
Observation: litmus is rapidly bleached

Chlorine is produced by:
- electrolysis of a solution of a chloride
- adding dilute hydrochloric acid to a solution containing chlorate(I) ions

Hydrogen chloride
Test 1:   place damp blue litmus paper in the gas
Observation: steamy fumes that turn damp litmus paper red
Test 2:   place the stopper from a bottle of concentrated ammonia in the gas
Observation: white smoke (ammonium chloride) is formed

Hydrogen chloride is produced by:
- the reaction between concentrated sulfuric acid and a chloride
- the reaction between phosphorus pentachloride and either an alcohol or a carboxylic acid

Sulfur dioxide
Test: place filter paper soaked in potassium dichromate(VI) solution in the gas
Observation: colour changes from orange to green

Sulfur dioxide is produced by:
- warming an acid with a solid sulfite
- burning sulfur
- reducing concentrated sulfuric acid

---

 e You must know the formulae of the ammonium ion, NH_4^+, and common polyatomic anions, such as carbonate, CO_3^{2-}, hydrogencarbonate, HCO_3^-, sulfate, SO_4^{2-}, sulfite, SO_3^{2-}, nitrate NO_3^- and chlorate(I), ClO^-.

 e The only hydrogencarbonates that exist as solids are those of group 1 metals.

Flame tests

To carry out a flame test, a *clean* platinum or nichrome test wire is dipped into concentrated hydrochloric acid, then into the solid to be tested and finally into the hottest part of a Bunsen flame.

The concentrated hydrochloric acid converts some of the unknown solid into a chloride. Chlorides are more volatile than other salts, so some of the unknown goes into the gas phase when heated in the hot flame. An electron is promoted to a higher energy level by the heat. It then falls back to the ground state and emits light of a colour specific to the metal present in the compound. The colours obtained in flame tests are given in Table 17.1.

Flame colour	Ion present in solid
Crimson	Li^+
Yellow	Na^+
Lilac	K^+
Yellow-red	Ca^{2+}
Red	Sr^{2+}
Pale green	Ba^{2+}

Table 17.1
Flame colours

ⓔ The flame test is the only test for a group 1 metal in a compound.

ⓔ Flame tests cannot be used on mixtures containing two of these ions because the colour produced by one of the ions will mask the colour produced by the other metal ion.

Precipitation reactions

It is important to know which ionic compounds are soluble in water and which are insoluble.

Soluble ionic compounds include:
- all group 1 salts
- all ammonium salts
- all nitrates
- all chlorides, apart from silver chloride and lead(II) chloride. (The solubility of bromides and iodides is similar to that of chlorides.)
- all sulfates, apart from barium sulfate, strontium sulfate and lead(II) sulfate. Calcium sulfate and silver sulfate are slightly soluble.

Insoluble ionic compounds include:
- all carbonates, apart from group 1 carbonates and ammonium carbonate
- all hydroxides, apart from group 1 hydroxides, barium hydroxide and ammonium hydroxide. Calcium and strontium hydroxides are slightly soluble.

The results of some precipitation reactions are shown in Table 17.2.

Solutions containing these ions are mixed	Result
Ba^{2+} and SO_4^{2-} (sulfate ions could be from solutions of group 1 sulfates, ammonium sulfate or from sulfuric acid)	• White precipitate of barium sulfate • On adding dilute hydrochloric acid, the precipitate remains
Ba^{2+} and SO_3^{2-}	• White precipitate of barium sulfite • On adding dilute hydrochloric acid, the precipitate dissolves with no fizzing
Ag^+ and Cl^- (silver ions are normally from a solution of silver nitrate)	• Chalky-white precipitate of silver chloride • Precipitate dissolves in dilute aqueous ammonia
Ag^+ and Br^-	• Cream precipitate of silver bromide • Precipitate is insoluble in dilute ammonia, but soluble in concentrated ammonia
Ag^+ and I^-	• Pale yellow precipitate of silver iodide • Precipitate is insoluble in both dilute and concentrated ammonia
CO_3^{2-} and any cation, other than a group 1 metal or ammonium (carbonate ions could be from solutions of group 1 carbonates or ammonium carbonate)	• White precipitate of the insoluble metal carbonate • On adding acid, the precipitate fizzes as it gives off carbon dioxide and disappears

Table 17.2
Precipitation
reactions

The ionic equations for these precipitation reactions will always be of the form:

cation(aq) + anion(aq) → formula of precipitate(s)

For example, the ionic equation for the formation of barium sulfate by mixing solutions of barium chloride and potassium sulfate is:

$$Ba^{2+}(aq) + SO_4^{2-}(aq) \rightarrow BaSO_4(s)$$

Tests for anions

Sulfate
To a solution of the suspected sulfate, add dilute hydrochloric acid followed by aqueous barium chloride. A white precipitate, which remains when excess hydrochloric acid is added, proves the presence of sulfate ions.

Sulfite
To the suspected sulfite, add dilute sulfuric acid and warm. If it is a sulfite, sulfur dioxide gas will be given off. This is tested for by placing a piece of filter paper soaked in potassium dichromate(VI) in the gas. A colour change from orange to green on the paper confirms the presence of sulfite ions in the original solid.

This test can be carried out on either an unknown solid or a solution of the unknown.

Carbonate
Add dilute sulfuric acid to the suspected carbonate in solid form. If it is a carbonate, carbon dioxide will be given off. This is tested for by passing the gas into limewater, which goes milky.

e The dilute hydrochloric acid can be added before the barium chloride. The formation of a precipitate proves the presence of sulfate ions.

Alternatively, heat the solid. Carbonates, other than sodium, potassium and barium carbonate, give off carbon dioxide.

> ℮ Both these tests work with hydrogencarbonates as well, so the initial conclusion should be that carbonate or hydrogencarbonate is present. However, the only hydrogencarbonates that exist as solids are group 1 compounds. If the unknown is found to be a group 1 compound, the test for hydrogencarbonate must be carried out. If the result is negative, the original compound is a carbonate.

Hydrogencarbonate

To distinguish between a group 1 carbonate and a hydrogencarbonate:

- add some of the unknown solid to almost boiling water. Hydrogencarbonates decompose with the production of carbon dioxide — fizzing occurs. Carbon dioxide can be tested for by passing it into limewater, which goes milky.

or

- add a solution of the unknown to a solution of calcium chloride. Hydrogencarbonates do not give a precipitate because calcium hydrogencarbonate is soluble; carbonates give a white precipitate of calcium carbonate.

or

- test a solution with pH paper. Carbonates have a high (alkaline) pH whereas hydrogencarbonates are almost neutral. On warming, if the solution is a hydrogencarbonate, the pH rises rapidly as the hydrogencarbonate decomposes to give a carbonate and carbon dioxide.

Chloride

To a solution of the suspected chloride, add dilute nitric acid until the solution is just acidic (test with litmus paper). Then, add silver nitrate solution. A chalky-white precipitate, which dissolves when excess dilute ammonia is added, proves the presence of chloride ions in the original solution.

Bromide

To a solution of the suspected bromide, add dilute nitric acid until the solution is just acidic (test with litmus paper). Then, add silver nitrate solution. A cream precipitate, which is insoluble in dilute ammonia but dissolves in concentrated ammonia, proves the presence of bromide ions in the original solution.

An alternative test is to add chlorine water to a solution of the unknown. Bromide ions are oxidised to bromine, which turns the colourless solution brown.

Iodide

To a solution of the suspected iodide, add dilute nitric acid until the solution is just acidic (test with litmus paper). Then, add silver nitrate solution. A pale yellow precipitate, which remains when concentrated ammonia is added, proves the presence of iodide ions in the original solution.

An alternative test is to add chlorine water to a solution of the unknown and then add a few drops of an organic solvent such as hexane. Iodide ions are oxidised to iodine, so the colourless hexane layer turns violet. The presence of iodine from the oxidation of iodide ions can also be detected by adding starch, which turns blue-black.

The result would be the same if bromine water had been added. Bromine is a stronger oxidising agent than iodine and therefore oxidises iodide ions to iodine.

Nitrate

To the unknown solid, add aluminium powder or Devarda's alloy and sodium hydroxide solution and then warm. Nitrate ions are reduced to ammonia, which can be detected by the gas evolved turning damp red litmus paper blue.

> **e** This test is only conclusive if ammonium ions have been shown to be absent. They also would give ammonia by reacting with the alkali present in this test.

An alternative is the 'brown-ring' test. Mix together solutions of the unknown and iron(II) sulfate. Then carefully pour some concentrated sulfuric acid down the side of the test tube. A brown ring forms where the concentrated acid and aqueous layers meet.

Tests for cations

Ammonium

Warm the unknown solid or solution with aqueous sodium hydroxide. Ammonium salts give off ammonia gas, which can be detected by placing damp red litmus paper in the gas. Ammonia turns the paper blue.

Group 1 and group 2 cations

Group 1 and group 2 cations are detected by the flame test (pages 212–13 and 317).

Magnesium compounds do not colour a flame, but are colourless in solution and give a white precipitate of magnesium carbonate when ammonium carbonate solution is added.

p-block and *d*-block cations

The detection of the ions of *p*-block and *d*-block elements is tested at A2.

Tests for organic functional groups

Test for a C=C bond

Add bromine water to the test compound. Compounds containing a C=C bond (unsaturated compounds) turn the bromine water from brown to colourless.

Test for a C–OH group

Add solid phosphorus pentachloride to the anhydrous compound. If the compound contains a C–OH group (carboxylic acids and alcohols), steamy fumes of hydrogen chloride will be produced. This can be detected by putting damp blue litmus in the gas. Hydrogen chloride turns the litmus red.

> **e** Do not state that this is a test for alcohols. If a compound gives a positive result for a C–OH group, you then have to determine whether it is a carboxylic acid or an alcohol. To distinguish between the two, add some of the test compound to sodium hydrogencarbonate solution. An acid will produce fizzing as carbon dioxide is given off; with an alcohol there is no reaction.

Test for a C–halogen group

Warm the unknown with aqueous sodium hydroxide for several minutes, being careful not to boil off a volatile organic compound. Allow the solution to

Before doing this test you need to show that the compound is not an ammonium salt, as ammonium salts give ammonia when heated with alkali.

A little ethanol is sometimes added to help dissolve the halogenoalkane and bring it into contact with the hydroxide ions.

cool. Add nitric acid until the solution is just acidic to litmus. Then add silver nitrate solution.

- Organic chlorides give a white precipitate of silver chloride, which is soluble in dilute ammonia.
- Organic bromides give a cream precipitate of silver bromide, which is insoluble in dilute ammonia but dissolves in concentrated ammonia.
- Organic iodides give a pale yellow precipitate of silver iodide, which is insoluble in both dilute and concentrated ammonia.

Alkanes

There is no test for an alkane. If all the above tests are negative, the unknown (limited to organic substances covered at AS) is probably an alkane.

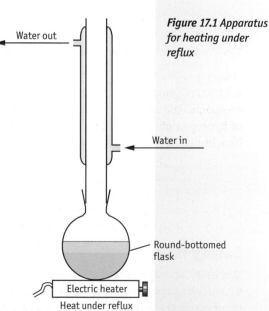

Organic techniques

> ⓔ You are expected to know techniques used in organic preparations. You must be able to draw diagrams of the necessary apparatus and know when to use each method.

Heating under reflux

Organic substances are volatile and the rates of many organic reactions are slow. Heating is used to speed up reactions, but if this is done in an open vessel, such as a beaker, the organic reactant and the product may boil off. To prevent this happening, a reflux condenser is used. The organic vapours that boil off as the reaction mixture is heated are condensed and flow back into the reaction vessel. As most organic substances are flammable, it is safer to heat the mixture using an electric heater or a water bath rather than a Bunsen burner.

The essential points of the apparatus (Figure 17.1) are:

- a round-bottomed flask
- a reflux condenser with the water entering at the bottom and leaving at the top
- the top of the reflux condenser being open
- the flask being heated using an electric heater, a water or oil bath or a sand tray

Reactions where heating under reflux is necessary include:

- oxidising a primary alcohol to a carboxylic acid
- oxidising a secondary alcohol to a ketone
- converting a halogenoalkane to an alcohol
- converting an alcohol to a bromoalkane by reacting it with potassium bromide and 50% sulfuric acid

The organic halogen compound is hydrolysed by the sodium hydroxide to give halide ions in solution, which then react with the silver ions to form a precipitate.

ⓔ Remember that an organic unknown may contain two functional groups.

Figure 17.1 Apparatus for heating under reflux

Water out

Water in

Round-bottomed flask

Electric heater

Heat under reflux

Distillation

Most organic reactions do not go to completion and there are often side reactions. This means that the desired product is present in a mixture. Organic substances are volatile, so they can be separated from non-volatile inorganic species, such as acids or alkalis, by distillation.

If there is a large enough difference in the boiling temperatures of organic substances present in a mixture, then distillation can be used to separate them.

The essential points about the apparatus are:
- a round-bottomed or pear-shaped flask containing the mixture
- a still-head fitted with a thermometer, the bulb of which must be positioned level with the outlet of the still-head
- a condenser with the water going in at the bottom and leaving at the top
- an open receiving vessel, such as a beaker, or an adaptor open to the air joined to a flask
- the flask being heated using an electric heater, a water or oil bath or sand tray

The mixture is carefully heated and the vapour that comes over at $\pm 2°C$ of the boiling temperature (obtained from a data book) of the particular substance is condensed and collected.

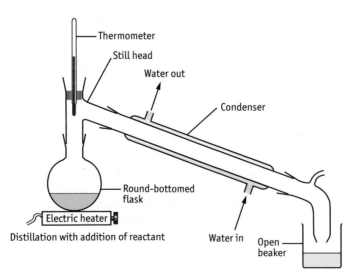

Distillation with addition of reactant

Figure 17.2 Apparatus for simple distillation

ⓔ Make sure that the top of the still-head is closed and that the Liebig condenser is drawn with a water jacket.

ⓔ You must be able to draw the two types of apparatus shown in Figures 17.1 and 17.2.

Safety issues

Questions are often asked about the safety precautions that should be taken in a particular experiment. Safety precautions and the reasons behind them include the following:

- Distillation and heating under reflux must be carried out in a fume cupboard if the vapour of one of the reactants or products is harmful, poisonous (toxic) or irritant.
- If a mixture is being heated under reflux or distilled, there must be some outlet to the air. If there is not, pressure will build up in the apparatus, which will then fly apart, spraying hot, flammable, and often corrosive, liquid around.
- Gloves must be worn when corrosive substances are used. Such substances must always be handled with care.
- The flask should never be heated with a naked Bunsen flame. This is because almost all organic substances are flammable and if the liquid being heated were to spill over or the flask to crack, a fire would result.

> ℮ It is assumed that laboratory coats and eye protection are always worn when carrying out practical work. Therefore, credit is not given for stating either of these as a specific safety precaution.

◀ This applies to halogens, hydrogen halides, halogenoalkanes and ammonia.

◀ This applies to concentrated acids and alkalis.

Yield

Yields are less than 100% because of:
- competing reactions
- handling losses during purification

℮ You must be able to calculate theoretical yield and percentage yield (pages 68 and 280–81).

Enthalpy change measurements

Enthalpy of combustion

The method for determining the enthalpy of combustion of a liquid, such as ethanol, is described on pages 87–89. Ways to increase the accuracy of the experiment include the following:
- The copper calorimeter should be first weighed empty and then when containing water. Alternatively, water could be added to the calorimeter using a pipette, *not* a measuring cylinder. If the volume of water is measured, the mass is calculated using the density of water, which is 1 g cm^{-3}.
- A screen should be placed around the calorimeter to maximise the transfer of heat from the hot combustion gases to the beaker of water.
- To ensure an even temperature throughout, the water in the calorimeter must be stirred continually.
- The temperature of the water should be measured for several minutes before lighting the fuel and for several minutes after putting out the burner flame.
- The temperature–time measurements are used to plot a graph from which the theoretical temperature rise is estimated by extrapolation. This reduces the error caused by heat loss from the beaker to the surroundings.
- The burner and its contents should be weighed before and immediately after the experiment, using a balance that reads to an accuracy of 0.01 g or better.

The calculation is carried out in three steps.

Step 1: heat produced by the combustion of the fuel $= m \times c \times \Delta T$

The mass, m, is the mass of water in the beaker, *not* the mass of fuel burnt. The specific heat capacity of water, c, is 4.18 J g^{-1} °C^{-1}, so the unit of the heat produced is joules, *not* kilojoules.

Step 2: amount of fuel burnt $= \dfrac{\text{mass before} - \text{mass after}}{\text{molar mass of fuel}}$

Step 3: enthalpy of combustion, $\Delta H_c = -\dfrac{\text{heat produced}}{\text{moles of fuel burnt}}$

All combustions are exothermic, so the value of ΔH_c is negative.

Displacement reactions and enthalpies of solution

In both displacement reactions and experiments to determine the enthalpy of solution, a solid is added to a liquid and the temperature change is measured.

The method for determining ΔH_r of reactions, such as the displacement of copper from copper sulfate solution by a more reactive metal, is described on page 84. The same method is used to find the enthalpy of solution of a solid.

Temperature–time graphs are necessary because the reactions are not instantaneous.

Errors can be reduced by:
- using powdered solid rather than lumps. This speeds up the rate of reaction, so there is less time for cooling.
- making sure that, for displacement reactions, enough metal is taken to ensure that the solution of the salt of the less reactive metal is the limiting reagent. For enthalpy of solution experiments, the water must be in large excess to ensure that all the solid dissolves.

A limiting reagent reacts completely, leaving an excess of the other reagent.

- measuring the temperature for several minutes before the start of the reaction and for several minutes after the reaction has finished. The measurements are used to plot a graph, which is extrapolated to find the theoretical temperature rise.
- continually stirring the contents of the expanded polystyrene cup
- placing a lid on the cup to prevent heat loss through evaporation
- weighing the cup empty and then, before the reaction starts, weighing it containing the solution. This gives an accurate value of the mass of the solution. The assumption that the density of a solution is 1 g cm^{-3} is not wholly accurate.
- measuring the volume of the solution using a pipette rather than a measuring cylinder, so that the amount (moles) of solute can be accurately determined. This is not necessary for enthalpy of solution determinations.

The calculation is carried out in three steps:

Step 1: heat produced by the reaction $= m \times c \times \Delta T$

The mass, m, is the mass of solution in the cup, *not* the mass of solute reacted. The specific heat of the solution is assumed to be the same as that of water (4.18 J g^{-1} °C^{-1}), so the unit of the heat produced is joules, *not* kilojoules.

Step 2: amount of solute reacted = concentration (mol dm^{-3}) × volume (dm^3)

(For enthalpy of solution determinations, amount of solute = mass/molar mass.)

> ⓔ The unit of volume measured by a pipette is cm^3. This must be converted into dm^3 by dividing by 1000.

Step 3: $\Delta H_r = \dfrac{\text{heat produced or lost}}{\text{moles of solute reacted}}$

If there is a temperature rise, ΔH is negative; if the temperature falls, ΔH is positive.

Instantaneous reactions

Neutralisation and precipitation reactions are instantaneous reactions. The method used for following both types of reaction is the same and is described on page 90.

Errors can be reduced by:

- using pipettes, rather than measuring cylinders, to measure out the volumes of the two liquids.
- making sure that one of the reactants is in excess. The value of ΔH can then be worked out using the amount in moles of the limiting reagent.
- for neutralisation reactions only, weighing the expanded polystyrene cup empty and after the reaction. This is a more accurate way of obtaining the mass of solution than using a pipette and assuming that the solution has a density of 1 g cm^{-3}.
- measuring the temperature of both liquids before mixing and averaging the two values
- stirring immediately on mixing the two solutions
- reading the maximum temperature reached

The calculation is carried out in three steps.

Step 1: heat produced by the reaction = $m \times c \times \Delta T$

> ⓔ The mass, m, is the total mass of the *two* solutions, *not* the mass of solute reacted. The specific heat of the solution is assumed to be the same as that of water (4.18 J g^{-1} °C^{-1}), so the unit of the heat produced is joules, *not* kilojoules.

Step 2: amount of solute reacted = concentration (mol dm^{-3}) × volume (dm^3)

> ⓔ The solute used in the calculation must be that of the limiting reagent.

Step 3: $\Delta H_r = \dfrac{\text{heat produced or lost}}{\text{moles of solute reacted}}$

If there is a temperature rise, ΔH is negative; if the temperature falls, ΔH is positive.

This not only gives an accurate value for the amounts (moles) of each reactant, but also gives an accurate value for the volume of the solution.

ⓔ There is no need to plot a temperature–time graph because the reaction is instantaneous, so there is no time for heat to be lost.

Titration techniques

Calculations based on titrations will not be asked in Unit Tests 1 or 2. Questions based on titrations may be asked in the Unit 3 test.

Preparation of a standard solution

In any titration, the concentration of one of the solutions must be accurately known. The method for preparing 250 cm^3 of such a solution — a **standard solution** — is as follows:

- Calculate the mass of solid needed to make a solution of the required concentration.
- Place a weighing bottle on a top-pan balance. Press the tare button, so that the scale reads zero.
- Add the solid to the weighing bottle until the required mass is reached.

> *e* The best way to do this is to remove the bottle from the pan and then add the solid, checking the mass until the correct amount has been added. This prevents errors caused by spilling solid onto the pan of the balance.

- Tip the contents of the weighing bottle into a beaker. Wash any remaining solid from the bottle into the beaker.
- Add about 50 cm^3 of distilled water to the beaker containing the solid. Using a glass rod, stir until all the solid has dissolved. In order to dissolve the solid completely, it may be necessary to heat the beaker.
- Pour the solution through a funnel into a standard 250 cm^3 flask. Wash the stirring rod and the beaker, making sure that all the washings go through the funnel into the standard flask.

> *e* You *must* use a standard flask, not a beaker, to measure the volume of the solution. The volume of a solution is not the same as the volume of the solvent used to make up the solution. Adding 250 cm^3 of water to the solid would not produce an accurate standard solution.

- Add more distilled water to the solution until the bottom of the meniscus is level with the mark on the standard flask.
- Put the stopper in the flask and mix thoroughly by inverting and shaking several times.

Performing a titration

The pieces of apparatus required are:
- a burette
- a pipette (usually 25.0 cm^3) and a pipette filler
- a conical flask

The chemicals required are:
- a standard solution
- the solution of unknown concentration
- a suitable indicator

To make 250 cm^3 of a solution of concentration 0.100 mol dm^{-3}, $\frac{1}{4} \times 0.100$ mol has to be weighed out. This is because 250 cm^3 is a quarter of 1 dm^3.

e The same method applies using standard flasks of other volumes, such as 100 cm^3, 500 cm^3 and 1 dm^3. Standard flasks are accurate to approximately ± 0.1 cm^3.

The method for performing an accurate titration is as follows:

- Draw a small amount of one solution into the pipette using a pipette filler and rinse it with the solution. Discard the rinsings.
- Using a pipette filler, fill the pipette so that the bottom of the meniscus in on the mark.

- Allow the pipette to discharge into a washed conical flask. When the pipette has emptied, touch the surface of the liquid in the flask with the tip of the pipette.

- Making sure that the tap is shut, rinse out a burette with a few cm^3 of the other solution and discard the rinsings.
- Using a funnel, fill the burette to above the zero mark and run liquid out until the meniscus is on the scale. Check that the burette below the tap is filled with liquid and that there are no air bubbles. Remove the funnel.
- Record the initial volume by looking at where the bottom of the meniscus is on the burette scale.
- Run the liquid slowly from the burette into the conical flask, continually mixing the solutions by swirling the liquid in the flask. Add the liquid dropwise as the end point is neared and stop when the indicator shows the end point colour. Read the burette to 0.05 cm^3.
- Repeat the titration until three concordant titres have been obtained.
- Ignore any non-concordant titres and average the concordant values, to give the average (mean) titre.

Titration calculations

These are described in Chapter 10 on pages 206–11.

Remember that the calculation is carried out in three steps.

Step 1: calculate the number of moles in the standard solution used in the titration.

> **e** Don't forget to convert the volume (pipette or burette) from cm^3 to dm^3.
>
> $$\text{volume in } dm^3 = \frac{\text{volume in } cm^3}{1000}$$

Step 2: use the stoichiometry of the equation to calculate the number of moles (amount) in the other solution used in the titration.

> **e** If the mole ratio is 1:1, make sure that you state that the number of moles in the second solution equals the number of moles in the first solution. If the ratio is not 1:1, make sure that the ratio is the right way up in the conversion:
>
> $$\text{moles of A} = \text{moles of B} \times \frac{\text{number of moles of A in equation}}{\text{number of moles of B in equation}}$$

Step 3: calculate the concentration of the second solution.

> **e** Make sure that you give your answer to the correct number of significant figures. If in doubt, give it to 3 s.f.

Worked example

A sample of 2.65 g of pure sodium carbonate, Na_2CO_3, was weighed out, dissolved in water and made up to 250 cm^3 in a standard flask. Some of this solution was placed in a burette and used to titrate 25.0 cm^3 portions of a solution of hydrochloric acid. The titres obtained are shown in the table.

The equation for the reaction is:

$$2HCl + Na_2CO_3 \rightarrow 2NaCl + H_2O + CO_2$$

Experiment	Titre/cm^3
1	22.35
2	22.40
3	21.85
4	22.50

a Calculate the concentration of the sodium carbonate solution.
b Calculate the mean titre.
c Calculate the amount (in moles) of sodium carbonate solution in the mean titre.
d Calculate the amount (in moles) of hydrochloric acid that reacted.
e Calculate the concentration of the hydrochloric acid solution.

Answer

a molar mass of sodium carbonate $= (2 \times 23.0) + 12.0 + (3 \times 16.0) = 106.0 \, g \, mol^{-1}$

$$\text{amount of sodium carbonate} = \frac{\text{mass}}{\text{molar mass}} = \frac{2.65 \, g}{106.0 \, g \, mol^{-1}} = 0.0250 \, mol$$

$$\text{concentration} = \frac{\text{mol}}{\text{volume}} = \frac{0.0250 \, mol}{(250/1000) \, dm^3} = 0.100 \, mol \, dm^{-3}$$

b The titre for experiment 3 is at least 0.5 cm^3 less than any other titre, and so is not used in the calculation.

$$\text{mean titre} = \frac{22.35 + 22.40 + 22.50}{3} = 22.42 \text{ cm}^3$$

c amount of sodium carbonate = concentration × volume

$$\text{amount} = 0.100 \text{ mol dm}^{-3} \times \frac{22.42 \text{ dm}^3}{1000} = 0.00224 \text{ mol}$$

ratio HCl:Na$_2$CO$_3$ = 2:1

$$\text{amount of HCl} = \frac{2}{1} \times 0.00224 \text{ mol} = 0.00448 \text{ mol}$$

$$\text{concentration of hydrochloric acid} = \frac{\text{mol}}{\text{volume}}$$

$$= \frac{0.00448 \text{ mol}}{\left(25.0 / 1000 \text{ dm}^3\right)} = 0.179 \text{ mol dm}^{-3}$$

Note that all volumes were converted from cm^3 to dm^3 by dividing by 1000.

Evaluation of error

There is always a built-in error when using apparatus such as balances, thermometers, burettes and pipettes.

The pipette may be stamped with the volume 25.0 cm^3, or the balance may read 1.23 g, but these quantities are all subject to some error. For example, the manufacturer of a balance may state that the mass of an object is accurate to + 0.01 g. Therefore, a reading of 1.23 g may mean a mass anywhere between 1.22 and 1.24 g.

If a container is weighed empty and then reweighed containing some solid, each weighing would be subject to a possible error of ± 0.01 g and so the mass of the solid could have an error of ± 0.02 g.

Worked example 1

An empty crucible was weighed, using a balance with a possible error of ± 0.01 g. It was reweighed containing some solid metal carbonate, MCO$_3$, and was then heated to constant weight. On heating, the metal carbonate decomposes according to the equation:

$$MCO_3 \rightarrow MO + CO_2$$

The readings were as follows:
- Mass of empty crucible = 14.23 g
- Mass of crucible + carbonate = 16.46 g
- Mass of crucible + contents after heating to constant weight = 15.61 g
a Calculate the mass of carbon dioxide given off.
b Hence, calculate the amount in moles of carbon dioxide produced.
c Calculate the mass of metal carbonate taken.
d Using your answers to **b** and **c**, calculate the molar mass of the metal carbonate and hence the molar mass of the metal, M.

e Assuming all the measurements of mass had an uncertainty of ± 0.01 g, calculate the uncertainty in the value of the molar mass of M.

Answer

a   mass of carbon dioxide = 16.46 − 15.61 = 0.85 g

b $amount = \dfrac{mass}{molar\ mass} = \dfrac{0.85\ g}{44.0\ g\ mol^{-1}} = 0.0193\ mol$

c   mass of MCO_3 = 16.46 − 14.23 = 2.23 g

d   amount (moles) of MCO_3 = amount (moles) of CO_2 = 0.0193 mol

   $molar\ mass\ of\ MCO_3 = \dfrac{mass}{molar\ mass} = \dfrac{2.23\ g}{0.0193\ mol} = 116\ g\ mol^{-1}$

   molar mass of M = 116 − [12.0 + (3 × 16.0)] = 56.0 g mol^{-1}

e   mass of CO_2 = 0.85 ± 0.02 g

The amount (moles) of CO_2 lies between 0.83/44.0 = 0.0189 mol and 0.87/44.0 = 0.0198 mol.

Since the ratio of MCO_3 to CO_2 is 1:1, the amount (moles) of MCO_3 also lies between 0.0189 and 0.0198 mol.

   mass of MCO_3 = 2.23 ± 0.02 g

   $maximum\ molar\ mass\ of\ MCO_3 = \dfrac{biggest\ mass}{smallest\ moles} = \dfrac{2.25}{0.0189} = 119\ g\ mol^{-1}$

Therefore,

   maximum molar mass of M = 119 − 60.0 = 59 g mol^{-1}

   $minimum\ molar\ mass\ of\ MCO_3 = \dfrac{smallest\ mass}{biggest\ moles} = \dfrac{2.21}{0.0198} = 112\ g\ mol^{-1}$

Therefore,

   minimum molar mass of M = 112 − 60.0 = 52 g mol^{-1}

The uncertainty of the molar mass of M is that it is between 52 and 59 g mol^{-1}.

e The element, M, could be Cr, Mn, Fe, Co or Ni, as their molar masses lie in the range 52 to 59 g mol^{-1}.

Worked example 2

A thermometer is labelled as having an accuracy of ± 0.2°C.

   In an enthalpy of neutralisation reaction, the temperature before the reaction was 17.4°C and after the reaction was 24.5°C. The mass of liquid used was 100 g. The specific heat capacity of the liquid is 4.18 J g^{-1} $°C^{-1}$. Calculate the percentage error caused by the thermometer and hence the error in the evaluation of the heat produced.

Answer

   temperature rise = (24.5 − 17.4) ± 0.4 = 7.1 ± 0.4°C

   percentage error caused by the thermometer = $±0.4 × \dfrac{100}{7.1} = ±5.63\%$

   heat produced = 100 g × 4.18 J g^{-1} $°C^{-1}$ × 7.1°C = 2968 J

   error in value of heat produced = ±5.63% of 2968 J = ± 167 J

Index

The periodic table

Group

Period	1	2												3	4	5	6	7	0
1	1.0 H hydrogen 1																		4.0 He helium 2
2	6.9 Li lithium 3	9.0 Be beryllium 4												10.8 B boron 5	12.0 C carbon 6	14.0 N nitrogen 7	16.0 O oxygen 8	19.0 F fluorine 9	20.2 Ne neon 10
3	23.0 Na sodium 11	24.3 Mg magnesium 12												27.0 Al aluminium 13	28.1 Si silicon 14	31.0 P phosphorus 15	32.1 S sulfur 16	35.5 Cl chlorine 17	39.9 Ar argon 18
4	39.1 K potassium 19	40.1 Ca calcium 20	45.0 Sc scandium 21	47.9 Ti titanium 22	50.9 V vanadium 23	52.0 Cr chromium 24	54.9 Mn manganese 25	55.8 Fe iron 26	58.9 Co cobalt 27	58.7 Ni nickel 28	63.5 Cu copper 29	65.4 Zn zinc 30	69.7 Ga gallium 31	72.6 Ge germanium 32	74.9 As arsenic 33	79.0 Se selenium 34	79.9 Br bromine 35	83.8 Kr krypton 36
5	85.5 Rb rubidium 37	87.6 Sr strontium 38	88.9 Y yttrium 39	91.2 Zr zirconium 40	92.9 Nb niobium 41	95.9 Mo molybdenum 42	[98] Tc technetium 43	101.1 Ru ruthenium 44	102.9 Rh rhodium 45	106.4 Pd palladium 46	107.9 Ag silver 47	112.4 Cd cadmium 48	114.8 In indium 49	118.7 Sn tin 50	121.8 Sb antimony 51	127.6 Te tellurium 52	126.9 I iodine 53	131.3 Xe xenon 54
6	132.9 Cs caesium 55	137.3 Ba barium 56	138.9 La lanthanum 57	178.5 Hf hafnium 72	180.9 Ta tantalum 73	183.8 W tungsten 74	186.2 Re rhenium 75	190.2 Os osmium 76	192.2 Ir iridium 77	195.1 Pt platinum 78	197.0 Au gold 79	200.6 Hg mercury 80	204.4 Tl thallium 81	207.2 Pb lead 82	209.0 Bi bismuth 83	[209] Po polonium 84	[210] At astatine 85	[222] Rn radon 86
7	[223] Fr francium 87	[226] Ra radium 88	[227] Ac actinium 89	[261] Rf rutherfordium 104	[262] Db dubnium 105	[266] Sg seaborgium 106	[264] Bh bohrium 107	[277] Hs hassium 108	[268] Mt meitnerium 109	[271] Ds darmstadtium 110	[272] Rg roentgenium 111							

Key:

Relative atomic mass
Atomic symbol
name
Atomic (proton) number

Elements with atomic numbers 112–116 have been reported but not fully authenticated

140.1 Ce cerium 58	140.9 Pr praseodymium 59	144.2 Nd neodymium 60	144.9 Pm promethium 61	150.4 Sm samarium 62	152.0 Eu europium 63	157.2 Gd gadolinium 64	158.9 Tb terbium 65	162.5 Dy dysprosium 66	164.9 Ho holmium 67	167.3 Er erbium 68	168.9 Tm thulium 69	173.0 Yb ytterbium 70	175.0 Lu lutetium 71
232 Th thorium 90	[231] Pa protactinium 91	238.1 U uranium 92	[237] Np neptunium 93	[242] Pu plutonium 94	[243] Am americium 95	[247] Cm curium 96	[245] Bk berkelium 97	[251] Cf californium 98	[254] Es einsteinium 99	[253] Fm fermium 100	[256] Md mendelevium 101	[254] No nodelium 102	[257] Lr lawrencium 103